World Energy Production and Productivity

Volume One in the
INTERNATIONAL ENERGY SYMPOSIA SERIES
A 1982 WORLD'S FAIR EVENT

World Energy Production

and Productivity

Proceedings of the
International Energy Symposium I
October 14, 1980

Symposium Chairman
JOHN C. SAWHILL
Deputy Secretary, United States Department of Energy

Edited by
ROBERT A. BOHM • LILLIAN A. CLINARD • MARY R. ENGLISH
Energy, Environment, and Resources Center
The University of Tennessee

Ballinger Publishing Company • *Cambridge, Massachusetts*
A SUBSIDIARY OF HARPER & ROW, PUBLISHERS, INC.

International Standard Book Number: 0–88410–649–7

Library of Congress Catalog Card Number: 81–12863

Printed in the United States of America

Library of Congress Cataloging in Publication Data
International Energy Symposium (1st : 1980 : Knoxville, Tenn.)
World energy production and productivity.

(International energy symposia series; v. 1)
1. Power resources—Congresses. 2. Energy policy—Congresses. I. Sawhill, John C., 1936–
II. Bohm, Robert A. III. Clinard, Lillian A.
IV. English, Mary R. V. Title. VI. Series.
TJ163.I5.I5635 1980   333.79   81–12863
ISBN 0–88410–649–7   AACR2

The 1982 World's Fair has taken as its theme "Energy Turns the World." It is committed to improving international understanding of the energy issues affecting the entire world. Through the International Energy Symposia Series, the 1982 World's Fair has set a goal of making a significant contribution to resolution of those global energy issues. I am pleased to dedicate these *Proceedings* as marking the first step toward achievement of that goal.

S. H. Roberts, Jr.
President
The 1982 World's Fair

Editors' Preface

The availability of energy affects the attainment or deferment of many human goals: economic development, environmental quality, and, most critically, world peace. In the world's less developed countries, energy shortages and higher prices impede progress toward a higher standard of living. In developed countries, changes in the cost and supply of energy necessitate significant alterations in lifestyles and lessen hopes for the future. Competition for energy resources within and among nations can thus be expected to grow, not abate, in the coming decades.

Because energy is essential to develop and secure a better way of life, it has become a limiting element in the equation of world progress and, as a result, a crucial factor in the continuing quest for world peace and equity. Clearly, a new phase in the politics and economics of energy use has been entered—one in which national interests, sectoral demands, environmental concerns, the need to alleviate human suffering, and the legacy left to future generations all compete for attention and action. These issues cannot be dealt with well in isolation; they require concerted consideration if they are to be resolved.

Purpose and Structure of the International Energy Symposia Series

The International Energy Symposia Series (IESS) is intended to provide such a broad-based forum for addressing these basic concerns about mankind's need for energy. Carried out under the auspices of an international energy exposition entitled the 1982 World's Fair, the IESS commenced with the convening of Symposium I on October 14–17, 1980, as a prelude to the Fair. Symposium I is to be followed by Symposium II held in November, 1981, and by a third Symposium held during May, 1982, when the Fair officially opens.

The IESS is designed to enable the deliberation and debate of alternative energy policies. Its broad theme, "Increasing World Energy Production and Productivity," is global in scope and provides room for analysis of both energy supply and energy demand. Set within the context of the World's Fair, a symbol of worldwide cooperation and sharing of knowledge, the IESS is intended to help rectify world energy problems by bringing many different viewpoints together in order to assess areas of conflict and reach areas of agreement. The IESS thus undertakes an evaluation of energy issues that includes but goes beyond solely technical considerations. It seeks to understand and resolve energy issues in light of the human condition as a whole. In doing so, it hopes to have a three-pronged impact—one which affects the work of scientific and technical personnel on energy issues, the public's understanding of the international energy situation, and policymakers' decisions on both a national and a global basis.

As structured by the Program Committee, the Organizing Committee, and the Symposia staff (see Appendix III), each Symposium in the Series is intended to complement the others, ultimately leading to a comprehensive set of energy proposals. The first Symposium has focused on defining the nature of the energy problem. Building on the results of Symposium I, Symposium II will serve as the analytical component of the Series from which general proposals and specific recommendations will be carried forward to Symposium III. Deliberations at Symposium III will be directed toward arriving at resolutions that will help rectify the world energy crisis.

This sequenced approach is a distinguishing aspect of the IESS, for the Series not only addresses issues but also provides a means of moving toward their resolution. At the end of Symposium III, a communiqué will be issued to the world's nations highlighting areas of agreement and conflict. The relationship of the three Symposia and the anticipated flow of ideas can therefore be shown as:

SYMPOSIUM I   →   SYMPOSIUM II   →
issue identification   issue analysis

SYMPOSIUM III   →   COMMUNIQUÉ
issue resolution

Organization of Symposium I

To fulfill its task of issue identification, Symposium I was organized into four sessions, each with the following individuals involved:*

- a *session chairperson,* to introduce those making presentations and to moderate the session's discussion period;
- two *authors*—leading energy authorities to present papers at the session;
- the *participants*—a group of energy experts from different countries and international organizations invited to the Symposium to discuss the papers and topics of each session;
- a *session integrator*—to summarize the papers and discussion of the session and to suggest issues for further consideration at Symposium II; and
- the *gallery*—the audience registered as attending the Symposium.

The ensuing proceedings result from the interaction of these personae.

As given below, each session had a predetermined topic to guide its presentations and ensuring discussion—topics that were selected by the Organizing Committee in the incubatory phase of the first Symposium. Within these broad guidelines, the sessions were intended to range freely. Each session thus took on its own flavor, distilling and devloping a few key issues. The general thrust of these issues is synopsized here, but for a more extensive review of the sessions, the summary chapters prepared by the session integrators should be consulted.

Session I—World Energy Productivity and Production: The Nature of the Problem

This session sought to establish definitions of the world energy problem within different national settings. Specifically, it was intended to address such questions as: How do individual countries view their respective "energy crises"? In particular, how do perspectives differ among energy-rich and energy-poor nations? Is energy consistently viewed as a pervasive world problem resulting in domestic turmoil and international tension, or is it actually a second-order problem, one subsidiary to more basic problems such as the need for economic development and environmental

---

*See Appendix I for a fuller review of the Symposium I program.

quality? Is it possible to establish a consensus on the various dimensions of world energy issues?

The Session I deliberations were multifaceted, but out of them a few central themes can be identified. For example, Mohammad Sadli, taking a humanistic and developmental approach to the energy problem, stressed in his paper that the oil crisis is having different effects on the domestic policies of different countries, including OPEC countries, and on the world's power balance. He emphasized that global cooperation between producers and consumers—developed and developing, North and South—is needed but "at the moment [this] prospect is a mere illusion."

Wolf Häfele's paper, on the other hand, adopted a more technically analytical approach to issues of energy supply and demand. In particular, he maintained that once labor and energy productivities are sufficiently high there is no reason why the world's energy problem should not be solved, and that nature has endowed us with enough resources provided we use them in a sophisticated manner. He noted, however, that nations must eventually switch from a consumptive to an investive mode of energy management but that conservation must not be stressed at the expense of skill and sophistication in national economies, for the latter must always stay ahead of the foreseeable increase in the cost, scarcity, and dirtiness of fossil fuels.

Summarizing Session I, Shem Arungu-Olende concluded that the overall energy problem is technical and substantive but moreover it is political. Its eventual solution lies in international cooperation with respect to financing, technology transfer, information flows, infrastructures, and so forth. Its resolution demands (1) systematic examination of energy supply and demand in the context of existing policies and (2) development of a coherent set of new policies for the more efficient production, processing, and use of conventional energy sources and for the transition to new energy sources. According to Arungu-Olende, the fact that the solution lies in a *combination* of sources must be kept in mind.

Session II—Improving World Energy Productivity and Production: The Role of Technology

The intent of Session II was to address the technological options open to the world community for dealing with the increasing social cost of energy. The two authors were asked to define the realistic limits of current technology and the prospect for technological change on the demand and supply sides of the energy equation. In addition, impediments to full utilization of current and future technological options were to be considered. Finally, an attempt was to be made to define a balanced long- and short-run program of basic and applied research and technical and commercial demonstration.

Within this rubric, the responses diverged widely. For example, John Deutch's paper emphasized that energy policies and programs will necessarily be different for the near, mid, and long terms but that an energy supply reliant on fossil fuels is unavoidable in the short and medium time frames. To develop the needed technology in this regard, he maintained that support should be given to joint government/private ventures such as the US Synthetic Fuels Corporation.

Amory and Hunter Lovins, however, maintained in their paper that the answer can be found along the "soft path," with conservation and renewable energy resources. They further maintained that the technology to make these approaches readily accessible is already available and already cost effective: the problem lies rather with the artificial price supports now being given to the "hard path"—to energy systems based on fossil fuels or nuclear power. Finally, they predicted that "by the end of the 1980s, energy will be the least of our problems as the linkages between the problems of water, soil fertility, energy, and the sustainability of systems in general converge as an integrated resources crisis of enormously greater scale and intractability." These themes were developed by the Lovinses in the ensuing discussions of both Session II and subsequent sessions.

In his summary of Session II, Hiroo Tominaga noted that in the near term, political/economic factors will be more significant than technological factors, but in the medium and long terms, technology should provide the alternative choices for political and economic actions—a position which appeared to be widely held at the Symposium and was one of its central points of agreement.

Session III—Towards an Efficient Energy Future: Critical Paths, Conflicts, and Constraints

This session focused on the critical choices required as nations strive for higher levels of energy productivity and on the normative context in which these choices must be made. Analysis was to take place at three levels: First, what are the conceptual issues that must be addressed? (Under this heading come questions of the marketplace's role, tradeoffs between equity and efficiency, conflicts with environmental goals, and the relationship of energy supply and demand to economic growth.) Second, what are the specific problems that must be considered? (This might explore, for example, the carbon dioxide question and its relationship to limiting the use of fossil fuels; problems of the proliferation, waste disposal, and safety of radioactive materials and the effects of these problems on the use of nuclear power; and limitations on the use of renewable resources because of cost, geographic distribution, and transportability.) And third, what are the important decisions that must be made? (Such decisions might include whether to emphasize centralized or decentralized supply, whether to concentrate on rapid replacement of the existing capital stock, and whether to share information among nations.)

José Goldemberg's paper, the first of the two at Session III, dealt with a specific instance where many of these issues are raised into question: Brazil's use of sugar cane for ethanol fuel production. From this inquiry, he concluded that technologies such as ethanol production from sugar cane or other biomass products are potentially adverse from an environmental standpoint but could bring jobs and energy self-reliance to developing countries.

B. C. E. Nwosu's paper, also written from the perspective of a developing country but more general in scope, maintained that the energy problem has not only economic but also political and military parameters which must be considered, including the reconciliation of the differing perspectives of the less developed countries and highly developed countries. He concluded that resolution of these differences lies in international cooperation and in the development of a just, participatory, and sustainable society that is global in scale.

In his role as integrator of Session III, David Rose first summarized the sense of the session and then went on to note that due consideration must be given to *all* the relevant issues, not just selective attention to a few, for the latter approach will lead only to partial, inadequate solutions. He also noted that it is important to specify the time horizon being used, in order to clarify debates on these issues. Finally, he stressed that institutional and other non-technical issues should be considered on a holistic basis, not simply an international one, and that the differences—from a global standpoint—between near- and mid-term supply increases and emergent long-term difficulties (e.g., coal use and the gradual build-up of carbon dioxide) must be taken into account.

Session IV—Alternative Policies for Improved Energy Productivity and Production

The purpose of this session was to present a comparative

review of alternative national initiatives dealing with issues of energy production and utilization. Using as its basis a positive analysis of what is currently being done worldwide, the session was intended to address these questions: What choices have been made and are being made by different countries? What options have been foregone or postponed, and why? What factors have most influenced critical energy decisions? On the basis of this review, can the world's nations be categorized according to which energy strategies appear most promising?

In the first paper of Session IV, David Sternlight, using the International Energy Agency's findings as his focal point, concluded that the overall objective of energy policy should be to achieve a supply/demand balance in energy services at market-clearing prices, such that a least cost solution results over time for consumers. To achieve this, there must be a rectification of the existing absence in most countries of major supply policies for renewable energy sources—sources whose utilization is now mainly in a research, development, and demonstration stage by the private sector.

Although multinational in scope, David Sternlight's paper concentrated primarily on the energy policies of industrialized, highly developed nations, most of which now import large quantities of oil. On the other hand, Sessions IV's second paper—that of Amulya Reddy—was written as a case study of India, a country that, while also oil importing, is still in the throes of a development effort aimed at raising its growing population to a higher standard of living. In this case study, an urgent situation was detailed: "The poor undermine the resource base in order to survive, . . . and the rich exhaust the resource base in order to preserve their affluence . . . ." To redress this situation, Reddy concluded that there must be a reduction of inequalities, both within and between nations, and that otherwise the energy crisis cannot be resolved.

As the final discussion period of the Symposium, Session IV's comments ranged broadly, dealing with the subject at hand and also with several others—in particular, the appropriate role of nuclear energy and the pros and cons of electricity as an energy mode. The session's integrator, Vaclav Smil, gave a general summary of the session and of his impressions of the Symposium by concluding that energy must be considered in an ecosystem framework and that "to maintain the richness of human civilizations we should approach our energy problems as complexifying minimalists rather than as simplifying maximalists." This view could be universally espoused at least insofar as it points out the enormous intricacy of the energy problem and its inextricable ties with other environmental and social issues.

Symposium I's Message

Fulfilling its function of issue identification, Symposium I achieved two things: it defined the energy crisis—on both a broad, conceptual level and a more specific, immediate plane—and it suggested areas in which possible solutions should be sought. In doing so, it laid the groundwork for Symposia II and III to follow.

With respect to the first achievement, that of defining the problem, Symposium I found that for the near and mid term the energy crisis is primarily a political and economic problem and only secondarily a technological one: in general, the needed technologies exist but the infrastructures to use them do not. In the long term, exotic technologies might offer alternative solutions, but they cannot be expected to make a substantial contribution for several decades. At the conceptual level, it was also agreed that the energy problem is unavoidably tied to other problems—in particular, problems attending the overuse or misuse of other natural resources and problems arising from disparities in standards of living both within and between countries.

These general points of agreement about the nature of the problem were, during the course of the Symposium, specified into a number of issues that need to be addressed. For example, the following questions were raised and discussed: How should the high demand for fossil fuel and its economic effects on oil-exporting and oil-importing nations be handled? What are the best methods for redressing the lack of technological sophistication in some nations? To what extent can energy conservation help offer a solution; to what extent might it impede that solution?

Although there was dissent on the precise elements of an overall answer to the energy problem, Symposium I did reach an informal but strong consensus about how that answer should be shaped. It was generally agreed that there was no *one* solution, hard path or soft path, centralized or decentralized; the answer instead lies in a *mix* of solutions appropriate to each nation and dependent upon the kind of economy, level of development, and array of energy resources the nation has.

It was, however, acknowledged that there are some issues which transcend national boundaries and which must be considered from a global standpoint if the appropriate mix is to be found. These supranational considerations include (a) energy resources such as nuclear power and biomass whose potential effects are of worldwide concern, (b) environmental conflicts and constraints concerning more than one nation, (c) the correlation between energy factors—both supply and consumption—and equitable economic distribution, and (d) appropriate global or multinational institutional arrangements to address these and other considerations.

Symposium I's message was not conclusive, but it did provide a tangible sense of mission. As shown below, it thus has helped to structure the analysis and resolution phases taking place in Symposia II and III.

Future Directions

Symposium II, to be held November 3-6, 1981, will allow analysis and discussion of the topics identified in Symposium I as of regional or global significance. The Symposium II program will be built around a format of work and plenary sessions. Seven concurrent work sessions are planned, each keyed to a position paper and case study: four of the sessions will be organized around world divisions according to national perspective (industrialized nations with market economies, industrialized nations with nonmarket economies, energy-surplus industrializing nations, and energy-deficit industrializing nations); the other three sessions will focus on issues of global importance (nuclear energy, biomass energy, and energy for rural development). The major areas to be addressed at each work session involve:

- an identification of the nature of the energy problem for the topic at hand,
- a preliminary evaluation of the options available and their technical and economic considerations,
- an analysis of institutional and societal factors affecting the problem,
- a review of pertinent education and information needs, and
- a final evaluation of the options available.

Plenary sessions will entail an opening session, two special topic presentations, an integrative session, and a closing address. As in the prior Symposium, Symposium II will include chairperson/integrators (this time in a combined role), authors presenting papers, a group of international energy experts as invited participants, and a registered gallery.

Based on the iterative process of Symposium II's analysis, discussion, and integration, general proposals and specific recommendations will be sent forward to Symposium III. Symposium III, scheduled for May 23-27, 1982, will have a format which differs from the earlier Symposia in that it will constitute a World Energy Congress with approximately one hundred delegates from fifty nations. Using the analyses and recommendations of Symposium II, this Congress will consider alternative policy proposals by dividing the policies into two groups: those on which there is agreement and those on which there is disagreement. The essence of these deliberations will be transmitted to the world community via communiqué, and a process will be developed for continued international dialogue on the areas of disagreement.

Editorial Considerations for Symposium I Proceedings

Although editorial changes have been held to a minimum in these *Proceedings*, some alterations have been necessary to insure consistency among the papers and the prepared or spontaneous remarks. In all instances, however, the overriding aim has been to retain the original tone and intended meaning. Insofar as this has been achieved, the editors extend thanks to the authors and other participants for their cooperation. Insofar as this aim has not been fully met, the editors request the understanding of those involved, who undoubtedly realize the complexity of the task and the rapidity with which it had to be executed.

One aspect of the *Proceedings* requires special explanation. The selected comments chapters, which record the sense of the formal discussion periods at the four sessions, presented particular editorial problems. Because informal dialogue external to the sessions was encouraged, some of the comments include indirect references to remarks made during the Symposium but outside the context of the discussion periods. Despite their possibly confusing nature, these references have been retained to preserve the integrity of the comments. In addition, it should be noted that participants and gallery members were encouraged to write in comments following the close of each session. Where pertinent to a particular point in the discussion, these written comments have been inserted in their appropriate places for continuity's sake; others of a more general nature occur at the end of the session's comments. In both instances, it should be remembered that because these written remarks were made after the fact, other discussants usually did not have an opportunity to respond to them, as they might otherwise have.

One final note about the selected comments chapters and Symposium I in general. The possibility of organizing each session's comments by topic was considered but not adopted; although this was done occasionally in a rudimentary way, especially in Session IV because of the broad-ranging nature of that session's discussion, the tactic was in general rejected as being both impractical and unfruitful. Instead, it was agreed that Symposium I's focus was to elicit and demonstrate a diversity of viewpoints, not necessarily to reach consolidated conclusions about specific subjects (the latter being the job of Symposia II and III). Thus, in Symposium I the free flow of ideas was of preeminent importance, and thus, in the selected comments chapters eclecticism became the dominant motif. It is hoped that both will serve the Symposia Series' purpose better than a rigidly structured approach would have.

Acknowledgments

Our sponsoring organization, the 1982 World's Fair, is to be commended for assuming the demanding task of hosting the International Energy Symposia Series. Through the Fair's unfailing support, this Series promises to make a much more substantial contribution to the body of knowledge about energy policy and prospects than would otherwise be possible. In addition, the editors would like to acknowledge the important support of the Symposium's cosponsors: the International Energy Agency and its parent organization, the Organisation for Economic Co-operation and Development; the United States Department of Energy; the Tennessee Valley Authority; and The University of Tennessee. The International Energy Symposia Series has been facilitated by a grant from the United States Department of Energy and by the support of its offices and officials. In particular, we are in debt to the continuous assistance of the Department of Energy's Former Deputy Secretary, John Sawhill, and its Chief of Public Presentations, John Bradburne. The International Energy Symposia Series is also cosponsored by a diverse group of foundations, organizations, and corporations concerned with the global energy problems that are associated with high levels of demand coupled with uncertain future supplies.

The *Proceedings* could not have been completed without the enthusiastic and professional efforts of the staffs of the 1982 World's Fair and The University of Tennessee's Energy, Environment, and Resources Center. Our editorial team included Jim Billingsley, who developed the book and cover design and layout specifications; Joyce Rupp Troxler, who prepared graphics; Carolyn Srite, who provided editorial assistance; Pauline Koehler and Karen Stanley, who proofed several drafts of this volume; and Rica Swisher and Peggy Taylor, who prepared manuscripts.

A special thanks is offered to Earl Williams and the staff of the Williams Company, who patiently worked with us to typeset and compose this volume on a very tight schedule. Finally, to those at the Ballinger Company with whom we worked—Carol Franco, Editor; Jerry Gavin, Production Manager; Robert Entwistle, Marketing Director; and Leslie Zheutlin, Marketing—we would like to express our appreciation for their professional support.

RAB, LAC, and MRE
Knoxville, Tennessee
May 1, 1981

Contents

Figures

Tables

Abbreviations

ASHRAE	American Society of Heating, Refrigerating, and Air Conditioning Engineers
bbl	barrel
BEPS	Building Efficiency Performance Standards (US)
BOE/D	barrels of oil equivalent per day
BPKM	billion passenger-kilometers
BTKM	billion tonne-kilometers
Btu/ft²·DD(°F)	British thermal units per square foot, divided by the number of degree-days in degrees Fahrenheit
DOE	Department of Energy (US)
DOF	development-oriented forecast
EEC	European Economic Community
EPA	Environmental Protection Agency (US)
FRG	Federal Republic of Germany
FY	fiscal year
GDP	Gross Domestic Product
GNP	Gross National Product
GWh	gigawatt-hour
ha	hectare

HDC	highly developed country
IAEA	International Atomic Energy Agency
IC	industrialized country
IEA	International Energy Agency (an agency of the Organisation for Economic Co-operation and Development)
IGT	Institute of Gas Technology (US)
IIASA	International Institute for Applied Systems Analysis
IIP	Index of Industrial Production
ILO	International Labour Organisation
IMF	International Monetary Fund
JET	Joint European Torus
JPSS	Just, Participatory, Sustainable Society
kcal	kilocalorie
kg	kilogram
kg/m²	kilograms per square meter
kJ/m²·DD(°C)	kilojoules per square meter, divided by the number of degree-days in degrees Celsius
km	kilometer
km²	square kilometer
km/l	kilometers per liter
kw	kilowatt
kwh	kilowatt-hours
kwh/capita	kilowatt-hours per capita
kwyr/yr	kilowatt-years per year
LDC	less developed country
LIC	less industrialized country
LNG	liquified natural gas
LPG	liquified petroleum gas
MCF	thousand cubic feet
MIT	Massachusetts Institute of Technology
mi/US gal	miles per US gallon

mmbd	million barrels per day
MTCE	million tonnes coal equivalent
MTCR	million tonnes coal replacement
NCE	noncommercial energy
NEDO	New Energy Development Organization (Japan)
NRC	Nuclear Regulatory Commission (US)
NTPC	National Transport Policy Committee (India)
odt	oven-dry tons
OECD	Organisation for Economic Co-operation and Development
OPEC	Organization of Petroleum Exporting Countries
QOL	quality of life
quad	quadrillion (10^{15}) British thermal units
R&D	research and development
RD&D	research, development, and demonstration
REA	Rural Electrification Administration (US)
REC	Rural Energy Center
RLF	Reference Level Forecast
SERI	Solar Energy Research Institute (US)
SFC	Synthetic Fuels Corporation (US)
T	one metric tonne—i.e., 1000 kilograms, or 2,205 pounds avoirdupois
TOE	tons of oil equivalent
TW	terawatt
Twyr/yr	terawatt-years per year
UNCTAD	United Nations Conference on Trade and Development
UNEP	United Nations Environment Programme

UK	United Kingdom
US	United States
USSR	Union of Soviet Socialist Republics
WAES	Workshop on Alternative Energy Strategies
watts/m²	watts per square meter
WCC	World Council of Churches
WGEP	Working Group on Energy Policy (a part of India's Planning Commission)
WOCOL	World Coal Study

The following terms and prefixes have been used to express quantities given in powers of ten:

Quantity	Term	Prefix
10^{15}	one quadrillion	peta
10^{12}	one trillion	tera
10^9	one billion	giga
10^9	one billion	giga
10^6	one million	mega
10^3	one thousand	kilo

Opening Address

John C. Sawhill
Deputy Secretary
United States Department of Energy

It is indeed a great honor to have been asked to chair this first of three Symposia to be held over the next year and a half. The Symposia Series leads up to the 1982 World's Fair, which will be here in Knoxville and will have energy as its theme. I would like to compliment the organizers of this Symposium for the outstanding job they have done in bringing together so many diverse interests from throughout the global community.

The theme of the Series is "Increasing Energy Productivity and Production in the World." That theme was chosen to provide for analysis and proposals directed at both the demand and the supply side of the energy equation. The Symposia Series is a forum for leading energy authorities to (1) evaluate current international energy policies, (2) explore prospects and possibilities for a transition to a high-energy-productive society, and (3) formulate and present to the world community constructive proposals for making that transition.

The activities of Symposium I will focus on the nature and extent of the world energy problem. The efforts here will set the stage for the second and third Symposia in the Series by addressing four fundamental areas:

- first, the nature of the world energy productivity and production problem;
- second, the role of technology in improving world energy productivity and production;
- third, critical paths, conflicts, and constraints towards an efficient energy future; and
- fourth, alternative policies for improved energy productivity and production.

1

Few would argue that energy is among the most important issues that will determine the course of human events. What you accomplish at this Symposium and at the two which follow could have a significant impact on international and domestic policies and on the ability of our countries—collectively and individually—to assure our economic health through energy security.

I am sure all of us have followed closely the events in Iran and Iraq over the last three weeks. We no doubt share great concern that the hostilities between these two countries be brought to an end. Clearly, there is the human factor—the loss of life and suffering being experienced by citizens of those two nations and citizens of other nations who by no choice of their own have been caught up in the conflict. There are the economic hardships that will affect those two countries regardless of the outcome of the war and when it ends.

But a more important message to all of us here is the message of vulnerability. We are reminded forcefully of how delicate is the thread that binds the international energy structure. For those of us in the United States and for most others in the industrialized world, the Iran/Iraq conflict has had negligible impact because of our high oil stocks and relatively small dependence on Iraqi and Iranian oil. Yet while many people focus on the industrialized countries and their economies in times such as these, there is no question but that this interruption will have a greater impact on some of the less developed countries of the world—countries that are proportionally more dependent on oil imports and less well equipped to cope with interruptions.

As the industrialized economies have struggled to deal with the impact of the rapid oil price increases of the last year and a half, the adversity experienced by the developing world is even greater. The developing nations—those most desperately in need of economic growth and development—instead face deepening trade deficits, greater accumulation of debt, and diminishing prospects for helping to meet the needs of their people.

There are no quick solutions to the energy problem and the interrelated economic, social, and environmental issues that must be considered in achieving adequate energy supplies. In the United States, two federal government agencies [the Council on Environmental Quality and the Department of State] recently completed a major study called the *Global 2000 Report* which looked at the world in that year based on a continuation of present policies. The picture it presents in the absence of change is indeed gloomy.

- Energy projections indicate no early relief from global energy supply/demand balancing problems.
- Oil production will lag behind demand growth.
- Political and economic decisions could level oil production before alternative energy sources are developed.
- Even firewood resources are dwindling, and demand will exceed supply by 25 percent. For hundreds of millions of people in developing countries, firewood is the principal energy source.

The *Global 2000 Report* looks broadly at many issues, including population, resources, and the environment, but raises the prospect of a long-term energy supply imbalance, noting that "a world transition away from petroleum dependence must take place but there is still much uncertainty as to how this transition will occur."

It states that many less developed countries will have increasing rather than decreasing difficulties in meeting energy needs. It points out that the world's finite future resources are theoretically sufficient for centuries but are not evenly distributed. I would note that while economic wealth also is not evenly distributed, it has in fact undergone some major redistribution in the last seven years. The oil-exporting nations have achieved substantial wealth as a result of their increased control over the production and pricing of oil.

These factors suggest certain responsibilities that nations must assume if the vision of the future is not to be a mirror image of the projections of the *Global 2000 Report*. My own view of the future is more optimistic, partly because of gatherings like this and the international recognition that change must and will occur. That change will be either by design or by circumstances. I believe all of us here would choose the former.

If that is to be the case, each of us must assume certain responsibilities:

- The industrialized nations must reduce their reliance on imported oil. They must use energy more efficiently and increase domestic production. They must use their technical skills and their financial resources to a greater degree in helping less developed countries exploit their indigenous resources.
- The oil-exporting nations must bring stability to their pricing and supply policies. They must be willing to increase their support to developing countries by recycling more of their surpluses to the developing world.
- The developing countries must participate in the interna-

tional energy and economic dialogue. They must build and strengthen their technical and financial infrastructures dedicated to meeting future energy needs.

The interrelated nature of these efforts is illustrated by a recent quote from Ghana's Finance Minister Amon Nikolai. He urged the United States "to get the economy moving—it makes life that much easier for all of us." Certainly, we know the importance of a strong economy, but we also know that, as a nation, our progress depends to a large extent on the actions of others. It is a lesson that was learned well in the 1970s.

At the beginning of that decade, energy in many forms was relatively cheap, relatively abundant, and reliable. But that era has ended. In the thirteen years leading up to the 1973 Arab oil embargo, world energy consumption grew at an annual rate of about 5.5 percent, while oil consumption rose about 7.5 percent annually. Those growth rates dropped dramatically after that. We project that in the 1980s world energy consumption will grow 2 to 4 percent and oil consumption will grow less than 1 percent, a reversal of a trend that saw oil consumption growing faster than total energy consumption.

As the world's growth rate in energy consumption declines, we are likely to see major shifts in the distribution of energy consumption between the developed and developing countries. In recent years industrialized countries accounted for nearly 85 percent of world energy consumption, but by 1990 that proportion is expected to decline to about 76 percent, as energy use in the developing countries grows. That will occur as the developing countries return to more traditional growth rates. The World Bank projects that total energy use in the developing countries will grow at an annual rate of about 6.2 percent, well above the 1975–80 rate of 3.7 percent although still somewhat below the 6.9 percent rate of the previous twenty-five years.

United States Department of Energy projections suggest that the oil-importing developing countries will increase their oil needs from about 5.3 million barrels per day in 1978 to about 7.2 million barrels daily in 1990. And while there are many uncertainties in any forecasts, total world oil production will not increase significantly, and it could even decline. That raises a fundamental question: How will developing countries meet their projected needs as they compete with other countries for valuable oil? One answer is through the efforts of industrialized countries to meet their own future needs by improving energy efficiency and increasing domestic energy production while helping other countries to do the same.

The United States has made significant progress in the last three years in developing and implementing a comprehensive energy program. Today it is projected that US oil consumption in 1990 will be in the range of 15.7 million barrels per day, a decline of over 14 percent from 1978. The United States has established an oil import target of 4.5 million barrels per day by 1990, which would be almost one-half the import level at the beginning of 1978. It is showing good progress toward meeting that goal. United States oil imports have declined 1.5 million barrels per day from a year ago and over 2 million barrels per day from the early 1978 levels.

Those numbers are important for two reasons. First, they show that the United States has in fact made a commitment to itself and to other countries to reduce its impact on the world oil market. And second, it means that that much more oil is available to other countries, including developing nations.

There are a number of reasons for this progress and for my confidence that the United States will be a major positive force in future world energy developments. Let me cite a few.

The first and perhaps the most important reason is that the United States has begun to price its domestic resources at world energy levels. By October of next year, US domestic oil prices will be totally decontrolled.

Decontrol of domestic oil prices is sending the right economic signals to US consumers, thereby leading to increased energy efficiency, and to US producers, thereby resulting in increased exploration for domestic oil and gas reserves while making alternatives to those fuels more competitive. The number of US rotary rigs in operation for the week ending September 29, 1980, reached 3,138, which is an all-time high, and domestic production of crude oil for the year to date [October 14, 1980] is 2 percent higher than a year ago, and in the latest four-week period, 2.6 percent higher.

Coal production in the United States in the first six months of 1980 was 10 percent above production a year ago. And coal provided over 50 percent of the nation's electricity in the first six months, which has not happened on an annual basis since 1968.

To a large extent these trends are the result of new economic realities, but the United States is doing much more to assure that these trends continue. It is taking steps to move its utility industry more aggressively from oil to coal. It is spending over 1 billion dollars this year for research and development into new technologies that will allow cleaner and more efficient use of coal. It is expanding the leasing of federal lands, and it is implementing a massive synthetic fuels program.

The United States has made important strides in bringing natural gas from Alaska to the lower forty-eight states. It has aggressive programs in the residential, commercial, industrial, and transportation sectors to improve energy efficiency.

Internationally, the United States has participated with other industrialized countries in making commitments to greater cooperation that will reduce its collective use of imported oil while strengthening its ability to respond to oil supply interruptions.

The United States recognizes its responsibility to use its vast domestic resources to help others in their desire to become less oil-import dependent. It knows that its coal can be an important asset to other countries, so it is looking actively at steps to improve its transportation and port facilities and to assure that it becomes a reliable source of coal to meet a growing world demand. I might note that private investment in US coal export facilities has increased significantly in the past six months.

The United States knows that its technologies can be beneficial to other nations, and it has taken steps to make those technologies widely available. But it also recognizes that for developing countries to achieve their goals will require significant expenditures. The fact that the cost of oil imports to developing countries has risen nearly tenfold in the last decade and is likely to double again to about $100 billion by 1990 represents an economic drain that is far greater proportionately than it is in the industrialized world. That fact presents a major challenge to all of us in the coming decade.

The World Bank has proposed the establishment of a separate lending facility affiliated with the Bank which would nearly double the $13 billion in loans now targeted over the next five years to help developing countries meet additional energy needs on their own. Bank Vice-President Ernest Stern recently noted that the Bank "wouldn't be proposing such a large expansion if we didn't consider there were sound investment opportunities in the less developed countries."

The source of funding for such proposals presents an interesting challenge. Mr. Stern noted that without OPEC money the affiliate would not be created. Clearly this suggests that the industrialized oil-importing countries, as well as the oil-exporting countries, must join together if the people of the less developed countries are to achieve their aspirations.

Achieving global increases in energy production and greater energy efficiency entails international cooperation on a scale that has not yet been realized. There are complex issues with which we are all familiar that separate nations. It is a difficult challenge.

I firmly believe that this gathering can make a major contribution toward meeting that challenge. Therefore, my charge to you in this first Symposium is to clearly identify and define major energy issues in the context of a global community in which we seek the betterment of all our citizens; focus national, regional, and global attention on these issues; and force debate in a cooperative and concerted manner. I believe that if you accomplish those objectives, you will have made an important contribution to the success of the next two Symposia and to the achievement of an energy future far different from that presented by the *Global 2000 Report*.

I am reminded of the American patriot Patrick Henry, who said, "I like dreams of the future better than the history of the past." By your participation in this Symposium, you have committed yourself to be a force in turning dreams into realities. I wish you well and look forward with enthusiasm to the results of your work.

World Energy Productivity and Production:

The Nature of the Problem

Introduction

McGeorge Bundy
Professor of History
New York University

Because time seemed short and because I thought it best not to delay the prompt presentation of the two main papers for Session I, I offered no substantive opening comment. Let me here instead make a few summary remarks on what I learned in that session.

The two papers are unusual in their force and significance, and still more unusual in their mutually reinforcing complementarity. I do not question Wolf Häfele's broad quantitative indications of the size of the challenge we face in energy production, if we are not to accept bitter disappointments to the real aspirations of men and women in all kinds of societies and all parts of the world. Nor do I doubt his general judgment that the problem is ultimately one of human attitude and consciousness. I also think Professor Häfele is right in placing great emphasis on productivity—especially if I may interpret that emphasis very broadly to include kinds of productivity that may be covered in the theory of general information to which he looks forward but that are not yet well recognized in economic theory or in prices. I even wonder whether somewhere in this kind of productivity there may be an instrument for a partial reconciliation of the abiding differences between the scenarios of Häfele and the hopes of those in agreement with the Lovinses. Finally, I respect, and regard as absolutely required for both understanding and action, Professor Häfele's insistence on examining these matters in worldwide terms. One looks forward to the day when it will be practicable to include China in this world.

I find a similar range and sensitivity in Professor Sadli's view from the developing world. Without surrendering his particular concern for the villages of Indonesia, he has been able to offer searching and sympathetic assessments of the positions of more than one kind of developing society, just as he has recognized reciprocal obligations of the developing and developed countries in any fruitful joint process of energy development. I find it more than a coincidence that in a symposium held in the heart of the Tennessee Valley, with its Tennessee Valley Authority, both the general theory of Professor Häfele and the specific concerns of Professor Sadli unite in commending the social value of electrification.

I find myself in equal sympathy with our integrator. Dr. Arungu-Olende correctly tells us that precisely because all this is hard and demanding, we must be about it now—and that the requirement is not merely for an approach where each society tends to its own future, but rather for a wider and more humane approach which still respects, because it must, the fact that we live in a world of separately governed nation-states. The difficulty is almost insuperable, I think, but not quite, because even where interests sharply conflict, as they do not only across borders but within them, I see no fatal necessity that the great games of energy must always be zero-sum. I prefer, both intellectually and morally, the aphorism made famous by John F. Kennedy, that a "rising tide lifts all the boats." We cannot, in this case, wait for the tide, but if meetings like this one can strengthen our determination to work to meet the common human need for energy by a common commitment—and to seek to reconcile our differences by understanding and respecting one another—if, in short, we can proceed in the temper of this meeting, then it will not be so bad that we have added one more to the number of energy conferences.

CHAPTER ONE

World Energy Productivity: The Nature of the Problem for Developing Countries

Mohammad Sadli
Professor of Economics
The University of Indonesia

All countries are affected by the energy crisis, whether they are big or small, rich or poor, oil exporters or oil importers. Oil is a vital input for modern and modernizing economies. It is a much-traded commodity internationally, with only a few large exporters, and hence its market structure is by nature not competitive. Europe and Japan, as well as most developing countries, are too dependent upon imported oil. When the international price increases many-fold in a short time, predictably whole economies become upset—not only national economies, but also the world trading and monetary system.

Developing countries feel themselves to be the hapless victims in a big market struggle between two giants, the industrialized countries and the oil-exporting countries. Although they are the victims of OPEC price decisions, they do not blame OPEC alone. They realize that the market power of OPEC also derives from the great and growing consumption needs of the West. These rich countries are also to blame if they are not able to effectively restrain their appetites for oil. The developing countries are several times the losers: in the end they have to make do with less aid from the industrialized countries; their plight is not fully assumed by the OPEC countries; and their exports become victims of the narrowing markets in the rich countries. The developing countries remain price takers in the big contest, with no bargaining power, while their pleas for clemency remain unheard. They are even penalized if there is a failure of effective conservation policies in major consuming countries like the United States.

Because of its important geopolitical aspects, the oil crisis is also changing the power balance in the world. In many developing countries it also has an effect on domestic policies. This is partly because of the fragility of many such political structures and the great and magnified impact the changing oil price has had on the fortunes of governments and ruling elites.

OPEC countries have their share of problems also. Problems of "too much, too soon" can be equally unsettling for elitist regimes, as events in the Middle East demonstrate. An oil bonanza of some duration can distort the domestic price mechanism with consequent misallocation of resources. This bonanza intensifies the duality in the economy.

LEAST DEVELOPED COUNTRIES

The adjustment problems to the energy crisis are different for the poor oil-importing developing countries, for the middle-income ones, and for the oil-producing and -exporting countries.

For the least developed countries the problems are the most difficult, both in the short run and in the long run. The productivity of such countries is low, especially in the large traditional sector. Foreign trade sustained by a limited modern sector is small except when there are large mineral developments. The mounting cost of oil imports has taken a significant portion of these export receipts—for some countries today even an overwhelming portion.

The directly productive units in the modern sector may be able to afford this expensive commodity, but transport costs and utility rates will go up, causing a lot of overall hardship. In such countries there is an expanding gray area between the modern sector and the traditional sector. It is sometimes called the informal sector. It consists of small entrepreneurs, working with limited capital implements on a shoestring budget. Examples of such activities requiring petrol are those of the small fisherman with his motorized boat, the owner/driver of a small truck or taxi, the village repair or handicraft workshop, and so forth. They are getting pinched between increasing working capital requirements and the stagnant purchasing power of their customers.

The era of cheap energy and expanding world trade has conveyed many benefits to these still poor but developing societies. Among other things, it has greatly enhanced people's mobility and productivity. With progressing urbanization and industrialization, workers begin to commute by bus or truck. Yet their daily wage is still low, often the equivalent of a dollar or two. It would mean great hardship to spend an important part of such a wage paying for increased fares. Hence such energy systems often require subsidization in the present transition period, which in turn puts additional burdens on governments and industries.

The availability of cheap kerosene was also a boon to these poor developing countries. It became the main source for lighting in rural areas and as a substitute for firewood. But even in the days before the oil crisis, the very poor households living below the poverty line had to use kerosene sparingly. Some governments fear that village people will now resort more to cutting forest trees for their cooking needs, thereby endangering the ecology of the environment.

At the macro, or balance of payments, level, the steep oil price increases are claiming a higher share of national export receipts. If the oil bill goes beyond one-third of such receipts, other essential requirements have to be sacrificed. Many developing countries still have to import food and clothing. Fertilizers are another major component of their import bills. And for their economic development they have to import capital goods, spare parts, and implements. Repayments of debts and interests are also noncompressible obligations for many of these countries. In the short run, faced with sharply increased oil bills, these countries have to curtail part of their consumption or their economic growth, or both. As consumption levels are often already at rock bottom, investments and other productive inputs usually are sacrificed. For instance, the country will import less capital equipment and fewer spare parts; perhaps less fertilizer also. By doing so, it soon gets trapped in a vicious circle. As its growth prospects diminish, its borrowing ability will also go down because lenders will doubt its repayment capability. The prospect for such countries of living on grants in perpetuity is not an attractive one, and not even a practical one.

These countries therefore must be internationally assisted for the period that is required to make structural adjustments and redesign their growth prospects. Because the short- and medium-term prospect for these countries to make a contribution to the international system is marginal, the motives for assisting them should first of all be humanitarian—call it "solidarity of mankind." International assistance and cooperation for mutual benefit is still not relevant for this category of countries. It may come later. On the other hand, the amount of balance of payment aid required for these countries is not great. The small low-income countries have a total oil import of less than 1 million barrels per day. Including the large countries, the total is still less than 2 million barrels per day. Some of the bigger countries around the Middle

East, because of their geopolitical importance, are to some extent already assisted by rich oil-producing neighbors on a bilateral basis. South Asian countries also have been able to profit from the manpower requirements of the oil-exporting Middle East countries by exporting labor and receiving significant remittances from these migrant workers. In a much greater plight seem to be the least developed countries of Africa, but also some small countries like the Maldives (whose fishermen are hard hit by the oil crisis) should not be overlooked.

How should this category of countries be assisted? In the short and medium term, their oil import requirements cannot be compressed much without greatly affecting the lives of their people. Most of their oil comes from OPEC countries. These OPEC countries are willing to give them a supply guarantee at official prices, so that they need not buy crude at spot market prices. This, however, may not be enough. These poor countries may not be able to pay the price. It should be possible to extend to a certain number of countries grants-in-aid by the international community, including OPEC countries. One problem will be where to draw the line. India has recently pleaded that, on the basis of its per capita income, it is entitled to receive such aid, if any will be given. Its present oil bill amounts to three-quarters of its total export proceeds, which is indeed killing.

Apart from helping the short-run problem of how to acquire the needed supplies of fuels, these low-income developing countries have to be assisted to improve their productivity and to make their economies more self-reliant. In other words, adequate development aid should be given, and the problem is how these necessarily limited assistances can be stretched to overcome the handicaps of having to live and to grow in a world where the cost of energy has risen enormously. These countries in the end have to be able to pay for themselves, and that means they have to be able to trade and to earn. For these countries, the first exportables are primary commodities rather than manufactured goods. Hence development means the need to increase agricultural productivity, for food crops and exportable commodities. In all this, how will the high energy prices affect their viability?

In many ways these countries will be at the short end of the deal. The high energy prices will cause higher prices for capital goods, building materials, fertilizers, and so forth. Hence the terms of trade for the factors of production they possess in abundance— that is, labor and perhaps land—will deteriorate. That is why it is so important for the international system to guarantee them fair and stable prices for their tradeable commodities. In the trading system today, they do not have the bargaining strength and the market power. For this reason the implementation of the United Nations Conference on Trade and Development's (UNCTAD's) Integrated Program for Commodities is so important for these countries.

There is a general and often-quoted admonition that developing countries in their further growth should adopt less energy-intensive technologies, development patterns, and styles of living, but what these exactly are in practice is still not clear. These developing countries also have a pronounced duality in their economies and societies. The problem of saving energy is much different in the traditional sector than in the modern sector. What are these countries' prospects for expanding their domestic energy production? Much could be done, and a little more energy production will go a long way because at their stage of development the needs are not great yet. But again, a combination of domestic efforts and policies and international assistance is required. Many such developing countries still have oil and gas prospects, but they receive only a very small fraction of the international spending for exploration and development. Most of the latter is spent in the developed countries, because these are the home countries of the international oil companies. And a second priority for these international companies is the exploration and development of oil and gas fields in the developing countries already well known for their hydrocarbon potential. These companies typically are after big deposits of an exportable nature. To search for small fields for local consumption in unknown territories entails too great a risk for them and is simply not attractive. On the other hand, the developing countries themselves have hardly the resources and capabilities to mount their own explorations.

Currently there are a few schemes to provide developing countries with international assistance money for survey and exploration. In OPEC circles the same idea is also being contemplated. Although a worthwhile objective, there are practical problems. The cost of (seismic) surveys is much less than for exploration and wildcatting. A few million dollars can mount a survey campaign. Hence a fund of a few hundred million dollars may go a long way in helping a number of developing countries. The assistance should be provided on a grant basis, or on an insurance basis, meaning that no repayment is necessary as long as no oil is found and lifted.

Exploration cost is much greater and entails a great risk. Tens of millions of dollars are required to explore one prospect, and these costs vary greatly depending on the terrain and geology. Hence an exploration fund must be big in order to assist a sufficient number of countries. Some OPEC officials, however, are not shy of thinking in terms of a few hundred million dollars or even a billion dollars or more. If such a fund could be conceived as self-sustaining, then perhaps the money could be raised, not only from OPEC sources but also from the industrialized countries. Part of the cost and risk could also be borne by the international oil companies. Probably all this is feasible. At least the idea merits further investigation. The receiving developing countries also should do their share. They have to be prepared to accept the international oil companies on a direct-investment basis. A production-sharing contract as practiced in Indonesia and other countries seems to be practical. The profit-sharing and incentive system should be liberal, in accordance with the likely prospect of the country. A profit split of between 75:25 and 60:40 with liberal cost recovery provisions will do. At higher levels of production, the profit split can be more advantageous for the host country. There is such a big margin between the international price of oil and the actual cost of lifting that a host government can fashion the profit split in such a way that it will attract the major international oil companies, or at least some independent companies. The role of these companies is very important because the best know-how to find oil is still with them. A host country can try to get cheaper help from other developing countries with experience in oil exploration, or from OPEC countries, but the capabilities of the established oil companies or the countries with long experience like the United States is no match for newcomers. Currently some big developing countries with indeed practical experience want to go it alone, at most buying the know-how from outside, but the fact remains that these countries have not been able to achieve self-sufficiency in their requirements. If such countries are still suffering from the financial pinch of the oil crisis, one wonders to what extent this predicament is not self-inflicted.

RURAL ENERGY FLOWS

In discussing energy problems of developing countries one cannot leave out the particular systems of many rural areas, especially in the lower income countries. Poor rural households cannot afford much kerosene. For cooking, and for warming in colder seasons or at higher altitudes, rural people use firewood, agricultural waste, or dung. These energy sources are often called noncommercial, because they largely are not acquired through the market but are collected near the farm or house, by members of the family, without incurring monetary expense.

There are several aspects of these phenomena of nonmonetary economics, as related to energy application. First, one cannot speak of demand, because in economics this means effective demand—that is, demand backed by purchasing power. It is better to speak of consumption or use. The amount of consumption of firewood by households is not related to income or prices but more to local availability. The poor people in the cities or district towns will not burn firewood because they will have to buy this wood and it may cost more than kerosene. Charcoal or coal briquettes for these more urban uses also may be cheaper than firewood because of lower transportation costs.

Local availability for these rural households means, first of all, availability in their house gardens or communal grounds, where there is often a lot with trees and bamboo. Since the need for one cooking a day is generally not much, the supply of fuelwood in the immediate surroundings is usually enough, although this is probably more true for lush tropical countries like Indonesia. Even here we have overcrowded Java where house garden plots become very small. The degree of monetization of village life in Java has also progressed more than outside Java and Bali. Where there are monetary incomes and cheap availability of kerosene, the energy sources of these rural households are also mixed.

From household studies in rural Java it is not clearly established that a growing population's need for fuelwood must encroach upon forest lands. The latter phenomenon is more attributable to the progressing hunger for land to grow food. But the ecological risk is there, and it depends on the availability of affordable substitutes, like kerosene in Indonesia and charcoal or coal briquettes in other countries. These ecological hazards are also increased by the expansion of a market demand for firewood by village and other traditional industries, such as brick and tile making, limestone burning, and a variety of food-processing activities.

The extent of these traditional energy uses in poor developing countries is often estimated as 50 percent or more of total energy consumption—that is, the amount of commercial energy consumed is less than half of the total amount consumed. This aspect has its own important consequence, for if income rises, especially among the rural masses, these traditional energy sources will be replaced with more commercial ones. This accounts for the high growth rate of commercial energy consumption. Poor countries consume less than a barrel of oil per capita per year. At affordable prices their oil consumption goes up at a greater rate than the growth of their national economies. Another substitution takes place in the transportation and agricultural sector, where road transport and tilling of land are increasingly mechanized. One wonders whether it is possible for these developing countries to expect income elasticities of demand for oil to be under unity.

Because the use of these traditional sources of energy is rooted in poverty, the underutilized working time of family members, and a lack of cash income, the need for conservation and economical use of such traditional sources of energy is more difficult to achieve. Any improvement of cooking tools and utensils that economizes on the use of fuelwood or even kerosene comes up against the cash income shortage. In other words, the people cannot afford to pay for the improved tools. An expense having the character of a capital investment is often beyond their means. When steel wheels and rubber tires were introduced to make bullock carts more efficient, this improvement often was frustrated by the same limitation. This capital shortage limitation should be considered when governments are looking for rural energy solutions in the direction of biomass, biogas, or solar heat applications that have intermediate technologies but also capital input requirements.

Over time, rural electrification may provide a partial solution to the problem of rural energy needs. Electricity is too expensive for cooking and heating purposes, but for lighting it will be cheaper than burning kerosene. If power and productive uses can be found or developed that will balance load requirements between day and night, then electricity can be generated and distributed with economies of scale, and the primary source need not be oil.

If the income level is still low, there is a dilemma, because the capital cost of the system is great while the average electricity use per household is still very small. Economies of scale cannot be applied, and the load factor will be low. But the alternative is probably dispensing a lot of kerosene at great opportunity cost, or otherwise the countryside will be pitch-dark at night.

That is why rural electrification, although in most cases probably still a luxury requiring subsidies, is promoted by governments in many developing countries. If there is a clear productive use requiring great amounts of energy, like the pumping of water for irrigation in India, then rural electrification can justify itself.

There is also the prospect that electricity can promote productivity in rural areas through cottage industries, handicraft, repair-shops, and so forth. Rural electrification in developing countries will be a very appropriate energy development objective, at some stage of per capita income, when the system can pay for itself. The question is whether or not low-income developing countries can afford to build it ahead of time, with subsidies in the initial period, assuming the social benefits are greater than the monetary costs and expecting external economies to come out of it. India, with a lower per capita income than Indonesia, has now about half of its villages provided with electricity. On the other hand, in Indonesia not even 10 percent of the villages have electricity yet. In the Philippines rural electrification is a big development project, and recently the connection of the one millionth household was celebrated. It is also a popular object of international aid, because it can improve the life of the people at the grassroots level. It is antipoverty. It can be connected to electricity-generating systems based on coal, hydropower, geothermal, and other nonoil sources. However, because of the requirement of minimum economies of scale, it may not lend itself well to sparsely populated low-income countries.

Rural electrification is only a partial answer to the growing energy needs of the rural area. Where petroleum products are expensive, renewable energy sources are a preferred alternative. Biomass, biogas, solar heat, wind power, and so forth are possibilities to be investigated. For countries that still have a favorable land-to-population ratio, biomass applications hold promise. There are experimentations to plant fast-growing trees, in village compounds and for reforestation. Alcohol production from sugar cane, sweet potatoes, or cassava is being researched for its economic viability. The possibilities of making economic use of tropical wildgrass, mangrove, water hyacinth, and so forth are intriguing propositions. At the moment, however, none of these has become a hard reality, and the problem of appropriately expanding and diversifying the energy basis of the rural area is largely still unresolved.

MIDDLE-INCOME COUNTRIES

Middle-income developing countries, such as Brazil and Korea at the upper end and the Philippines and Thailand at the lower end, have more or less the same problems as their poorer cousins, but because of their higher productivity they have more technical flexibility and resilience and greater hope for survival and growth. But they also need a lot of international assistance.

For these countries, the oil crisis is most of all a balance of payments problem. In a country like the Philippines, for instance, fuel imports constituted about 6 percent of total foreign exchange receipts before the oil crisis, but now they are over 20 percent, and this figure has been climbing since 1974. Calculated against merchandise exports (the new exports to pay for growth), the figure is about 40 percent. Countries of medium size and (lower) middle-income status, like the Philippines or Thailand, consume approximately one-quarter million barrels per day, or about 90 million barrels per year. If the price doubles within a short time, it means that several billion dollars more have to be financed, or oil imports have to be reduced with a consequent lowering of economic growth. This decline of growth is inevitable at periods of sudden oil price jumps, such as have occurred twice in the seventies now.

The reduction of oil consumption is also difficult for these middle-income countries, especially the lower middle class ones. They are in the early phases of industrialization when energy, specifically oil, is used to a large extent for productive purposes. Industry, mining, transportation, and some agricultural uses comprise over 80 percent of total use. The national income elasticity of oil consumption is normally greater than one. The switch to forms of energy other than oil is not a practical option in the short and medium run. Many of these countries' modern sector production capabilities have been developed since the sixties on the basis of low-cost oil, and other potential energy resources such as hydro-power and coal have remained relatively undeveloped.

The industrialized countries, one can say, have similar problems of financing the very much increased oil bill. In absolute terms, their requirements are much greater. So what is so special about these middle-income developing countries? There are two major differences. The borrowing ability of the industrialized countries is greater. They have better access to the money and capital market, because the financial intermediaries are the property of and are controlled by the industrialized nations themselves, and the recycling of the Middle East surpluses goes through these institutions. Second, because the middle-income countries cannot easily repress the growth of their oil imports, their situation can become aggravated in the short and medium term. The industrialized countries have been more successful in their conservation efforts; so much so that their oil imports are declining, or at least the income elasticity of demand is reduced to less than unity—for instance, 0.6 or 0.7. On the other hand, these middle-income countries still enjoy economic growth rates that are about twice those of advanced countries.

The borrowing capability of middle-income countries is currently constrained by their already high levels of indebtedness compared with their exports and other receipts. These dynamic middle-income countries have been great borrowers in the past to finance their respectable growth rates, and now they face additional borrowing requirements to finance their oil bills. Fortunately, during much of the seventies these countries have enjoyed strong balance of payments positions, thanks to favorable trade terms for their major export commodities and expanding merchandise exports. These prospects look much more dim now, with protection rampant in the industrialized countries and a possible prolonged recession or low-growth period shadowing the horizon.

Meanwhile the adjustment process to higher-priced fuels is a painful domestic process. It can erode the already somewhat shaky foundations of domestic political regimes if cost-push inflation jolts the living standards of the fixed-income earners and the urban poor. In South Korea and Thailand these cost-push inflationary pressures have already upset governments, and in other countries political stability has been impaired. Political instability induced by the oil crisis has a lot to do with the suddenness and the magnitudes of changes that are required, causing sudden bursts of inflation and deprivation for the common people.

In the end, more balance of payments loans to pay for the increased oil bill are not the answer. Such will, by themselves, not restructure the economy. What is needed is financial, technical, and trade cooperation from the industrialized countries. These middle-income countries have an entrenched interest in world trade. In the end it is again world trade, in a restructured and expanded form, that must provide the situation. The middle-income countries, apart from still being commodity exporters, particularly the lower middle-income group, increasingly are becoming exporters of manufactured goods. The growth of these exports depends on free access to the markets of the developed countries, based on a reformed division of labor and comparative advantage. If the low-income countries have a major stake in the UNCTAD's Integrated Program for Commodities, the middle-income countries have a similar stake in the implementation of the Kennedy and Tokyo Rounds of trade liberalization. Although the recent outcome of the Tokyo Round is not ideally suited to the interests of the developing countries, it is at least a good start to arrest the drift towards protectionism.

In the meantime these middle-income countries need financial assistance to help them in restructuring their economies. Such credits are medium-term to longer term for programs and for specific projects. Fortunately, the World Bank is already opening the possibility for such assistance. The Philippines, for instance, has recently received a $200 million loan, for fifteen years with a five-year grace period, at 9.25 percent annual interest. This loan is to be used in three areas; namely, energy development, industrial restructuring (e.g., aiding the conversion of the cement industry from imported fuel to locally available coal and the modernization of textile manufacturing plants), and export financing, including the establishment of a dozen new export-processing zones. The country is also required to overhaul its domestic industry protection system, that is, lower tariff barriers.

Country ceilings of private commercial banks should also be overcome if the international community has confidence in the outcome of these restructuring efforts. These ceilings are mainly reflections of institutional limitations of the private banks. The liquidity is there and the long-term prospects of these middle-income countries remain good, especially if restructuring policies could be implemented under the auspices of the World Bank and the International Monetary Fund (IMF).

OPEC COUNTRIES

OPEC countries call themselves developing countries, and most of them are members of the Group-of-77 in UNCTAD. But they are far from homogeneous. Among them are low-, middle-, and high-income countries, but a high per capita income is not sustained by high manpower productivity in agriculture and industries. A relevant and important distinction is between low absorbers and high absorbers of the oil revenues. For example, Indonesia is a high absorber and some of the Middle East countries are low absorbers. These high and low absorbers have different problems. The high absorbers have an opportunity to speed up their domestic development, and typically their oil revenues will not create trade surpluses. Countries like Indonesia continue to borrow internationally.

The high absorbers will probably produce as much as they can. Although they know that oil in the ground may be worth more, in real terms, later than now, they like to work from the assumption that oil revenues, properly invested, have a higher rate of return, can accelerate the transformation of their economy, and can provide much needed employment.

Yet, even these high absorbers cannot escape the pitfalls of an economy dominated by oil revenues. Domestic inflation is persistently double digit, as in the other OPEC countries. Although economic growth rates are respectable, better than 7 percent per annum, this high inflation hits the working class, the small or landless farmers, and the small enterpreneurs. Hence unequal income distribution tends to be aggravated, adding fuel to social unrest in a somewhat longer time span. A strong balance of payments position coupled with high domestic inflation works against the interest of the upcoming domestic industries, which feel the pinch of foreign competition. These OPEC countries are becoming affected by what in IMF circles is called the Dutch disease. Indonesia tried to devalue its currency in an effort to protect and stimulate its nonoil economy, but devaluation creates additional inflation, and hence the plight of the fixed-income earners and small farmers is aggravated. The oil economy accentuates the economy's duality, and one of the question marks is whether or not exchange rates and other economic parameters should reflect and serve this duality. The finding of a single equilibrium rate is more difficult, because which sector of the economy should such a rate serve—the oil and urban sector or the agriculture and manufacturing sector?

High absorbers may not have an explicit oil production conservation objective, but the fiscal and incentive system may produce such an effect. The incentives for exploration and production will be greater if the oil companies can count on a greater profit margin or a greater cash flow. When Indonesia in 1976 increased its government take from 65 percent to 85 percent of profits, the oil companies' reaction was to curtail exploration expenditures. Now, with higher crude prices and an unchanged tax system, the exploration budgets are restored. Profit and cash flow margins also decide to what extent small deposits will be developed and secondary recovery initiated.

A nagging problem for the high absorbers is the determination of domestic fuel prices. Because the domestic cost of oil is relatively low—that is, free of taxes or royalties—governments tend to set domestic prices below international prices, often well below. The motivation is domestic politics. Raising fuel prices every year by significant margins is a very unpopular exercise. If non-OPEC developing countries have no alternative, OPEC countries can afford to pay the subsidies. But such policies will work against objectives of oil use conservation and energy resource diversification. With low fuel prices, the growth in demand can easily go beyond 10 percent per year, and the specter is raised of declining exports. Large subsidies also undermine the growth of public savings and thus the long-term capability of the government to accelerate development by public investments.

Developing countries, whether they are oil-producing and -exporting or not, in the long run cannot afford to underprice oil products in their domestic markets. Such will only weaken their development system, and the history of subsidies provides ample proof that subsidies are being enjoyed not by the target groups only. The problem is, what political pressures could induce governments to take the correct policies? Can extensive and continuing public education help this process?

For the low absorbers, the conservation principle becomes more compelling with increasing oil prices, so much so that at the moment there is an expectation that the large OPEC countries will demonstrate a "backward-bending supply curve." Such behavior will increase the uncertainties for the West. Although in the abstract this phenomenon is believable, there are many influences working on the decisionmaking process in these OPEC countries. The determination of where their national interest lies is also not always a simple and unequivocal matter. Why is it that Saudi Arabia so far has not espoused this principle? The popular explanation is that the political decisionmakers identify their national interest with that of the West, but to what extent is this perception of the national interest stable? On the other hand, why did the other Arab countries increase their production when exports from Iran were down because of the effects of the revolution there?

CONCLUSION

The prospect for greater stability or predictability in the international oil market is currently still very dim. In the next five years, while supply and demand are still precariously balanced, this prospect may not improve. No structural excess supply of significance can be foreseen with enough certainty. On the supply side, the OPEC phenomenon of the backward-bending supply curve can make itself felt, while no new significant increases in oil production in non-OPEC countries can be expected. On the demand side, the conservation and diversification efforts of the industrialized countries, although promising, cannot guarantee results in time. The prospects for economic growth, or lack of growth, and recessions for the industrialized economies still continue.

Can international cooperation and agreements bring relief to these troubled times? Actually, two alternative courses present themselves for the world: further conflicts and uncertainties, leading possibly towards deepening of the crisis; or the willingness at some stage to negotiate, to agree, and to manage the global system cooperatively between producers and consumers, between developed and developing countries, between North and South.

At the moment the latter prospect is a mere illusion. The industrialized countries in their recent Summit of the Seven in Venice have only indicated a hesitant interest towards a North-South encounter at the highest level, as proposed by the Willy Brandt Commission. The West, still fighting a stubborn inflation and recession problem with the further prospect of only low growth, is not in a mood to grant major concessions concerning aid and reforms of the international system. On the other side, the unity among OPEC members and among the Group-of-77 is far from good. They cannot speak with one language and cannot present a common practical proposal for negotiation. OPEC wants the price of oil assured against the risk of inflation and currency fluctuations, so that it can gradually increase in real terms until it finds a long-term equilibrium position related to other energy sources. But an international agreement on the movement of oil prices is not enough without agreements on available supplies, and this will pose another set of negotiation difficulties. OPEC is also committed to the other developing countries to link the oil problem to the demands of the Third World for a reform of the international economic order.

Although the nonoil developing countries presently have no great bargaining power over these affairs, world diplomacy cannot leave them alone. On the other hand, the major Arab countries regard the oil as their prime political weapon—a weapon which has to serve their conceived national objectives, such as the solution to the Palestinian problem, their national survival, and so forth. It is hard to imagine, under such circumstances, whether and when an international agreement could be arrived at concerning oil, energy, and world development. But the alternative for us is to go through a period of great and prolonged uncertainty with a lot of hardship for most of mankind.

Energy conferences like this one now abound, everywhere in

the world. I have been to two, recently, and actually I was invited to proceed from here to another one. The sun never sets for these energy conferences. One wonders what real value they have, because the presentations and arguments tend to grow stale and sound repetitive. The small consolation is that these meetings may over time contribute to better understanding of the problem and to public political education, upon which governments can formulate policies and act rationally, nationally as well as internationally, so that before the century is over the world can be restructured and an era of new prosperity can begin, shared by the greatest number of people.

CHAPTER TWO

World Energy Productivity and Production: The Nature of the Problem

Wolf Häfele
Deputy Director
International Institute for
Applied Systems Analysis

It certainly can be stated that the energy problem has been studied intensely and widely during the last years. It is now appropriate to stop for a moment and ask the question: What, after all, is the nature of the problem? What are the categories, the terms to deal with the problem? Having participated in these studies I would like to reply that the nature of the problem turned out to be quite different from what seemed apparent in the beginning. It seemed to be a substantive resources and engineering problem. In contrast to that, I would now state that the nature of the energy problem is, however, political and abstract; it is ultimately a problem of human attitude and consciousness.

In order to unfold and explain this statement it might be helpful to look first into the problem of energy demand, in order to arrive possibly at some insights which in turn might be helpful when looking into the problem of resources and energy production. In doing so, we turn first to Figure 2-1.

In Figure 2-1 a number of terms are introduced that are commonly known but nevertheless helpful to reconsider. An energy chain starts with the use of primary energy, such as coal, crude oil, or uranium. Usually such primary energy cannot be used directly; it must first be converted into a more convenient form. Electricity, gasoline, warm water, or steam are examples for what is labeled secondary energy. Upon such conversion, losses appear. When electricity is generated, only 30 or 40 percent of the primary energy will result in secondary energy. Such secondary energy must be transported and distributed, which also leads to certain losses. It becomes final energy when it reaches the user. Electricity at the plug or gasoline at the filling station are examples of such final

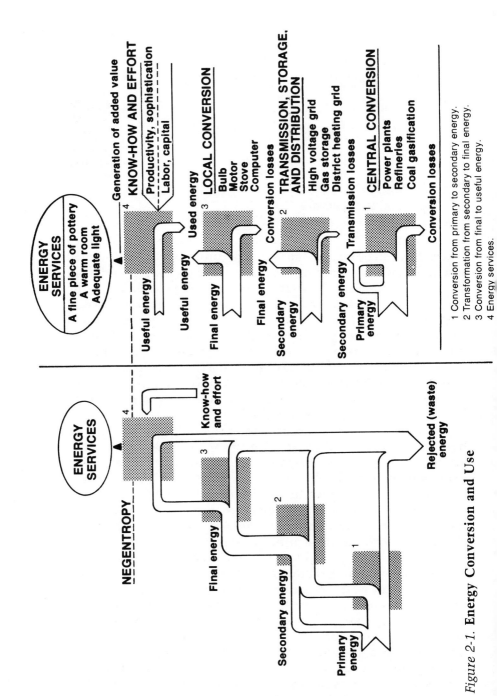

Figure 2-1. Energy Conversion and Use

energy. In addition, using final energy practically always means energy conversion. Gasoline is used in a motor to be converted into mechanical energy, or electricity is used in a bulb to be converted into photons for lighting purposes. Again, such conversions entail losses, and thus only a fraction of the final energy becomes useful energy. Obviously, all energy losses and the useful energy—when degenerated after use—add up to the original amount of primary energy according to the law of energy conservation.

While the distinction between primary, secondary, final, and useful energy is important, it is well known and somehow trivial. However, when rigorously looking into the function of useful energy, the analysis becomes less trivial. Namely, what is useful depends to some extent on the situation of the user. For example, at night it might be an electric lamp for illumination. In periods of cold weather it might be warm air for comfort heating in a home, or it might be kinetic energy such as in an automobile. Each of these is a useful energy "service." Yet, when we turn on a light to read a book, for example, most of the illumination is in a sense wasted, because other things than the book are illuminated too. Similarly, warm air in a house heats not only the people but other objects in the room as well, whose heating may be unnecessary. Much of the kinetic energy in an automobile is wasted as friction in the brakes, or in other actions that are not intrinsic to the process of transportation. What should be examined, then, is not final energy, not even useful energy, but energy services—for example, legible books, human thermal comfort, transportation. These services are not a mode of energy. They can be truly consumed, but they do engage energy. And such engagement of energy only results in a service if, at the same time, a capital stock, labor, and skill also are being engaged. Only when these are taken together can the service in question come into being.

This can be further illustrated by the example of a potter. He wants to earn a living by producing pottery. For this purpose he has equipment (a potter's wheel, a motor for driving the wheel, and spatulas); raw materials (clay); an energy source for running his motor; and his own labor and skill. Here we shall consider the potter's equipment to be his capital stock and the clay to be his resource. It must also be noted that in order to engage the services of his capital stock the potter must employ resources, energy, his labor, and his skill. Thus, his capital stock may be considered a stored form of such inputs. Over time, the potter's output of pottery per day improves. He produces less scrap, he uses less labor, and his skill increases. The amount of sustained influx of kinetic energy per output of pottery decreases; energy is substituted in part by skill. Encouraged, our potter buys a more sophisticated potter's wheel but keeps his original motor. He also buys more sophisticated spatulas and other tools. His output improves further, and, in turn, the amount of sustained influx of kinetic energy per output of pottery decreases even further. The services of his capital stock and his skill are substituting for energy services and possibly for his labor as well.

Now, there are four points that must be stressed after having gone through this analysis:

1. If services, coming from the engagement of energy, resources, equipment, skill, or labor, are at least partly substitutable, they must have a common denominator. Such common denominator has something to do with the respective information content of these sources for services. The notion of information is used here in a more general sense than, for instance, in Shannon's information theory. A theory of such general information does not exist yet, and in the absence of such theory, man has invented a most pragmatic approach to evaluating such information content: prices. The free interplay of market forces is a mechanism for the evaluation of the relative information content of energy, resources, equipment, skill, and labor. If we remain within the category of mere energy services, the analysis can be carried a step further. When energy uses take place the energy becomes degraded, and the entropy increases accordingly as friction processes, heat losses, and other degradations take place. The relation between thermodynamic entropy and information in the more specific sense of information theory is well known and established. It might be enough in this context to mention the names of Boltzman, Shannon, and Prigogine. Now, if the notion of general information points to a common denominator for services from sources of all kinds, it is clear that the nature of services can only be abstract and subtle in the sense of such general information. A further point, then, is that for such general information there is no law of conservation. And indeed, the services can be consumed while the energy engaged does observe a law of conservation. Thus, the nature of consumption is also abstract; it can be misleading to consider consumption in terms of liters of a certain liquid, or in terms of tons of a certain raw material. This not only would be simplistic but can be truly misleading. For instance, when dealing with consumption in such simplistic terms, one might indeed be led to believe that the world is running out of resources, while in reality it is not. Later we will deal with this point more explicitly.

2. Having understood that the engagement of energy requires a

simultaneous engagement of resources, equipment, skill, or labor, it then becomes obvious that an analysis of energy demand necessarily requires a context just because of this simultaneity. Such a context comprises the degree and sophistication of the capital stock in question, the degree of skilled labor, and many other factors. This is the reason why energy demand analysis—and that is often analysis of conceivable energy conservations—is so complex and so controversial, too. Indeed, purely on grounds of natural sciences and engineering it cannot be argued that 2, 4, or 8 kilowatt-years per year (kWyr/yr) per capita are a plain necessity. Houses, for instance, can be insulated in such a way that they consume not 4,000 liters per year of fuel but only 2,000 liters or 1,000 liters or even 200 liters per year. Or cars can be made to use not 15 liters per 100 kilometers but only 10 or 8 or 5 or less. A diesel locomotive requires for the same transportation service only about one-fifth the requirements of a steam locomotive. Energy services can in these cases be replaced by services coming from know-how. In other cases they have been replaced by services coming from capital and labor. Whether 2 or 4 or 8 kWyr/yr per capita is a reasonable target value when energy strategies are being conceived is essentially an economic, institutional, and above all a political question. In the heated debates on energy conservation, it is the feasibility and desirability of institutional and political measures that matter.

3. In this sense it might be helpful to look into Figure 2-2. What is plotted there is the per year ratio of energy required per dollar of Gross National Product (GNP). GNP is shown as a function of the Gross Domestic Products (GDPs) per capita for seven world regions—regions which have been identified by the International Institute for Applied Systems Analysis (IIASA) and which, when taken together, are globally comprehensive. They are explained in Figure 2-3. The triangular points refer to historical data. It is interesting to note a wide discrepancy between Regions I and II—that is, North America and the Soviet Union—and the rest of the world. Particularly in the Soviet Union there was obviously in the past a tendency to replace services coming from capital stock with services coming from energy. The round points refer to scenarios that have been prepared and analyzed by IIASA. The point that shall be stressed here is that we at IIASA found it prudent not to assume in our scenarios too drastic jumps in the considered energy-to-GDP ratio. Therefore we assumed a certain convergence toward values close to 0.5 watts per dollar* per year.

---

*Note: Here and elsewhere in this article, all dollars are in 1975 US dollars.

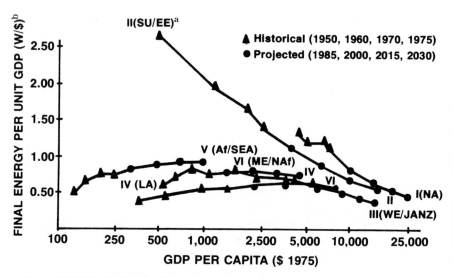

Source: W. Häfele, "A Global and Long-Range Picture of Energy Developments," Science 209 (July 4, 1980), p. 177, Copyright 1980 by the American Association for the Advancement of Science.

[a] See Figure 2-3 for an explanation of initials used here.
[b] Watts per dollar.

Figure 2-2. **Energy Intensiveness in Different World Regions, High Scenario**

But this has, of course, the economic, institutional, and political implications that were mentioned above.
4. The term productivity is well defined in economic analysis. It refers to the ratio of GNP per working hour. To illustrate the point envisioned here, it is sufficient to consider a Cobb-Douglas equation in its simple classical form. It relates capital stock and labor to the GNP per year as follows:

(1) $Y = A \cdot K^{\alpha} L^{l-\alpha}$
    Y = GNP, K = capital stock, L = labor
    α is a constant whose value is between zero and one, and A is a proportionality parameter.

We then find the ratio considered here as follows:

(2) $\dfrac{Y}{L} = A\left(\dfrac{K}{L}\right)^{\alpha}$

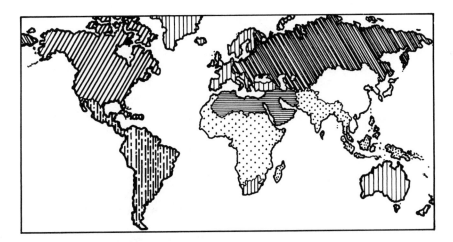

REGION I (NA) North America

REGION II (SU/EE) Soviet Union and Eastern Europe

REGION III (WE/JANZ) Western Europe, Japan, Australia, New Zealand, S. Africa, and Israel

REGION IV (LA) Latin America

REGION V (Af/SEA) Africa (except Northern Africa and S. Africa), South and Southeast Asia

REGION VI (ME/NAf) Middle East and Northern Africa

REGION VII (C/CPA) China and Centrally Planned Asian Economies

Source: Adapted from W. Häfele, "A Global and Long-Range Picture of Energy Developments," *Science* 209 (July 4, 1980), p. 175, Copyright 1980 by the American Association for the Advancement of Science.

Figure 2-3. **The IIASA World Regions**

Economic analysis has long recognized that increases in Y can only partly be explained by increases in K and L. A major share of increases in Y can only be explained by other influences than K and L. They are all aggregated in A, commonly expressed as

(3) $A = A \text{oe}^{\lambda t}$

due to what is called technical progress. This is, of course, of immediate importance for the productivity considered; that is, the ratio Y/L. Technical progress has indeed led to increases of productivity. During the years between 1965 and 1975, values between 2 and 3 percent per year have been observed, and it was such increase of productivity that was mainly the basis for the prosperity of these years.

The organizers of this Symposium have now coined the term "energy productivity." Therefore it is natural to refer to the ratio Y/E instead of the traditional ratio Y/L, the inverse of the quantity given in Figure 2-2. So we may state that at IIASA we expect, in the long run, energy efficiencies of $1/0.5$—that is, $2/watt per year. While the absolute value is indeed of interest, the more important fact is that we expect the energy productivity to increase steadily. A reasoning parallel to that given in equations (1) through (3) could formally describe such an increase of energy productivity when appropriate production functions are taken into account. Again, there are a great number of influences that become aggregated in this process.

Above we have introduced the notion of services. In the language of economic analysis this means value added or the GNP (or gross regional product) as considered here. As we have seen above, energy services are one element of such services. They are at least partly substitutable, and accordingly the elasticity of substitution, especially for energy, plays a major role in recent econometric analysis. Energy services are interwoven in the whole spectrum of influences that lead to an accomplished gross product. Energy productivity is therefore definitely not the same as thermodynamic efficiency. Thermodynamic efficiency is a term which is vigorously defined within the realm of physics and needs no interweaving with any other notions outside of physics. It is indeed very important to distinguish between energy productivity and thermodynamic efficiency. Their evolution in time may be parallel or not. For instance, it has often been argued that electricity should play a declining role in modern energy strategies because of its relatively low thermodynamic efficiency upon conversion from primary to secondary energy. But the fact is that in the past the growth rates of electricity were clearly above the growth rates of energy uses when averaged over all forms of final energy. This points to the high energy services that go along with the use of electricity, which remain, however, invisible when one concentrates only on thermodynamic efficiencies. Cleanliness, deliverability, controllability are examples of such aspects of the energy services from the use of electricity. We may recall here our observation on the ultimately abstract nature of the notion of energy services and their relation to the notion of information. We therefore must explicitly contradict the simplistic views that primarily concentrate on thermodynamic efficiencies. It can well be argued that electricity should play an increasing role in modern energy strategies. What really matters is that energy productivity is in line with the general problems of productivity.

It is in the light of these insights that we now should reflect on the problems of energy production and, as the organizers of this Symposium see it, of world energy production. At IIASA we in the Energy System Program Group have recently concluded a major study entitled *Energy in a Finite World* (to be published by Ballinger Company in 1981), where we dealt with this problem in great detail by the method of careful scenario writing. There the point of emphasis was internal consistency of these scenarios as well as global comprehensiveness. The approach was to concentrate on the factual basis of the energy problem by assuming that all societal, institutional, and political problems could be solved fairly successfully, and thus we arrived at some sort of an upper limit for what can be reasonably expected for the next fifty years. Or to put it negatively: it was not the intent to focus on societal, institutional, and political problems of one or a few nations and to consider what is likely to happen or what is desirable to happen during a shorter period—say, the next twenty years or so. Therefore, the scenarios developed at IIASA must be read carefully and not taken as straightforward predictions that are optimal in some way. At IIASA we ourselves consider them as tools to become aware of the important aspects and parameters of the worldwide energy problem. Recently these scenarios have been described in a magazine article (W. Häfele, "A Global and Long-Range Picture of Energy Developments," *Science* 209, No. 4452, July 4, 1980), and here I shall explicitly refer to them. It is, however, not my intention to repeat these descriptions. Instead, we may start right away with Table 2-1, which briefly summarizes a few main parameters of the IIASA high and low scenarios. This quickly leads to Table 2-2, which illustrates the supply scenarios in terms of global primary energy, disaggregated by source.

In fifty years from now—that is, for the year 2030—the scenarios envision a total primary energy need, in terawatt-years per year (TWyr/yr), of 22.39 TWyr/yr (low scenario) and 35.65 TWyr/yr (high scenario), respectively, as compared with 8.21

Table 2-1. **Two Global Scenarios: Definitions of Main Parameters**

	High scenario	Low scenario
Assuming, by 2030:		
Population		
factor increase	2	2
total (10^5)	7,976	7,976
Gross world product		
factor increase	6.4	3.6
growth	declining rate	
average growth (% /yr), 1975-2030	3.4	2.4
Final energy demand		
factor increase	4.0	2.5
2030 annual rate (TWyr/yr)	22.8	14.6
Results in, by 2030:		
Primary energy		
factor increase	4.3	2.7
2030 annual rate (TWyr/yr)	35.65	22.39
average per capita (kWyr/yr)	4.5	2.8

TWyr/yr of commercial energy by the year 1975, the base year of the scenarios. The striking feature of these scenarios is their fossil nature. Nonfossil energy assumes not more than roughly one-third of the total needs. This is essentially due to the limitation of build-up rates. New technologies, nuclear or solar, just take too much time to assume a higher share, even when societal and institutional factors are not particularly inhibitive. Let us then look into the cases of oil, coal, and gas, to be discussed in that order.

The use of oil increases strongly despite the fact that its relative share decreases. This means a steady shift from conventional, relatively easily accessible, and clean oil sources to more and more unconventional, relatively difficult, and dirty oil sources. Apart from secondary and tertiary recovery schemes this means that unconventional locations, such as deep offshore and polar regions, as well as unconventional kinds, such as heavy crudes, oil shales, and tar sands, come into the picture. What is surprising is that the resources of this kind are large, by far larger than the conventional resources of, for instance, the Gulf region. Siberia, Canada, the United States, and South America must be mentioned here. Or, in other words, the geopolitical implications of such a transition will be significant. They have been discussed elsewhere by the author ("The Energy Problem and Its Long-Range Geopolitical Implications," *Foreign Affairs*, Winter 1980–81). Here it shall only be stressed that it gradually becomes questionable whether

Table 2-2. **Two Supply Scenarios, Global Primary Energy by Source, 1975-2030** (terawatt-years per year)

Primary Source[a]	Base year 1975	High scenario 2000	2030	Low scenario 2000	2030
Oil	3.62	5.89	6.83	4.75	5.02
Gas	1.51	3.11	5.97	2.53	3.47
Coal	2.26	4.94	11.98	3.92	6.45
Light water reactor	0.12	1.70	3.21	1.27	1.89
Fast breeder reactor	0	0.04	4.88	0.02	3.28
Hydroelectricity	0.50	0.83	1.46	0.83	1.46
Solar[b]	0	0.10	0.49	0.09	0.30
Other[c]	0.21	0.22	0.81	0.17	0.52
Total[d]	8.21	16.84	35.65	13.59	22.39

[a]Primary fuels production or primary fuels as inputs to conversion or refining processes; for example, coal used to make synthetic liquid fuel is counted in coal figures.
[b]Includes mostly "soft" solar—individual rooftop collectors; also small amounts of centralized solar electricity.
[c]"Other" includes biogas, geothermal, and commercial wood use, as well as bunkers used for international shipments of fuels; for 2000 and 2030, bunkers are not estimated.
[d]Columns may not sum to totals because of rounding.

the costs and prices that go along with such uses can be afforded. In the first place, this refers most straightforwardly to the direct production costs. While the production costs in the Gulf region are still on the order of $1–2 per barrel of oil or less, we considered at IIASA production costs of up to $25 per barrel; otherwise the large resources necessary for meeting oil demands as described in Table 2-2 may not become accessible. We are talking of costs, not of prices. Such costs will also heavily depend on environmental abatement measures. So far we probably have no understanding of what it means to harness 5-7 TWyr/yr—70–98 million barrels per day!—of oil which is more and more unconventional in nature. In many cases this necessitates secondary and tertiary measures, such as regional developments, and implies large societal impacts. It means a size of industrial business that is hard to imagine today. These things should be kept in mind as well when we talk of costs.

The case of coal in the IIASA scenarios is even more striking. In the high scenario as much as 12 TWyr/yr is envisioned for the year 2030; for the low scenario the corresponding number is 6.5 TWyr/yr. These figures must be compared with the 2.3 TWyr/yr of the year 1975. In the scenarios it becomes apparent that this implies among other things a large world trade in coal. Western Europe will become a large importer of coal, quite in contrast to the situation of the past and today. In the scenarios it is assumed that such imports will come primarily from North America and the Soviet Union. The implications are tremendous. As in the case of oil this leads to the problem of costs in their most general sense.

For the case of gas we had assumed that no worldwide transportation would be feasible, a very conservative assumption which will probably prove to be wrong. But even for the conservative case of our scenarios, the increase in gas consumption is significant. It will be 5 TWyr/yr for the high scenario and 3.5 TWyr/yr for the low scenario as compared with 1.5 TWyr/yr in the year 1975. The cost increase and the environmental and other secondary and tertiary impacts may be less drastic than in the case of oil and coal, but they could be significant enough.

The question that we are driving at is: Can man afford such monetary and nonmonetary costs (and prices)? We wish to maintain that this is a question of productivity and energy as well as labor.

In order to develop this idea, let us consider from an engineering point of view a given economy as a huge black box. Inside that box, production factors such as labor, capital, and skill are at work, but energy must be provided from the outside. Then an energy productivity of $2/watt per year as mentioned earlier simply implies that one barrel of oil relates to roughly $440 of GNP. Even oil prices as high as $50 per barrel (if we think of monetary and nonmonetary implications) would then appear still very reasonable, according to an engineer's judgment. But if the energy productivity was only $0.55/watt per year, as has occasionally been the case in the past when the per capita income was somewhere at $1,000 per capita (see IIASA Region II), only $100 of GNP would correspond to the one barrel of oil in question. Then indeed the $50 per barrel that is required to operate the box comes dangerously close to the total output of $100. Obviously, such a consideration is not sufficient. In reality the GNP, the box's output, is allocated almost entirely to pay the wages for labor, the interest for capital, and to make savings. It requires all the skill of the politicians to balance the societal, institutional, and political forces that influence these allocations. But by going through the exercise of the engineer's box we have still seen that high energy productivities can sufficiently ease the problem of high energy prices.

Figure 2-2 now tells us that in the past, beyond the maximum which seems to appear somewhere at $2,000 per capita, increasing energy productivities seem to go along with an increasing GDP per capita. As we have observed above, this is largely due to an ever-increasing labor productivity. Parallel trends in energy and labor productivities are not surprising. The common denominator is skill and sophistication. As we have seen above, this was one of the sources of the services. The point can now clearly be stated: progress in skill and sophistication in economies must always be larger than the foreseeable degrading of the quality of fossil fuels as expressed in their increasing costs (and prices). Or in other words: the figures of Table 2-2 ask for more general progress, not less.

This points to a problem which really concerns many of us deeply. An almost exclusive concentration on energy conservation measures and related changes of lifestyles could very easily result in reduced productivities. Overemphasizing the service sector of an economy could lead to such effects. And this could increasingly lead to a situation where man cannot afford the dirty fossil fuels that were mentioned above. Conditions of energy demand and supply could diverge, not converge, when energy conservation measures are stressed in a too simple-minded way.

The uses of fossil fuels considered here mean a consumption of fossil fuels. The information contained in these fuels is consumed; what results is a state of less information when carbon dioxide and water are dissipated into the environment. It must be argued that such a consumptive mode of using resources will at least eventually lead to an impasse. The reasoning given above covers the next fifty years, but time progresses further. What about the situation in the second half of the next century?

It is then fundamental to observe that apart from the consumptive mode of resource use, there is also an investive mode of resource use. This distinction is fundamental for the nature of the energy problem, as investive uses of resources could lead to a satisfactory solution of the energy problem forever, at least in principle. This point can best be illustrated by the example of the use of fissionable material. Nature has endowed man with a certain finite amount of fissionable material. How much of this fissionable material is available for man is really unclear. We know much more about fossil resources than about fissionable resources. At IIASA, we have concluded that a figure of 20 million tons of natural uranium for the world as a whole is a good estimate for the time being. This relates to something like 300 TWyr of thermal power when used in presently existing reactors, that is, in the burners. And this means a consumptive mode of utilization; the fissionable material is consumed. By contrast, the use of fissionable material in breeder reactors is principally different in nature. In these reactors the fissionable material is applied without being consumed. Once a certain inventory is invested it becomes possible to burn the fertile material, not the fissile material. And this is the investive mode of using fissionable material. The material is used as an investment. As far as the actual situation of breeder reactors is concerned, it is correct to observe that the consumptive mode of resource use is only transferred from the fissile to the fertile material. But there is so much fertile material available that this actually means a qualitative change. It can be shown that instead of 300 TWyr something like 300,000 TWyr come into reach, which for all practical purposes is an infinite amount. Explaining, in Knoxville and Oak Ridge, Tennessee, the features and the potential of nuclear breeding is more inappropriate than carrying coal to Newcastle. So I abstain. The distinction between consumptive and investive modes of using resources should be clear.

The point now is that this distinction can be applied also to other energy forms, and above all to solar power. The use of solar power also requires huge investments. Quite apart from the physics of photovoltaic cells or other means of absorbing solar power, a certain amount of materials is required at the ground when solar power is to be used at a truly large scale. Estimates range from 10 to 100 kilograms per square meter (kg/m^2). As indicated in Tables 2-1 and 2-2 we are interested in dozens of TWyr/yr of energy production capacity, worldwide and in the long term. Let us assume for the moment that we speak of 10 TWyr (thermal power). At 50 watts per square meter ($watts/m^2$) this translates into 200,000 square kilometers (km^2). With 50 kg/m^2 this means 100 billion tons of materials. For comparison, today's world production of steel and concrete is close to 700 million tons, each. One can play with the numbers, but in any event the amounts of materials that must be invested are huge. We must also speak of investive uses of resources when we speak of large-scale solar power.

Using other renewable energy sources leads to similar considerations. One might take the case of wood. There, a power production density of 1 $watt/m^2$ is an indicative number for future commercial uses of this energy source. Ten TWyr/yr then translates into 10 million km^2, roughly the size of the Sahara. Without trying to suggest such a scale of wood use for an actual energy strategy, it becomes nevertheless clear that also in this case the one-time wood inventory to be made is significant.

As mentioned before, this is neither a paper on future energy strategies nor a briefing on IIASA's energy scenarios, but a paper on the nature of the energy problem. It might therefore be sufficient to capture the problematic situation of investive uses of resources by pointing again to the costs. It is estimated that for investive uses of resources, that is, nuclear and solar power, as well as power from renewable resources, an average figure of $3,000/kw thermal would be indicative. Let us then assume that the average citizen of the world consumes only 3 kwyr/yr, or, in short, 3 kw (thermal). This is a rather low figure. Today's world average is 2kw, while in the United States the per capita consumption today is 11 kw. The first part of this paper outlined the kinds of problems encountered when asking for an appropriate target figure of the per capita consumption of energy. So let us assume just for the moment the above figure of 3 kw. This then means a capital stock of $9,000 per person for energy purposes alone. The present share of energy-related capital stock is close to 25 percent. Let us favorably assume that this share is increased to 33 percent, thus relatively increasing the importance of energy. Then the total capital stock per person would be $27,000. This is indeed a high investment when compared with today's capital stock as given in Table 2-3, particularly for the developing countries.

Table 2-3. **Energy Consumption and Capital Stock**

	Capital stock per capita (1973 $)
Developed market economies	8,500
Developing countries	380
Centrally planned economies (excluding China)	2,700
World	2,000

Source: W. Ströbele, Untersuchungen zum Wachstum der Weltwirtschaft mit Hilfe eines regionalisierten Weltmodelles, Dissertation (Hannover: Technische Universitat, 1975).

As shown in Table 2-3, the developing countries' average capital stock today is only $380, in contrast to the $8,500 of the developed market economies and even to the $2,700 of the centrally planned economies. If we look at the world average of $2,000 in stock and envision the above-explained target figure of $27,000, then we realize that an exclusive reliance on the investive uses of resources is a project of the distant future, probably more than a hundred years from now. Only strongly enhanced labor productivities can get us to this point. Only when the GNP per capita becomes high throughout the world—when people become rich, not poor—can investments be made that lead to capital stocks of that order. And the higher the energy productivity becomes, the sooner can the mode of investive uses of resources be envisaged, as investive use alone can lead to a satisfactory, long-term, and stable solution of the energy problem. Mankind will therefore have to rely on the consumptive uses of resources for quite some time. But as we have seen above, this too leads to the problem of enhanced labor and energy productivity.

One may therefore give a simple answer to the question on the nature of the energy problem: once labor and energy productivities are sufficiently high, there is no sufficient reason why the world's energy problem should not be solved. Nature has endowed us with enough resources, provided we use them in a sophisticated manner. If we do not, we will probably fail. It is therefore compelling to strive for the necessary progress: scientifically, technically, economically. All this is ultimately a societal problem, a problem of attitude and consciousness. We must be ready to go forward, not backward, into the open and not hide away from the challenges. After all, we should not be too surprised, for this has always been the situation mankind has found itself to be in.

CHAPTER THREE

Selected Comments

JOY DUNKERLEY

Professor Sadli this morning asked what good is served by meetings of this sort, and I think he answered his question perfectly in his own brilliant analysis of the position that developing countries find themselves in at the present time. A particularly interesting feature which he emphasized was the heterogeneity of developing countries—these countries differ in their levels of economic development and in their energy resource endowments. It follows from this heterogeneity that each country has different sorts of problems, and also that international assistance must differ depending on the particular problems of the country receiving assistance. This is a point that is often ignored in policy analysis, and I think Professor Sadli did us all a great service by pointing out the implications of variation in the needs and resources of developing countries.

AMORY B. LOVINS

I am sure we all appreciate Professor Häfele's lucid discussion of energy services and his long-term global perspective, but based on discussions with some members of the Energy Group of the International Institute for Applied Systems Analysis (IIASA) and with expert energy modelers who reviewed the model in detail, I have some serious problems with the IIASA energy model. First, that it is what in the trade is called a "goulash model"—that is, it strings together many different types of models (input-output, econometric, accounting, linear programming) designed by different people for different regions for different purposes. The output from such a combination not only doesn't have predictive value; it is essentially meaningless. It is just as if, when you look through a red filter in order to see things in a particular light and then add a green filter and a blue filter and so on, ultimately you've filtered out so much reality that only darkness remains.

Second, the only way that these submodels can be used in combination is to have the operator subjectively adjust the output of one to match the needs of the next and to compensate for the deficiencies of each. It is these adjustments that largely determine the output. I therefore think this model should be thought of not as an objective model but as a simulation game, acting out a set of preconceptions. Third, there's a marked asymmetry in the criteria, deployment rates, and level and degree of analysis applied to hard and soft technologies. The transition to coal- and nuclear-powered central electrification is structurally assumed. It is not objectively derived or economically justified, and I think it is unrealistic to emphasize a brittle, homogeneous, and globally coordinated energy system based on huge transnational flows of capital and fuels while putting far less emphasis on energy productivity gains and on locally adapted resources designed for resilience in the face of discontinuities in the next fifty years.

Finally, by constraining energy productivity gains to a level several times below the economic optimum which we will describe in our paper [chap. 6] tomorrow morning, the model comes up with efficiency gains much smaller than those actually achieved in most industrialized countries since 1973. In other words, the model calls for a supply system that would imply very high energy prices, and these should elicit far greater efficiency gains than are shown, so the model's implicit long-run price elasticity of demand is implausibly low. This appears in Professor Häfele's first slide, in which the change of energy-to-GDP ratio had a similar slope historically, when real energy prices declined, and for the future, when we expect steeply rising real prices.

WOLF HÄFELE

I think you and I have disagreed on these points since 1976, and we continue to do so here. I emphasize again that, in contrast to what you imply, I have never pretended that there is an objective way to predict the future for fifty years. Concerning a specific point of disagreement: while you may find our assumed conservation measures too weak, others consider them to be far too strong. We have discussed it over and over again, so at this stage I think it would be best if we simply agreed to disagree.

HERMAN FRANSSEN

I found both presentations extremely interesting, but what is usually left out when people make long-term forecasts (and by long-term I mean beyond the next five to ten years) is: How do we get there? What does it mean in terms of the changes that our societies will have to undergo? Whether we take Professor Häfele's scenario, which is a plausible scenario, or Mr. Lovins's scenario, which is also a plausible scenario, the problem is that both scenarios would involve massive changes in our current consumer habits. It would mean large shifts in the way society uses resources over the next ten to twenty years, when the transition that these authors project is going to be made. I would appreciate it if Professor Häfele would throw some light today—as I am sure Mr. Lovins will tomorrow—on how he thinks this kind of shift from consumer goods to capital goods will be made.

WOLF HÄFELE

I should emphasize that I lectured not on *the* energy problem in all its detail, but more simply on the *nature* of the energy problem. If I had had to lecture on the former, I would have taken a somewhat different approach. But I find it important that one first have a conceptualization of a strategic nature, that is both globally comprehensive and long range, because it can strongly influence the immediate steps one takes today. For example, consider both the synfuel program starting in this country and similar efforts elsewhere. There are essentially two approaches to coal liquefaction or gasification: allothermal processes and autothermal processes. Currently, practically all the schemes being envisaged are autothermal—that is, the coal being liquified is the source not only of carbon but also of hydrogen and process heat. Allothermal schemes, on the other hand, use hydrogen and process heat provided by some external source. Now the importance of that distinction can only be fully realized if one looks at the global, long-range implications of the two different approaches.

Also—as a second example—the preparations and investments needed in Europe to achieve a certain desired degree of energy conservation or participation in new coal mines or oil technologies can be extensive and complex. Most importantly, they take time, and to properly understand such time constraints again requires that one have a long-range appreciation of the nature of the energy problem.

Thus, to properly address your question, "how do we get there from here?" would require at least another lecture, but one which would necessarily have to be preceded by something similar to what I have just presented.

MARCELO ALONSO

My question is addressed to Professor Häfele. But first I would like to say that it is very stimulating to note that both speakers this morning have emphasized that the so-called energy problem is not just a problem related to technology, to resources availability, and to financing for the production of energy, but it is essentially in the short term a geopolitical problem that to a great extent depends on the worldwide distribution of the resources, combined with different political and ideological goals of governments having access or no access to energy, and that is beyond the technological problem. The technological problem can be solved, we hope. Many of the solutions already exist, but it is the geopolitical aspect that I find extremely difficult to solve in the near future. And if you watched television this morning—watched how Abadan was being destroyed by Iraqi guns—you see what I mean by that.

Now, my question to Professor Häfele is in relation to his scenario, which I find extremely interesting. I have been following very closely the development of his work, for which I have a great respect. But I just want to ask a particular question, because at this moment we are living in a world that is consuming energy at a rate of about 8 terawatts with an average rate of consumption of 2 kilowatts per capita. But that is entirely meaningless, as we see when we compare the rates in the industrialized countries (ICs) with those in the less developed countries (LDCs). In North America, energy is used at the rate of 10 kilowatts per capita, while in Latin America, which is a fairly developed area in relation to the other underdeveloped areas, energy is used at the rate of 1 kilowatt per capita, so we already have a difference of 10 to 1 in the use of energy per capita. Let us now assume that in the year 2000, we are going to use not 30 terawatts, which is the maximum considered in the IIASA model, but just 22 terawatts, which is the more conservative level that you put in your chart. In this model, if I recall well, the average world consumption would be assumed to be about 4 kilowatts per capita. However, in the ICs a substantial increase to about 15 kilowatts per capita is assumed, whereas in Latin America the most we could reach is about 4 kilowatts per capita, so a tremendous difference between the ICs and the LDCs would still exist.

So my question is: How, through the use of that model, can we really reorient the present energy trends to effectively disassociate from an oil-based economy? How can we really help that? And second, how can we talk the more industrialized countries into—rather than increasing their per capita consumption—being a little more conservative; even decreasing their per capita consumption to make room for those who need more energy because they are still very low in their energy per capita consumption?

WOLF HÄFELE

In our scenarios we generally imply that the ratio of energy consumption in the developed countries to energy consumption in the developing countries is improving somewhat. The present situation is not only 1 to 10 but in some cases as high as 1 to 50. Specifically, in North America energy consumption is on the order of 10 or 11 kilowatts per capita, while in Central Africa and Southeast Asia it is 0.2 kilowatts per capita, not 1.0 kilowatts per capita. We have indicated that, in our high scenario, this ratio of 1 to 50 improves by a factor of 3. In 2030 it would therefore be only 1 to 15. Thus, while idealistically all sorts of assumptions are expressed, realistically one must understand the implications of the possible growth rates and the interconnections among the growth rates in different regions. And these did not allow an improvement by more than a factor of 3 in that ratio. That means that in the high scenario, the 4 kilowatts per capita as a worldwide average would be based on somewhat smoother distribution between the extremes; that is, it would be less uneven than it is today. In order to accomplish that minimum goal, which is by no means an ultimately satisfactory situation, we do have to go deeper into unconventional resources, and as Dr. Alonso rightly points out, these have a different geography and thereby also different geopolitical implications, which must be understood and which will be of a radically different nature than today, when most of the oil is coming from only one part of the world. All this means stresses and strains, and by presenting these scenarios we hope to provide general orientations so that as these stresses and strains occur on the political side, it will be more easily understood what efforts towards resolving these stresses and strains could be meaningful. Common to all such efforts would be a relative decrease in the energy consumption in the developed countries.

JOSÉ GOLDEMBERG

I would like to quote from Raymond Aron, the political commentator, and then ask Professor Häfele a question. The quote is the following: "I don't like to predict the future because the courage and audacity of men keeps changing it all the time." The point I would like to ask is the following: In your paper you mentioned that if you were to cover the Sahara Desert with forests—that means 10 million square kilometers of forests—you could produce 10 terawatts. Now that is not so absurd, because you need 1 hectare of land to feed a person. In approximately 1 hectare of land you can produce the energy that a person in a modern society needs. So what's so wrong about it? Why do you so underrate, in your predictions, solar energy and particularly biomass energy sources?

WOLF HÄFELE

I never suggested having trees in the Sahara, although I did say that 10 terawatts per year based on a 1 watt per square meter production density would mean 10 million square kilometers— but not in the Sahara. Still, I am not discouraging the idea. It is a possibility and should be used to the largest extent. In our book [Report of the Energy System Program Group of IIASA, *Energy in a Finite World*, Ballinger, 1981], where we examine more carefully the upper limits of different supply options, we feel that 5 or 6 terawatts is an upper limit of world biomass production for energy purposes. This is a large number, and if Brazil can make a beginning, then more power to them. If you ask me why we don't have a larger share for biomass resources in our scenarios, it is because we feel there is a great ponderousness in the system—that introducing large-scale biomass exploitation takes time and cannot be accomplished so quickly.

P. C. ROBERTS

I want to speak for a moment about Professor Häfele's treatment of energy intensity, which you can speak of in kilowatt-hours per dollar or joules per dollar. Now, at the Systems Analysis Research Unit in the Department of Environment in the United Kingdom, we have been studying this quantity for the last three years or so, and I made the point that it derives from two separate quantities. The first one is energy efficiency, which I will define as the quantity of energy required to produce one unit of a good or service. So, for example, it might be one passenger-kilometer, or one ton of cement, or something like that. Now in general, energy efficiency improves over time. You can find, for nearly all the quantities that you want to investigate, that efficiency as defined in the way I've just stated improves with time.

The other quantity that is included in energy intensity is the pattern of goods and services. What happens over time is that the pattern of goods and services changes as we get richer. In general, we have in the past moved towards higher energy intensity quantities. For example, travel: travel is of very high energy intensity, so the more we travel, the higher the energy intensity of the pattern of goods and services. However, there is no underlying reason for a particular change in the direction of higher energy intensity. For example, all information services have low energy intensity. I am speaking of telephone, telegraph, television, books, phonograph records—all these things have quite low energy intensity. What has been happening in recent years is that although travel and things like that have increased, information services have also been increasing, and what you see if you measure solely joules per dollar is a combination of these two.

If you stay with joules per dollar, you cannot investigate this additional factor, which is the change in the pattern of goods and services, and which is a matter of culture—the way in which our culture moves. There is no underlying reason why we should move in one direction or the other. Now my question to Professor Häfele is: Has he or his team at the International Institute for Applied Systems Analysis separated these two factors and studied them separately, as we have?

WOLF HÄFELE

The answer is "yes," but I couldn't report on that in thirty minutes. There are four reasons for increased energy demand—namely, increases in population, economic growth in the sense of more of the same, the improvements in technology, and changes in the structure of the economy. Of these four factors, the latter is the most difficult one. In the Federal Republic of Germany a Parlimentary Inquiry Commission on future nuclear energy policy has, based both on analyses covering the next fifty years and on public hearings exploring more immediate concerns, tried to identify the future of structural changes by the year 2000. The problem is a difficult one. One must be careful, for example, that efforts to export energy demand to other countries not amount to exporting crucial parts of the domestic economy. The task is to be globally comprehensive and consistent—and not just for the analysis of energy demand, but also for the balance of payments problem. By that I refer to the fact that individual nations might pursue individual policies (e.g., going heavily into microelectronics) that when taken together are globally undesirable—we can't have everyone in microelectronics. Because of such international implications and dependencies, assessing future structural changes in the economy is exceedingly complex. It requires a variety of assumptions, and what we have done is to identify these clearly and to make sure that they are in fact consistent with one another. It is, as you say, not an easy problem, but it is an important one.

PIERRE JONON

I would say that I quite agree with the presentation of Professor Sadli this morning. But I think that an important idea in this presentation is the developing country's need not only to develop energy sources but to increase at the same time the efficiency of using tools, such as heating systems, cooking systems, and so on—and we met this problem in France with our overseas islands, such as Guadeloupe, Martinique, or Reunion Island. In these countries, we decided a few years ago to adopt the same selling price of electricity that we have in Europe, in spite of the fact that the production cost in these places is about three times higher than in Europe. The result was a very rapid increase of electricity consumption, the rate being now higher than 20 percent per year. Our problem is now: How to develop the production plans at such a rate? So, we are considering if it would be possible to use a part of the investments in this country to develop the efficiency of using tools—putting, for example, more insulation in houses.

MOHAMMAD SADLI

I think all countries, whether they are developed or developing, should meet the same requirements, or use the same recipes, for energy conservation, diversification, and also trying to develop more energy—more oil and other sources—more of the same, in principle. In my paper I only indicated some of the problems, some of the complications. Usually—Professor Häfele also pointed this out—there's a substitution problem. If you want to save energy, you have to put in more capital. And then poor developing countries may have a problem, because as you know, capital is expensive. Labor is not expensive, so if you can find the right substitution, where you can substitute more labor for energy, then you are on the right track technologically. But this is still very difficult, and we don't have the technological capability to investigate the right options and the right direction. In principle, however, I agree that everybody should meet these same three requirements.

WOLF HÄFELE

I would like to strongly second that statement. We have gone to pains to take these national differences into account, not only in terms of energy consumption per unit of Gross Domestic Product (GDP) when you are building up an infrastructure, but also in view of the energy conservation measures that can be expected under different conditions. The principal shortage in the developing countries is a capital shortage, and if the largest shares of their GDPs are going just for consumption, leaving only a little bit for investment, certain routes that would appear viable were more capital available are no longer viable. It has indeed been a major point of our studies to reflect these situations properly.

DAVID J. ROSE

First, a comment about Professor Sadli's excellent paper calling for international collaboration and outlining very well the complexities of the problem—and the discussion following, having to do with the use of biomass energy resources, which Professor Sadli pointed out was a great danger for the developing countries. I think this has been very much underestimated all through history, even from Plato talking about the deforestation of Greece and the Mediterranean littoral, up to the present day with the deforestation of India, Nepal, and other places. Read Roger Revelle's excellent article in *Science*—the July 4, 1980, issue. This delusion about the inexhaustibility of biomass extends right up to the present, where all those who bought wood-burning stoves for use in the state of Maine now discover that wood is being cut there faster than it grows, and now ask me "what will we do?" And my advice is: "sell the stoves, or better still, don't buy them in the first place."

A question for Professor Häfele: in his excellent presentation and his discussion of energy conservation, he points out the need for much more sophisticated approaches to the use of energy. I think that is happening, and I'd like his comment on what he thinks about some of the extensions of what is sometimes called availability analysis, thermoeconomics or other names, where the questions are now being asked: What is it for? What is the purpose? What do you need? What are the tradeoffs between energy and capital? Do you see a hopeful trend growing here?

WOLF HÄFELE

The problem of deforestation in poor, developing countries is well recognized. You see it everywhere if you fly out over those countries. What I wanted to address was the cause of that deforestation. Is it the need for fuel, for firewood, or is it the need for agricultural land, or to hunt for game? The distinction is probably important, because if the worry is biomass for fuel, you know you can find certain solutions in fast-growing species, for example. You can cope with the problem. If the risk stems from agriculture and food production, then you are in a different ball game, and it may be more difficult.

MOHAMMAD SADLI

I don't know whether I got the question of Professor Rose right. I understood that he is asking whether there are methodological advances in the art of understanding substitutability on the end-use side. Is that it?

DAVID J. ROSE

The question was availability analysis; that is, how much energy is available from given sources, or how much is needed for accomplishing specific energy-dependent tasks—transportation, or making things, or keeping warm. This new field developing is sometimes called availability analysis, thermoeconomics, and so forth, and I was wondering about your opinion of this developing art, which I am sure you have seen.

WOLF HÄFELE

We anticipate significant developments emerging in that area. However, if one addresses the question from that side, the analogy I would offer is that it is like dealing with the leaves of a tree instead of the trunk of the tree. In the past, our methods have dealt with the trunk of the tree, but now we have to go to the individual leaves. At IIASA we do that with large-scale input-output analysis in conjunction with other approaches based on linear program methods using shadow prices.

MIGUEL S. USSHER

This is addressed to Professor Häfele on his excellent presentation. Obviously the scenarios have to be based on models, and these models on equations. I will refer to the third equation in Professor Häfele's paper. He talks of a parameter called technical progress, which is extremely difficult to define. I thus would ask three questions:

- First, how were the factors that were involved in equation (3) related to "technical progress?"—I mean the variables that shape those factors, and how they were calculated.
- Second, are those factors different for different groups of countries?—If so, what were the criteria used?
- Third, and finally, how were the changes in "technical progress" foreseen over a fifty-year period?

We have run models and found extreme difficulties in trying to guess what technical progress is when fed into models.

WOLF HÄFELE

Thank you for these most pertinent and important questions, indeed. In answering them, let me emphasize that we see three different kinds of mathematical models, and this distinction is important. The first kind of mathematical model appears in natural sciences and most prominently in physics. There, the mathematical model and reality are identical. It is the beauty and the privilege of that field that mathematics is identical with reality. The second field of mathematical models includes the usual econometrical models—that is, the intelligent processing and evaluation of time series—for forecasts over a period of four, five, six, or seven years at the most. And there's a third kind of mathematical model, namely: coherent, consistent scenario writing.

With scenario writing, it is the focus not on a future but on futures—the plural—that is so important to understanding the alternatives. For this reason out study always addressed futures—the plural. There, the objective of the mathematical modeling is not to predict one reality but to describe several conceivable future realities. We have gone to some pains to focus on only those parameters for which a certain evaluation or assumption is meaningful. This deemphasizes the input-output technique because it is very data intensive, and follows more an aggregated, mostly macroeconomic fashion where feasible. Now this necessarily leads to questions of productivity. There is no way of evaluating productivity in the future, just because it is such a soft element, influenced so strongly by the spirit and ingenuity and habits of people. We did assume, generally, a somewhat decreasing rate of increases in productivity, to be on the safe side. Should it turn out that there is an outburst of spirit in some part of the world making productivity better, the better for the reality.

As for the second question—we did assume different values for the different world regions. We have gone to some effort, within the limits that are available to us, to communicate with local groups and to understand their specific situations. In different degrees, this was successful for North America and for the Soviet Union. We also have this sort of communication with Southeast Asia, with Latin America to an extent, and with Africa to an extent, though of course these assumptions are still open to debate, which means the whole exercise is designed in such a way that if we learn more about these parameters, we can repeat the calculation. The changes in productivity that we assumed are therefore ultimately based on informed judgment, though they must be mutually consistent—and there the mathematical modeling is the tool for assuring consistency.

HANS H. LANDSBERG

I would like to direct my question to Professor Sadli, whose paper I found extremely informative and precise. I would like to zero in on his characterization of what is sometimes known as the "big bargain," which is usually trilateral and, in keeping with our chairman's caution, has nothing to do with the Trilateral Commission. The question here involves his characterization of hopes for such an outcome as a mere illusion. I would agree with him, but I think he doesn't go far enough, and I wonder whether he would comment on this. Of course, it is thankless to speak of big ideas with anything but breathless admiration, and I admire his courage in not having done so.

Beyond the lack of agreement on each side of the prospective participants in the bargain, isn't there also a problem that he begs: namely, that any oil-pricing formula has to be followed by some kind of oil allocation scheme. That must be by type and by destination; it must have a priority system like all rationing systems, with various special considerations of this or that category; it must have some room for contingency planning; and so on and so forth. More generally, I think, it must be open to accommodate technological change, economic fluctuations, and other modifications. Let's make no mistake about it: the supplier has to *supply*, but the consumer also has to *take* at given prices. And finally, I think the scheme would have to extend to competing sources. You cannot do all this for one source that to a large extent is substitutable—that is, can be replaced by other energy sources.

From the foregoing, what we would end up with, I would think, is a worldwide, detailed, supply and demand planning system—implementation control system—for energy sources throughout the world. I have grave doubts that to envision this is anything more than an illusion, and maybe a dangerous one, because we may spend a long time working away at it at the expense of other problems. Before we do so, let us just imagine that such a system had been in effect for the last twelve months—how would it have worked? And also, let's say, in terms of even a purely national effort of the same kind?

In conclusion, I would like to ask Professor Sadli (a) whether he agrees that the issue goes further than lack of agreement on all sides and (b) whether he sees some modest steps that may move us some way in that direction.

MOHAMMAD SADLI

I see two aspects of the problem: one is the political aspect, the other is the more technical aspect, and both are very complex. I remember, in OPEC circles, that we had an effort to try to find out whether technical matters such as production programming within OPEC member countries were technically possible. That question, for instance, was never solved or resolved, even after a few years.

You're right—there are many aspects of technical problems that beg for solution or resolution in the form of: "Will arrangements between countries or the global system ever be subjected to planning systems?" The answer is—well, if things continue as they are going now, this is near solution. The other aspect is the political aspect. The will to come to that long-term solution even on a piecemeal basis is questionable—for example, could the seven countries who form the nucleus of OPEC come to an agreement to freeze oil prices for a number of years, just to give the world respite from inflation and so on? This would be a political decision that actually would not solve the problem in the long run, because there are other problems also—relationships with developing countries and so on. However, if you cannot solve the problem in an integral way, you can try to do it on a piecemeal basis—step-by-step. This may lead in the end toward the desired solution, through a political process.

When I wrote my paper for this Symposium, I was already influenced by the events in the Middle East, when Iraq and Iran were not fighting. Even between OPEC members it was already very difficult to come to an agreement. Now these two countries are not even talking to each other. So, you know, at the moment all global solutions for the energy problem, for the development problem, are merely theoretical. But on the other hand, we (and I am very optimistic that sooner or later the world), through recognition of problems and through crises, may eventually see the light—although perhaps on a piecemeal basis.

MÅNS LÖNNROTH

I have a short question which is perhaps more like a comment. The question is really related to the last two interjections, and I would like to ask Professor Sadli and perhaps also Mr. Landsberg the following: Isn't the alternative to some kind of an overall agreement between the OPEC countries and the oil-consuming countries not so much a piecemeal situation but rather regional arrangements in the sense of some kind of regional arrangement around the Caribbean Sea with Mexico, Venezuela, and the United States, Canada, and so forth; some kind of regional arrangement around the Mediterranean between the Common Market European countries and some of the Arab oil countries; and some kind of regional arrangement in the Far East? These arrangements, by the way, could very well be mutually agreed upon in part and coerced in part.

My second point is related to both Sadli's and Häfele's papers. Sadli's paper is essentially, I think, a political paper; Häfele's essentially more an engineering analysis. I would guess that one of the things that we know the least about is the nature of the problem of increasing energy productivity in developed countries. I think that what we see right now in terms of the relations between the United States, Western Europe, and Japan—with respect to the steel industry, car industry, petrochemical industry, manmade fiber industry, and so forth—are exactly the types of problems we will see with respect to increased energy productivity; namely, the conflicts over trade and industrial restructuring. But more can be said about that later.

MOHAMMAD SADLI

While waiting for the global solution on energy and development, of course one can proceed wherever possible with regional solutions, and this is happening—in Latin America, in Southeast Asia. But of course it is a partial solution and limited in extent, often an accommodation or a solution to a problem between South-South, whereas the big resolution needed is between North and South.

HANS H. LANDSBERG

There is nothing wrong with trying to find common goals, but if you just want to look at the United States and Mexico in such an enterprise, you'd have to deal with the enormous Mexican population growth—in finding markets, so to speak, for Mexican people in the United States. You'd have to deal with the Mexican farm export; you'd have to deal with the US/Mexican water supply problem. You cannot get into the energy problem with Mexico and the United States, for instance, by just talking about energy. When you bring in Canada, there are different problems, but the same kind of difficulties will turn up. That doesn't mean one should not be trying. But I think just identifying what is common between any two countries illuminates the degree of difficulty.

MOHAMMAD SADLI

When I said regional solution, I actually meant it in quotation marks. But I don't think they are solutions—they are more like going back to slightly more control by the more powerful of the less powerful. That is actually what I was referring to.

JOHN S. FOSTER, JR.

I just have a quick observation and then a question for Professor Häfele. First—the observation. It's clear we are getting some better understanding of the nature of our energy situation and what the future might hold as each year goes by. None of us likes the scenarios for the future—each for different reasons. None of us can know the future—none of us. We don't know how much solar or nuclear or synfuels we'll have by 2000 or beyond. So it seems to me a useful thing that could come out of this conference is to describe the experiments we could run to help determine the answer to the complex technical, social, political, and economic interactions. It does little good to argue endlessly about things just on faith; these are things that we can do something about. Arguing won't do it.

Second—the question for Professor Häfele. In your description of a range of situations that might exist in the year 2030, you indicated a lot of use of gas and oil. Question: Would you say just a few words about what percentage of that gas and oil is natural; what percentage of the natural fuels would come from discoveries; to what degree are we depending on discoveries? And if it's all from new discoveries, how comfortable are you with that situation?

WOLF HÄFELE

I address myself to the second question, and it is a most salient one. To first give a specific answer: by the year 2030, depending on the geography (that means no uniform numbers throughout the world), the contribution to the world's oil supplies coming from reserves known today is essentially zero. Excluding for the moment the centrally planned economies, the numbers associated with our high scenario indicate only 4 percent of the oil supply in 2030 coming from not-yet-discovered conventional sources outside Region VI (that is, the Arab countries). Finding these sources will of course require drilling and other exploratory activity. Some 40 percent of the year-2030 oil supply is attributed to unconventional resources whose existence is known—that is, Athabasca or Colorado or Eastern Siberia. And some 30 percent is synfuels, with the remainder in oil imports from Region VI. These numbers are only roughly indicative. The synfuel share is large in Western Europe and in North America. It is smaller in other regions—South America, for instance. But reaching these synfuel shares requires action, and I am extremely happy to see this starting now. Our analysis indicated that to achieve the shares in 2030 that I just mentioned, you have to have synfuel production at a significant scale shortly after the year 2000. Furthermore, the numbers for conventional and unconventional oil both require drilling. If you look at the drilling densities all over the world, there's only one place that is well drilled, and that is the United States. Although the findings are particularly high in Western Europe, I must say that there is essentially no drilling there because Western Europe is considered to be a garden. But it might be very expensive now to maintain that garden without drilling. Other areas haven't had much drilling either—and I enjoyed that question because it points to a fairly obvious necessity: we have to act and not wait passively for something good to happen.

VACLAV SMIL

I will be brief and crisp. Let us assume that Professor Häfele could transport himself fifty years back to the year 1930 and do his analysis at that time. In 1930 Chadwick had yet to discover the neutron, and consequently there was no thought of fission reaction—there were 17 million draft animals working in the United States; big Saudi Arabian or Libyan or Mexican oilfields were not yet even discovered; Russia did not have a kilometer of natural gas pipeline; China was a miserable and torn country; and even John von Neumann was not dreaming about computers in 1930. Now, how would Professor Häfele, under those circumstances, forecast fifty years ahead, for the year 1980? I just could not do it. I like to look back, and when I do I see the great limitations of any very long-range forecast. You should look back.

WOLF HÄFELE

As a matter of fact we looked back one hundred fifty years. We looked into the replacement of wood by coal and of coal by oil and by gas. There are stunning regularities, and the message there is that the energy infrastructure is particularly ponderous. Worldwide it takes roughly one hundred years for a new energy source to move from a market share of 1 percent to a market share of 50 percent. It is for this reason that we have gone as far as fifty years but not further. Beyond fifty years you can hope for the unexpected, but for considering changes in the energy infrastructure on the terawatt scale within the next fifty years, it is prudent to count on what you have and to leave the unexpected for the time beyond fifty years. We have taken great pains to make that argument in detail in the book [Report of the Energy System Program Group of IIASA, *Energy in a Finite World*].

VACLAV SMIL

You are absolutely right that you cannot bring in new energy sources at a terawatt scale within a few decades, but unforeseen, and I think often largely unforeseeable, developments will bring tremendous changes in the ways we are using our energy resources, such as the changes brought by massive electrification of households, the adoption of tractors and chemicals in agriculture, ubiqitous applications of computers, the rise of China as a major power, and so forth. And most of these changes, and above all their eventual combined effect, can never be anticipated over a fifty-year span.

WOLF HÄFELE

Well, you can conceptualize that. I really must say—the alternative is not to do anything at all. Let me compare that to a personal situation. When you are parents and you are raising children, you develop your concept of what their educational curriculum should be, in spite of not knowing the future—and many things can happen. If you really reflect on it, you realize that you conceptualize because you do not know the future. And what you do in your private life we have to do as a society. If it turns out to be better than expected, then let's all be happy about it. But I think of this as a responsible human position: simply to take an account of what you can reasonably expect.

JANE CARTER

"Between the idea and the reality ... falls the Shadow" (T. S. Eliot). Between Professor Sadli's "world of internal contradictions" and Professor Häfele's scenarios for 2030 lie a wide range of technical, political, and social options. To examine these, as we must, we need to take into account not only the unpredictability of the nation of change but also the discontinuity and lumpiness of change, which may combine at the end of the day to invalidate the scenario or the plausibility of objectives. This is essential background to the examination of specific issues and courses of action, and I agree with Drs. Foster and Franssen that identification of such options and issues would be a valuable outcome of this Symposium.

WILLIAM BAIRD[1]

Virtually any study of human history indicates that human behavior is in response to social pressures. Current population growth, resource and food shortages, and large differences among developed and undeveloped world communities make it not only difficult but dangerous to build complex mathematical models that extrapolate existing data to produce scenarios for future possibilities. No existing mathematical model would have predicted that Iran and Iraq would go to war against their own economic self-interest and destroy much petroleum at its source. Similarly, no existing model for computer analysis can accurately predict human response to the social pressures in which access to nonrenewable energy resources is central. While fossil and nuclear fuel resources are finite, the range of human behavior is infinite, therefore unpredictable.

HERMAN FRANSSEN

For a person like myself, who spends half of his time on very short-term forecasts and the other half on longer term (10-year) energy demand and supply estimates, a fifty-year outlook with very specific growth figures by sector and fuel boggles the mind. Looking back at what all of us, including myself, were saying in the early 1970s about the energy outlook in the 1980s (only ten years away), we would like to shove all those forecasts under the table because we would be ashamed to show their views today. The world economic and political situation continues to change dramatically, and therefore it is very difficult to say anything really sensible about the long-term energy future. We probably all agree that too much dependence on imported oil from regions that are in a great deal of turmoil is detrimental to the future of the world economy. Let me just add one more thing about the uncertainties of projecting energy supply and demand. We are in the process of updating our International Energy Agency (IEA) short-term forecast of energy and oil demand and supply for 1981 and I don't know yet what is going to happen in the first quarter of 1981. The possible effects of events in 1980–81 will have important consequences for the long-term outlook.

NAIM AFGAN

I would like to express my point of view regarding the Session I papers. As was pointed out, the energy problem can be seen as either a short-term or a long-term problem. I would like to support the idea Professor Sadli expressed that the short-term problem is political and has to be treated as a complex problem, especially as it relates to the developing countries. Because developing countries do not only need energy; they need more support for their total economic progress. So in my opinion, we cannot concentrate on energy problems to the exclusion of all others, for these other problems are even more severe than the energy problem in some of the less developed countries.

I would agree with those who think that the long-term energy problem is primarily a technical and scientific problem, for our ability to satisfy our future energy need will be strongly dependent on our future progress in finding better solutions than we know now. As Professor Häfele pointed out, the future strategies that will be built into our scenarios will depend on the future levels of energy technology achievement we will reach in the future.

DEE ASHLEY AKERS[2]

The Session I tendency to deal with the world energy problem in terms of productivity, technology, and costs is a natural reflection of current energy crisis thinking in which supply-delivery problems obscure the basic relationship between natural resources and the social system. Enlightened governments have recognized for decades that resources that are essential to human existence—especially those which derive their basic value from nature—belong to the public and their use is regulated in the public interest. In this perspective, the use of the marketplace as the only information tool in determining how the resource is used is an abdication of political responsibility. The resulting gap between extraction costs and selling price makes little sense in the public business of promoting the beneficial use of a necessary energy resource, and adds to the problem rather than its solution. From an international point of view, the only common denominator in the energy problem, besides a need, is access to the resource; any international analysis of the problem that ignores this consideration lacks a common interest base. Democratic resolution of the world energy problem may not be possible, but prospects of resolution are dimmer if this common interest base is not at least factored into the statement of the problem.

PAUL DANELS

Regarding the geopolitical perspective outlined by Sadli and the techno/economic approach outlined by Häfele, the externalities and their roles in determining the outcome of each model need to be addressed. Furthermore, an analogy should be drawn between the energy problem of less developed countries—that problem's impact and nature, and approaches to it—and the energy problem of disadvantaged Americans. Much of Sadli's paper sounded like a discussion of the problems of low-income minorities in the United States. And finally, the peculiarly urban aspects of the energy problem should be elicited and developed.

FERNANDO ALTMANN ORTIZ

I really have no questions but a general comment on what is being said at this Symposium. In his opening address, Dr. Sawhill told us that 85 percent of the world's energy is absorbed by 25 percent of the global population. This is completely unfair, I must say. And Professor Sadli told us about the problems of the undeveloped countries, but there is no solution given in his wonderful and beautiful paper. Professor Häfele gave us the scope of the world energy for the next fifty years, but are we going to survive the next fifty years? By "we," I'm speaking of the 85 percent of the world population that is using only 15 per cent of the global energy.

I think we have to realize that this energy problem is not a crisis of prices; it's a problem of looking for today's and tomorrow's solutions at the same time. Let me tell you, very briefly, something about Costa Rica. In Costa Rica, we produce electricity by using all hydropower, yet we have not touched 93 percent of our hydro potential. We are working on geothermal; we are working on biomass and biogas; we are doing some oil exploration with Mexican aid. In 1970 Costa Rica spent 2 percent of its total exports buying energy oil. Today it spends 40 percent, although its oil need is only 20,000 barrels per day. Though we had experienced an increase in oil consumption, in the last eighteen months we were able to bring down our oil need by 14 percent. If big countries like the United States or the Federal Republic of Germany or Japan—the big oil consumers—would accomplish half of Costa Rica's oil demand reduction, we would have a beautiful chance to survive, if we take into consideration that 10 percent of the oil used daily in the United States equals the daily oil needs of the 85 percent of the world population.

Tomorrow's solution will take years to come: it will take a change in our way of living. But there is only one solution for today—the political solution, which necessitates having the understanding and cooperation of all nations. On this point, people try to blame OPEC. I am not trying to defend OPEC, but please bear in mind that OPEC countries are selling nonrenewable energy—as oil is. And they are obtaining money in return, which is very easy to renew. After they finish with their oil, some of the OPEC countries will have nothing for the future.

Today's solution is the political solution. My country promotes a regional agreement with Mexico and Venezuela. This agreement will give us the opportunity to arrive at practical solutions. The agreement did not come easily, but it was feasible. It was signed between two oil-export countries—one OPEC country, which is Venezuela, the other a non-OPEC country, which is Mexico. This has provided a broad base from a political point of view. The countries that have the right to participate in this wonderful agreement—an example of cooperation and solidarity among countries—are very different. We have countries that speak English, like Barbados and Jamaica. We have countries with different races, religions, political ways of living—we have leftist countries, military government countries, and democratic countries (as my country is). The agreement was signed on August 3, 1980.

I would like to help international cooperation by putting this agreement—this paper agreement, a very small one—in the hands of the Symposium, because it represents a pragmatic solution to our problems. Such solutions are the only way that the aforementioned 85 percent of humankind can survive the energy crisis. And I will say that this agreement offers a unique opportunity for developed and undeveloped countries participating at this Symposium, especially since it is taking place in the country with the greatest energy consumption in the world. I therefore feel that this paper should be part of this Symposium in order to be copied in other countries and other areas.

Joint Declaration of the Presidents of Mexico and Venezuela

The President of the United States of Mexico, José López Portillo, and the President of the Republic of Venezuela, Luis Herrera Campíns, meeting in San José by invitation of the President of Costa Rica, Rodrigo Carazo Odio, set forth the following Joint Declaration:

The Governments of Mexico and Venezuela,

Reaffirming the close bonds of friendship and cooperation that have existed traditionally between Mexico and Venezuela as well as between themselves and Central America and the Caribbean;

Convinced that acts of cooperation jointly and severally liable between developing countries are indispensable in order to reach their objectives of economic and social progress in an atmosphere of peace and liberty;

Conscious that all countries should contribute to the realization of a new international economic order based on justice and equity and, in this context, find concrete solutions that may regulate and rationalize the production, distribution, transport and consumption of energy;

Reaffirming their conviction, in consonance with the common position adopted by the Group of 77, to continue the struggle to revalue raw materials in the international market, to diversify the sources of energy and rationalize their use on a worldwide scale as well as to ameliorate, in general, the present unjust relations between the industrialized world and developing countries; and

Taking into account the intention of both parties to give priority to the provision of petroleum to other developing countries and considering, likewise, that independently of other bilateral and multilateral actions already in effect or to be undertaken it is opportune to accomplish jointly concrete measures of a regional character that may contribute to continuing alleviation of the pressing problems of net hydrocarbon-importing countries of Central America and the Caribbean.

Mexico and Venezuela now put into effect the following

ENERGY COOPERATION PROGRAM
FOR THE COUNTRIES OF CENTRAL AMERICA
AND THE CARIBBEAN

I

Mexico and Venezuela resolve to attend to the net internal petroleum consumption of imported origin of the countries of the area, designating for this a total volume of up to 160,000 barrels daily, and to contribute to corresponding official financing.

II

In such manner, the net internal petroleum consumption of imported origin for each one of the countries benefiting from the present Program will be satisfied equally by Mexico and Venezuela.

III

Supplies will be provided in accordance with commercial contracts that Mexico and Venezuela may establish separately with the Governments of the countries benefiting from the Program.

IV

The supplies that Mexico and Venezuela may sell within this Program will be governed by the policies and usual commercial practices of each, including those relative to conditions of disposal and sale prices in their respective international markets.

V

Mexico and Venezuela, by means of their official financial entities, will grant to the benefiting countries loans for 30 percent of their respective petroleum purchases with terms of 5 years and annual interest rates of 4 percent. However, if the resources derived from these loans are designated for priority economic development projects, particularly those related to the energy sector, said loans may be converted to others of as long as 20 years duration with annual interest rates of 2 percent.

VI

The conditions stated in this Program will be applied on the basis that the benefiting countries will continue to fulfill efforts to rationalize the internal consumption of hydrocarbons and promote domestic energy production.

VII

In the manner in which circumstances may permit, it will be intended that the petroleum transport objective of this Program be effected on ships operated by the Naviera Multinacional del Caribe (the multinational merchant marine of the Caribbean).

Without precluding that the Program be extended to other countries of similar economic conditions, this same Program will begin with the volumes provided at present to the countries specified in the Appendix; it will have a duration of 1 year from this date and will be renewable annually by previous mutual accord. Mexico and Venezuela will arrive gradually at the proportion of supply that corresponds to them in conformance with the present Program during the course of the first trimester of 1981.

The Presidents of the United States of Mexico and of the Republic of Venezuela sign two identical copies in Spanish of the present Joint Declaration in the presence of the President of the Republic of Costa Rica, in the City of San José, on the third day of the month of August in one thousand nine hundred and eighty.

JOSÉ LÓPEZ PORTILLO
President of the United
States of Mexico

LUIS HERRARA CAMPÍNS
President of the Republic
of Venezuela

Appendix to the Energy Cooperation Program for Countries of
Central America and the Caribbean
Barbados
Costa Rica
El Salvador
Guatemala
Honduras
Jamaica
Nicaragua
Panama
Dominican Republic

NOTES

1. William Baird; Teacher; West High School; Knoxville, TN.
2. Dee Ashley Akers; Director, Government Law Center (Technological Law); University of Louisville; Louisville, KY.

For identification of the other discussants, please refer to the list of participants in Appendix II.

CHAPTER FOUR

Summary

Shem Arungu-Olende
Senior Technical Officer
United Nations Conference on New
and Renewable Sources of Energy

The theme for the 1982 World's Fair in Knoxville, Tennessee, "Energy Turns the World," is both challenging and fitting. Challenging in that the world's economic activities, of which energy is a part, are complex. The world's economy has become interdependent; at the same time the economy has been going through a most difficult stage in its history. Energy resources are unevenly distributed and the major consumers have become increasingly dependent on sources from outside their borders. World trade in energy is perhaps the single largest in volume. Energy issues are, therefore, geopolitical and thus difficult to resolve. They affect national security and economic well-being, and consequently they are sensitive and touch on a country's very survival; because of all this their resolution is hard.

It is a fitting theme in the light of the current energy situation characterized by high oil prices; by the realization that the main sources of energy in current use, notably oil and natural gas, are depletable and will soon run out; and by the need to develop new and renewable sources of energy, matched only by our inadequate knowledge of the extent and potential of these resources. The theme is also fitting in that the world is beginning to realize that widespread development and use of energy resources have major direct or indirect environmental and social consequences that are only beginning to be appreciated.

The subject matter of the Energy Symposia Series is "Increasing Energy Productivity and Production in the World." Its basic goal is evaluating current international energy policies and exploring prospects for possible transition to greater energy productivity, with a view to developing equitable and efficient solutions to the present national and international problems of energy supply and demand. This, as rightly stated by Symposium I's chairman, is indeed a very ambitious objective. But it is incumbent on us all, experts and decisionmakers alike, from developed and developing countries, to draw on our experience and knowledge in order to try to meet this objective.

The focus of Symposium I is on defining the nature and extent of the world energy problem, thereby setting the stage for the subsequent Symposia. In this context, particular attention has been given to:

- World Energy Productivity and Production: The Nature of the Problem
- Improving World Energy Productivity and Production: The Role of Technology
- Towards an Efficient Energy Future: Critical Paths, Conflicts, and Constraints
- Alternative Policies for Improved Energy Productivity and Production

The present discussions relate to "World Energy Productivity and Production: The Nature of the Problem." Two papers were presented, one by Professor Mohammad Sadli and the other by Professor Wolf Häfele.

Mohammad Sadli, while concentrating on the energy problems of the developing countries, states nevertheless that the energy crisis—caused largely by a series of sharp increases in the price of oil—has affected all countries, for oil is a vital input for modern economies; it is also a much-traded commodity with many net importers and only a few major exporters. Furthermore, because of a massive flow in funds and accumulation of excessive reserves by certain countries arising from the price increase, the monetary system has been adversely affected. The crisis of the last decade and subsequent developments are changing the power balance in the world, he adds.

The main problems facing developing countries are those of adjustment, which vary according to the category of the developing country, that is, whether least developed, middle-income, or oil-exporting. The least developed countries are the most acutely affected: their productivity is low and oil imports claim a large portion of their export receipts, so that other equally important requirements such as investment are sacrificed. On top of all this, many have to import food, fertilizers, and other manufactured goods. In the rural areas, for example, kerosene has been a significant source of energy, but with runaway prices, it is feared that consumption will shift back to firewood, which will in turn result in more cutting of forests with attendant negative impacts on the ecology.

Concerted action is called for to assist these countries, first, to enable them to meet their short-term adjustment problems of acquiring much-needed supplies of fuel. Some of these countries have already received assistance from developed as well as oil-exporting countries and international organizations such as the World Bank. But all this aid has not fully responded to the question of the best mechanisms for putting available aid into action.

The countries should also be helped to raise their agricultural productivity, increase their food production, increase their exportable commodities, and thus make them more self-reliant. Since the tight oil import bills have eroded their balance of payments and terms of trade, the international system should guarantee for them stable and fair prices for their exportable commodities; this calls for the restructuring of the international aid, trading, and monetary systems. The countries themselves should in turn review their domestic policies and in the process institute energy savings in the public sector.

Many least developed countries have the potential for expanding domestic energy production, but they lack adequate resources to mount their own exploration programs. Much could be done, and this could make a major difference. But in order for the effort to bear fruit, a combination of domestic policies and international assistance is required. Sadli indicates that several schemes of assistance have been suggested, including (a) funds on a grant basis for seismic surveys and (b) larger high-risk funds to cover exploration costs. These merit further investigation, he stresses. But in order for these to be successful, developing countries will have to be more flexible in their cooperation with developed countries, in particular multinational corporations whose participation is deemed important because of their know-how and experience in exploration.

The middle-income developing countries have more or less similar problems to those of the least developed countries. However, the latter have more flexibility and greater resilience. Their major problem is related to the balance of payments. Many may experience difficulty in reducing oil consumption; nor is it practical for them to switch, within a short time, to alternative sources of energy. In general, they have limited access to borrowing facilities; in any case, their borrowing capacity is further constrained by their high levels of indebtedness. The impact of a high oil bill can further erode the somewhat shaky political foundations of many, so says Sadli. They require international assistance to adjust to the reality of the new situation.

The adjustment will be over a medium-term period, declares Sadli. In this context, what is needed by middle-income developing countries in the form of assistance are not merely loans to pay increased oil bills, but, even more important, financial, technical, and trade cooperation from industrialized countries and multinational institutions to help them restructure their economies. Their long-term hope rests on a rehabilitated and expanded world trade, for they increasingly are becoming exporters of manufactured goods in addition to traditional commodities. While some help to this end has been forthcoming, it requires expansion. There is scope for developing energy resources in these countries, but exploration risks are high. Financial assistance should therefore be made available to them. Development of such resources as coal and hydro should be cost-beneficial in the light of the current energy situation.

For the oil-exporting countries, the oil crisis has meant prosperity with mixed blessings. For most, the windfall profits have distorted income distribution and wealth, thereby giving rise to potential political instability in the years ahead. The sudden influx of funds will also create additional inflationary pressures; investment is not uniformly allocated, so that while the oil and modern sectors enjoy a boom, agricultural and traditional sectors receive hardly a cent and thus remain unchanged. But there is divergence of opinion on the type of long-term action strategy required. Many would like to distribute the use of the new wealth over a long period of time and thus maximize their utility between generations; some are aware that they are producing more than is prudent; but, suffice it to say, they are all under international pressures to produce more. Others, on the other hand, believe that it is in their interest to produce as much oil as possible and invest the surplus funds in factories, dams, and so forth, thereby creating external economies.

While oil-exporting countries are advocating or rather justifying higher prices as a means of encouraging conservation in the developed countries, they are themselves not living up to these standards. The price of their domestic fuel is invariably much below the international level, leading to undesirable waste; in fact domestic consumption in many of these countries is growing at a much faster rate than Gross National Product (GNP), certainly much faster than the average for the developing countries taken together. This, stresses Sadli, should be curbed—through frequent adjustment of the price of petroleum products.

In general, the oil-exporting countries have had difficulties with the developed countries regarding the nature of the oil-pricing problem. Oil-exporting countries feel that oil prices should be high enough to encourage conservation and the development of alternative sources of energy. Furthermore, they view the oil-pricing problem as part of that of the international economic system related to trade, aid, finance, and investment, and they feel that it must be tackled in the context of the establishment of a new international economic order. But much closer to home, they desire more long-term security for their national wealth and finances; hence their stake in safeguarding the international system.

For all developing countries, there is the special problem of rural areas, a problem which is more acute in the least developed countries. Rural inhabitants are faced with ever-increasing prices of kerosene on the one hand and dwindling traditional sources of energy (e.g., cow dung) on the other hand. They also have limited purchasing power and thus cannot afford better equipment with improved energy efficiency and economy of use. The need for conservation thus becomes more difficult to respond to. Their lot must be improved.

Sadli feels that rural electrification may provide a partial solution to this problem, but at exorbitant costs because of the nature of the rural market with its widely dispersed, relatively small load centers. However, he is hopeful that rural electrification can promote productivity in rural areas. The question is then whether developing countries can afford to build rural networks ahead of time with subsidies in the initial period, in anticipation of social benefits that would in the long run outweigh monetary costs. Sadli also advocates the development and use of renewable resources to meet the requirements of rural areas, especially as hydrocarbons become too expensive; these renewable resources, however, have not reached a point of widespread practical commercial application at costs that are well within the reach of the rural population, he asserts.

Sadli draws attention to the need for an international dialogue and for open-ended summits. In the dialogue, each side must be willing to give concessions. The sacrifices are big and tradeoffs not even. But vision is required for an international agreement that gives less emphasis on purely national interests. In other words, global development is the ultimate answer that will lead to desired changes in the market in international trading and monetary systems. But he cautions that great expectations about immediate results should not be entertained, for the problems are multifaceted and their convergence is not in sight. Their resolution will require a lot of patience and change in outlook. He is, however, hopeful that time and the urgency of the situation will triumph in the end.

Wolf Häfele takes a more generalized approach in addressing the energy production problem. He concludes that while the energy problem has been intensively and widely studied during the past years, the exact nature of the problem, its categories, and so forth have not been examined in depth. He further concludes that the nature of the energy problem is political and abstract and that it is ultimately a problem of human attitude and consciousness.

He explains this by first looking into the problem of energy demand in order to gain some insight into the nature of the energy supply problem. The demand for energy is met, he states, not usually by the primary energy directly, but rather by more convenient, final forms of energy that have been transformed through conversion. In the conversion process, losses occur. The final forms are further converted into useful energy. He affirms that what is useful energy in turn depends on the type of use and the condition of the user. Consequently, what should be examined is not final energy, not useful energy, but energy services. Energy services, while they engage energy, are not a mode of energy. It is important, Häfele declares, to note that such an engagement of energy only results in a service if capital stock, labor and skills, and other resources are engaged simultaneously.

He then introduces a number of concepts and in the process stresses several points. First, if services from the engagement of energy, resources, and skill or labor are at least substitutable, they must have a common denominator, which has something to do with the respective information content of these sources for services. A theory of such general information does not exist yet. But, in its absence, a most pragmatic approach to evaluating information content has been invented: prices. Consequently, the free interplay of market forces is in effect a mechanism for evaluating the relative information content of energy, resources, equipment, and labor or skill. If the notion of general information points to a common denominator for services from sources of all kinds, then the nature of such services can only be *abstract*, in the sense of such general information, he asserts. He further demonstrates that the nature of consumption is *abstract* and submits that it is consequently misleading to consider consumption simply in terms of, say, liters of petroleum or tonnes of coal.

Second, because the engagement of energy requires simultaneous engagement of resources, equipment, and skill or labor, analysis of energy demand requires a context, comprising the degree and sophistication of the capital stock in question, the degree of skilled labor, and many other factors. This explains why energy demand analysis, incorporating conceivable conservation measures, is both complex and controversial. Energy services can, under certain conditions, thus be substituted for labor, know-how, capital, or other resources. Consequently, whether one particular per capita consumption target or another is set—when conceiving energy strategies—becomes essentially an economic, institutional, and, above all, a political question. Third, it is these institutional, economic, and political implications, he states, which perhaps help explain the wide disparity of the energy intensiveness of the different regions of the world.

He finally introduces the notion of energy productivity, which he arrives at by analogy with the concept of economic productivity. Productivity in economic analysis refers to the ratio of GNP per working hour. Classical economic analysis relates GNP to capital stock and labor. Economic analysis has long recognized that increases in GNP can only be partially explained by corresponding increases in capital stock and labor; a major share of the increases in GNP is due to other influences, called technical progress, that have led to increases in productivity. By analogy, energy productivity can be viewed as the ratio of GNP per unit of energy consumed in a given time. By the same token, technical progress is expected to lead to increased energy productivity.

He stresses that a clear distinction should be made between thermodynamic efficiency, on the one hand, and energy productivity, on the other. Thermodynamic efficiency is narrow. The two do not necessarily run in parallel, as is clearly borne out in the case of electricity. Because of its low thermodynamic efficiency, electricity has been expected to play a declining role in modern energy strategies, but electricity growth rates have, on the average, been higher than total energy, due to electricity's higher productivity value as an energy service.

Häfele then discusses energy supply problems, drawing on the scenario development experience of the International Institute for Applied Systems Analysis (IIASA). According to their low and high scenarios, estimated consumption requirements fifty years from now are expected to be three to four times the current levels. The striking feature about these scenarios is that the bulk of the requirements would have to be met by fossil fuels, with only one-third contributed from nonoil sources, in part because new technologies such as solar or nuclear simply take a long time to contribute a larger share, even when social and institutional factors are not particularly inhibitive. In the scenarios, the use of oil increases strongly, despite the relative decrease in its share, implying a steady shift from conventional to more unconventional, relatively difficult, and dirty oil, obtained largely from enhanced recovery and from offshore and polar regions as well as other sources such as oil shale and tar sands. Production costs are expected to rise more than twenty times; part of this directly relates to environmental pollution abatement measures. Coal and natural gas are also expected to play a major role, implying major shifts in export/import relationships and significant increases in cost as well as environmental and other impacts. The geopolitical implications of such a transition will be significant. Furthermore, the development of such resources will by themselves imply large societal impacts. All this leads to the disquieting question: can we afford the monetary and nonmonetary costs as well as the prices that go with such uses?

Häfele maintains that this, in the final analysis, is a question of energy as well as labor productivity. He demonstrates first that low energy productivity compounds the problem of increased oil prices and cost of production, while high energy productivity can sufficiently ease the problem. He then draws attention to the observed parallel between increasing energy productivity and increased Gross Domestic Product (GDP) per capita in some regions of the world, notes that increased GDP is largely due to increased labor productivity, and goes on to affirm that the common denominator is skill and sophistication, which is one of the sources of services. He then draws the conclusion that progress in skill and sophistication must always be larger than the foreseeable degrading of the quality of fossil fuels as expressed in their increasing costs and prices; that is, more general progress is the desirable feature.

The above conclusion points, according to Häfele, to other important problems: an almost excessive concentration on energy conservation may very well lead to reduced productivity, and overemphasizing the service sector could lead to such effects. He maintains that overall this could lead to a situation where mankind cannot afford the dirty fossil fuels referred to above; that is, conditions of energy demand and supply could diverge, not converge, when energy conservation is expressed in a too simplistic manner.

Häfele then introduces the concept of consumptive and investive modes. The uses of energy considered over the past fifty years—wherein fuels, largely fossils, are consumed and where, in the process, part of the information contained in the fuels is lost—are defined as the consumptive mode of using resources. He asserts that this mode leads to an impasse. In the period beyond, it is imperative to anticipate the investive mode, one which will, in principle, lead to a satisfactory solution to the energy problem. In the investive mode, initial investment is high, but thereafter adequate supplies of fuel are assured. For example, once a certain inventory is initially invested in the breeder reactor, fertile material, not fissile material, is burned; this fertile material is created within the reactor without further consumption of the fissile material. He mentions similar examples for such renewable sources of energy as solar or wood: the initial investment would be high indeed, but thereafter the production of energy could continue for a long time without any further investment. He envisages an investive mode based on breeder, solar, and other renewable sources for the period beyond fifty years, but demonstrates that the required initial investment will be staggering, leading to as high as $27,000 per capita in capital stock in contrast to a current average of $2,000 per capita ($380 for developing countries, $8,500 for developed countries, and $2,700 for centrally planned economies).

Häfele concludes that only higher energy and labor productivities can lead us to the investive mode and sustain us there. According to him, then: once labor and energy productivity are sufficiently high, there is no reason why the world energy problem should not be solved. Nature has endowed us with the resources, and it is up to us to strive for the necessary progress—scientific, technical, economic, and so forth. All of these are ultimately a social problem, a problem of attitudes and human consciousness. The task before us is one of survival: it is a challenging task, but then such has always been the situation in which mankind has found itself.

Observations and additional comments I should like to make are as follows. What comes from the above discussions is that we are dealing with a multiplicity of problems called the energy problem. Sometimes reference is made to the energy crisis, which in such a context is used interchangeably with the energy problem. The energy crisis—in particular, the one that gripped the world in the wake of the sharp oil price increases in 1973 and subsequent years—is, in my view, but a manifestation of the energy problem. And I might add that it was a blessing in disguise; it came at the most opportune time and has jolted us into focusing, more than ever before, on the nature of the energy problem and its wider implications.

The energy problem has been with us for a long time—from the dawn of civilization. But the nature of the problem has shifted; the problem has become complex as the society and its institutions and organizational systems have themselves become complex; so that at the present time the energy problem, whether viewed from individual, national, regional, or international perspectives, is not only complex, but has manyfold manifestations. This perhaps partially explains the differences in perceptions of the problem discussed below. Suffice it to say, the problems will not simply disappear from the face of the earth. Indeed, they are in all likelihood going to become more complex in the future and therefore more difficult to handle, especially as conventional sources of energy become scarcer; as other problems, discussed below, grow more heightened; and as uncertainties during the transition period become compounded.

It is, therefore, incumbent on us to try to grapple with the energy problems as they stand now, and further to try to anticipate the kinds of problems that are likely to come in the future so that we can devise ways of resolving them, to the extent possible using regional approaches, but better still, a global approach—the latter, as discussed below, will require changes in our attitude, in our way of looking at the problem. That it is difficult to adequately resolve the energy problem needs no emphasis, for apart from the fact that the problem does not come alone, there is an added dimension, which is that resolving the energy problem involves the next-to-impossible task of balancing society's conflicting values.

In order to resolve the problems, it is important that, first and foremost, they be clearly defined. In this context the evaluation of viable and practical energy policies is an important step. Energy policies must, to the extent possible, strive to be in harmony with other national policies; this is a difficult and challenging task, for the objectives of energy policies sometimes appear to be at odds with other equally important objectives—for example, those related to protecting the environment and improving public health. The resolution of the problems demands a systematic examination of energy supply and demand in the context of existing policies. It also demands an articulation of a coherent set of policies for improved production, processing, or conversion of conventional sources of energy, for the development of new technologies for conversion and more efficient use of existent sources of energy, and for the transition to new sources of energy and improved ways of using these sources.

The strategies, the policies, should be able to accommodate and harmonize long-, short-, and medium-term considerations. Needless to say, current decisions about investments, developments, are strongly influenced by expectations about the future; they will in turn influence future decisions. A coherent view of the problems and their future solutions must therefore be developed; otherwise goals and programs for the short term may frustrate—indeed be in direct conflict with—those for the long term and vice versa.

But defining the energy problem is a formidable task. There are always different perceptions of the energy problem, perceptions that vary according to the region, depending on whether one is in industrialized nations like the United States or the Western European countries, centrally planned nations like the Soviet Union or the Eastern European countries, or developing nations like those in Africa, Asia, or Latin America. There are also differences of perceptions within each region. For example, within the developing countries much depends on whether one is referring to a least developed country, a middle-income country, or an oil-exporting country; similarly, there are disparities among industrialized countries, depending on whether one is, for example, in the United States or Britain. And so on within a country—for example, a government body or department, a private oil company, environmental organizations—all have different perceptions of the energy problem. It is important to appreciate the differences in perceptions of the energy problem, for they affect the kinds of solutions that may be envisaged.

All this leads to the important question: is there a universally accepted definition of the energy problem, a definition that we can use as a starting point in our attempt to solve the problem? This is a difficult question, and I am not sure that we can adequately respond to it yet. We must, however, strive to do so before the Symposia Series is over, so that we can arrive at some general agreement on the nature of the problem, for only then can a realistic solution be anticipated.

Energy problems have not come alone; rather they have been accompanied with other equally important problems—for example, inflation, recession, instability of financial systems, and so forth. The last decade has been marked by economic difficulties and grave strain on the international financial system; the combined impact of these has been especially acute in the developing countries, particularly oil-importing developing countries, where the problems have manifested themselves in the forms of higher prices of energy and food, non-energy-related inflation, a general slowdown or near breakdown of economic activities, mounting debts, crippling balance of payments problems, unstable raw material prices, worsening unemployment, and a general deterioration in terms of trade. The situation for these developing countries is bleak, to say the least. Needless to say, then, the energy problems can only be resolved if seen within the main economic and political context.

There is, in the developing countries—least developing, middle-income, and oil-exporting alike—an emerging general commonality in their view of energy problems. According to this view, energy problems and ways of resolving them must be seen within the context of the malaise of the "old" economic system and the need to restructure existing world economic and trade relationships and thereby move into the era of new international economic order. This partly explains the apparent difficulty that oil-exporting countries seem to have with oil-importing developed countries regarding the pricing of crude oil; the former view the oil-pricing issue as an integral part of the move into the new era, one which embraces the concept of sovereignty over a country's natural resources.

From the deliberations, it is clear that the energy problem is technical and substantive as well as political in character. The solution of the energy problem should incorporate both technical and political considerations. It is the political nature of the problem that makes it more difficult to solve. The technical dimension should embrace the development of more efficient technologies for the development and utilization of energy resources; for an increased resource base, more effort should be directed at enhanced survey, development of coal and new and renewable sources of energy, and better conservation.

Since energy use is pervasive, the energy problem affects every individual, every nation, and every economy, including the stability of international financial systems, and because of the interdependency of world economic systems, the energy problem is also an international problem. The international nature of the problem is further underscored by the fact that (1) energy resources are unevenly distributed, with the least populous regions, which have meager consumption levels, commanding the largest share in energy resources (oil in particular), and (2) energy as a commodity has perhaps the largest volume in trade, involving large transfers of financial resources from one country to another.

The fact that the problem is international makes it even more difficult to resolve. For however nicely we might couch the nature of the problem, and however we might ourselves want to look at the problem as an international one requiring an international approach to solve it, we cannot move away from the fact that people will continue to view the problem from their own national perspectives, to try to resolve it from purely national standpoints and thereby forestall an international strategy.

But before we could resolve the problem from an international perspective, it would be important to redefine our values and identify the criteria we consider most important and those factors we consider important for our survival. The problem is, however, that different countries view their survival problems differently. Therein lies one of the issues to which we must address ourselves: once we have identified certain areas, certain issues that are common to some of the problems, can we proceed to identify issues that we feel are common enough and that will thereby generate international dialogue and, we hope, lead to internationally agreed-upon solutions and resolutions? Another important question is: Can we conceive of a way to develop the kinds of value systems that will reflect our mutuality of interest, not only at the present time but in the future, knowing that this mutuality of interest will tend to conflict with our present perceptions of our national interests?

I am a little bit pessimistic about all this, but we could make a start. We could lay out some of the issues that we consider important, including those we think are sensitive but which all the same should be addressed. This is a difficult task, one which requires painstaking effort. Nevertheless, with determined but cautious effort and with some care in handling the sensitive issues, a lot of ground can be covered. This has been our experience at the United Nations, where we are currently organizing a United Nations Conference on New and Renewable Sources of Energy, to be held in Nairobi, Kenya, from August 10-21, 1981. Here we have been grappling with the problems of energy on an international and global basis; while the Conference is confined to new and renewable sources of energy, some of the issues that we touch on are relevant to those facing us in the Symposia Series. The first assessment is that the problems are international in character and need an international approach. Second, in order to resolve the problems, it is important to shift away from emphasizing purely technical considerations and to move more toward policy orientations and how to resolve policy differences in order to arrive at mutually acceptable solutions. By this, it is not meant that the importance of technical considerations should be downplayed; rather that they must be seen in the context of international policy concern.

In the task before us we should first and foremost strive to develop or agree on certain agenda for international cooperation. This international cooperation should be arrived at not only by resolving what we see as the problems that have been created by the oil crisis in the last decade, but also by looking at issues that relate to finance, transfer of technology, information flows, infrastructures, and so on.

These comments are made in full realization of the fact that the time for international cooperation has not arrived and may not be in sight in the foreseeable future. For a lot of effort has been devoted to, and much emotion directed and attuned to, the safeguarding and protecting of vital national—and in some cases regional—interests and goals. Energy issues have been and, it would seem to me, will continue to be viewed in this perspective, and no major changes in attitude are envisioned, not in the foreseeable future. And yet, as noted, a start in the direction of viewing problems, including energy problems, in terms of global perspectives is imperative. The nations of the world have become interdependent. Perhaps what is required—something that could accommodate and better encourage identification with international issues without scuttling national and regional ones—is to redefine national goals to take cognizance of the emerging world situation. This situation is characterized by said interdependence; by the advances in communications technology, making any part of the world within easy reach in minimal time; and by the internationalization of technology in general. With this new situation the entire world could be perceived as a nation, so that the desirable goals, development, and value systems could evolve accordingly. The problems to be solved would thus be of one "nation," with existing, heterogeneous national parts forming subsystems whose particular problems have to be addressed and harmonized to achieve the overall "national" goal.

In examining the energy problem and ways of resolving it at the global level, we will have to give due consideration to the fact that disparities in consumption have existed and continue to exist and that these have created some element of discontent in the political sphere. In the ensuing discussions it is noted, for example, that currently 15 percent of the world population—those in the industrialized countries—consume 85 percent of the energy. This disparity is further heightened by the uneven distribution of energy resources, with the largest known resources, in particular crude petroleum and natural gas, occurring in parts of the world with low levels of consumption, while relatively smaller resources have so far been discovered in countries with the largest consumption levels. It so happens that the higher per capita consumers have the higher technological capacity. These disparities will undoubtedly lead to conflicts in perceiving the nature of the problems and how to tackle them. Already many developing countries are apprehensive, justified or not, that the combined industrial, economic, and technological might of the developed countries will inevitably lead to the latters' continued control of energy resource development and production. Needless to say, then, if the current disparities are not corrected the world must be prepared for the eventuality of instability, fueled and caused by discontent.

The energy productivity and production problem must be seen in the context of the overall energy problem of which it is part. Once the real nature of the energy problem has been established, it becomes more straightforward to perceive, define, and, we hope, eventually resolve the energy productivity and production problem. Let me make the following comments in regard to the latter problem.

First, when addressing the productivity and production problem, we must keep in mind the fact that we are dealing with many energy sources, both conventional and new, and that the energy requirements will be met not by any single source of energy but rather by a mix of all the available energy resources. This is a point that is often lost; people then end up advocating one source of technology as opposed to another for solving the productivity problem.

Second, in resolving the productivity and production problems, certain constraints must be identified and reduced or overcome, as the case may be. These include financial and investment requirements; research, development, and demonstration; education and training; information on our available technologies and our research, development, and demonstration; adequate infrastructures; social acceptability; environmental problems and issues; and resources availability (e.g., land and water, etc.).

On yet another note, I should like to emphasize that when we talk about the kinds of investments in energy required, now and in the future, due account must be taken of the fact that from the point of view of the country, or for that matter the region or the world as a whole, investment will not be required in energy alone; there will also be investment requirements in other very important and crucial sectors, and some of these too are capital intensive. So, when full account is taken of the investments required in all these services (including energy) in, say, the next fifty years, the result will be staggering, to say the least. And I am not sure that existing institutions or the thinking within them can fully comprehend the enormity of the task.

It seems to me, then, that one of the issues that we must address in the Symposia Series is: Do we have institutions that are well geared to the kinds of investments referred to above? In other words, do they possess the capability to anticipate and plan for the magnitude of investment required for both energy and other services? If not, what measures must be taken to ensure adequate investments as required? When addressing these questions it must be borne in mind that faster growth rates are expected in the developing countries, not only in energy consumption or investment in energy services, but also in investment in other services required to accelerate their development.

In the final analysis, then, energy problems relate to politics, to political will and political vision. The problems are global in nature and require first and foremost a global approach to their resolution; this calls for flexibility, foresightedness, common sense, and resistance to narrow parochial or nationalistic approaches. Even then, the required changes and solutions cannot be achieved overnight; it is a long and arduous process involving complex negotiations, in which each side recognizes the economic and political realities facing the other. The mutuality of interest in solving the energy problem is evident. This should be a strong guiding factor in striving for a lasting solution.

In closing, I should like to echo Häfele's words: that we have to move forward, not backwards. I might add that in moving forward, we are going to meet numerous hurdles, difficult and complex issues, that we cannot afford to run away from. Let me also repeat Sadli's point: that we have to start and look afresh at some of the suggestions that have been put forward regarding the international approaches to resolving world problems—not just energy problems but all the problems, for it seems to me that until we can resolve the problems of international economic strategies, of international political differences and problems, we are not going to solve the energy problem.

Many issues have been raised in the course of the above discussions; they require further examination. Some of them, as they are formulated above or in modified form, could conceivably be the topics of discussions in the subsequent Symposia. The following additional suggestions may also warrant further consideration.

Some thought should be given to the development of mechanisms for better distribution of income and wealth in those countries with surplus funds; international assistance would be required to this end. This suggestion does not at first sight seem directly related to the issue of the nature of the energy problem. Nevertheless, it is important, for any move that ensures or assures the stability of oil-exporting countries is welcome. Needless to say, the region supplying the bulk of petroleum is unstable enough; anything that leads to further deterioration should be avoided. Related to this is the need to assist these countries in building stronger institutions and infrastructures that can better withstand the sudden changes in the volume of money. On a more general perspective there is need to build, in both developed and developing countries—particularly the latter—infrastructures and institutions that can better withstand the strain of sudden changes, part of which arise directly from the impact of increased energy prices.

There is also an urgent need to improve international financial systems and thereby evolve a system that is more robust and flexible. Furthermore, there is, as noted, a need to identify and appraise steps for the development of practical approaches to cooperation in the development of energy. Moreover, there is a need to appraise the best ways of responding to the special problems of developing countries, especially the least developed of developing countries. In the last context it suffices to bear in mind that the problems of the developing countries are different depending on resources endowment, level of economic and industrial development, and so forth. Consequently, envisaged responses to these problems will of necessity be different, but the range of these responses and their order of priority should be determined, and concerted action towards this end attempted.

When talking about future energy demands and the proper technologies to meet these demands, it is important to make as accurate a projection as possible. Forecasting techniques must be improved and must be made on the basis of realistic assumptions that reflect the complex issues involved. Understandably, one cannot be overly precise, especially on issues regarding the future—issues that may be subject to unforeseen changes which may not necessarily replicate the past and the present. The use of scenarios is welcome, for this way, certain trends which provide important insights into the possible future are highlighted. But scenarios alone are not enough. More detailed appraisal is required for the purpose of making decisions and defining policy options.

Due care must be taken when discussing issues touching on efficiency of energy use. We have been reminded that a clear distinction must be made between thermodynamic efficiency on the one hand and energy productivity, by which is meant energy efficiency, on the other hand. I might add that a more general concept of energy efficiency incorporating economic efficiency of energy use also deserves attention. In fact, one may go further and include in the latter the social dimension when considering efficiency of energy use. This point can best be illustrated when discussing energy efficiency in the transport sector. Here one is forced to realize that energy efficiency is not, and cannot be, the sole criterion for selection between one mode of transport and another, because figures often quoted for different modes of transport may not include the energy content of capital investment in say, highways, railroads, fleets of trucks, and rail cars. There is, therefore, need to carry out in-depth studies on the total energy content of different transport modes.

Conservation is an important consideration deserving the attention of the countries, developed and developing alike. Some people in fact go so far as to argue that conservation, especially in the developed countries, is not only crucial but, as a matter of fact, a desirable policy option, if only to ensure that more energy would thus be made available to the developing countries whose requirements for energy are expected to multiply in response to envisaged faster rates of industrial and economic development. While this argument is plausible, historical facts tend to preclude the type of altruistic attitudes and tendencies presumed by such positions. This observation not withstanding, the need for efficiency of use cannot be overemphasized. But conservation is achieved at some cost, a point that is often overlooked when talking about the subject. Strategies for conservation that are appropriate to different countries and that reduce the costs to a minimum, or to acceptable levels, while maximizing the benefits, need further examination. In this context the impact of such a strategy on the other services associated with energy services— labor, skills, other resources, and so forth—require more detailed examination.

In the course of the session's discussions a definition of energy productivity was arrived at by analogy to economic productivity, and several conclusions were drawn from the definition. In the analogy, the concept of Gross National Product was used. Now, it may well be said that the extent to which GNP adequately represents economic conditions or activity in a country, especially in a developing country, has not been established beyond reproach. Furthermore, the nature of the relationship between GNP and energy production or energy consumption is not fully understood and is the subject of further intensive studies. This calls into question the definition of energy productivity arrived at by an analogy that relies heavily on the concept of GNP. It is therefore important and imperative to look further into the question of energy productivity, in order to develop or define a concept that can withstand the passage of time. A clear understanding of the concept of energy productivity is compelling in the context of the present and planned Symposia Series.

Improving World Energy Productivity and Production: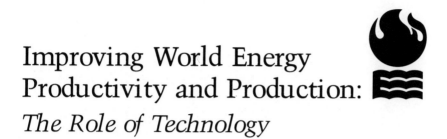
The Role of Technology

Introduction

Lin Hua
Director of the Second Bureau
State Scientific and
Technological Commission
People's Republic of China

The topic of this session is the important role of technology in improving world energy productivity and production. The papers to be presented here—the first by John Deutch and the second jointly by Amory and Hunter Lovins—address the technological options open to the world community in dealing with the increasing social cost of energy. Professor Deutch and the Lovinses have been asked to help us define the realistic limits of current technology and identify prospects for technological change. They have been given a difficult and complex assignment.

It is my hope that in our discussions today we will avoid defining the role of technology too narrowly. Clearly the choice of technology in a specific country—in fact, a country's energy policy itself—depends on many factors, including geography, climate, culture, economic structure, and so forth. For an analyst to advocate a single solution or course to international and national energy problems seems foolhardy. Within a country, the choice cannot be only soft technologies or hard technologies or only supply solutions or conservation. Instead, that choice must represent a balanced path selected from all the alternatives that are appropriate for that particular country. It should not amaze us, therefore, that when we meet as we have today, ostensibly to discuss energy technology in a global context, the discussion often turns quickly to economics, politics, ethics, or other seemingly nontechnical concerns within our respective countries. Ultimately, these are the factors that control our choices.

Still, while recognizing the importance of these other concerns, we must not avoid the more narrow, purely technical issues, some of which are extremely complex while others are rather mundane.

In seeking a balanced path we must develop agendas for research and development; use existing technologies to their fullest; address the difficult issues of international sharing of information; and never abandon the hope that workable solutions will ultimately be found. We must insure that research capable of pushing back the frontiers of understanding is pursued and that the necessary scientists and engineers are trained in our universities.

Let me close this introduction by making two final points. First, during this session we must come to understand better the relationship between technology and economics. I know that each paper to follow addresses this question; however, perhaps I am thinking along different lines from our authors. Specifically, in today's world how many nations can *afford* to allow economics alone to dictate their choices of energy technologies? Second, we must maintain the perspective that energy is only one of our resources and one of our problems. In particular, let us not lose sight of the relationship of energy to our land, water, and forests and to the need for international justice.

CHAPTER FIVE

Energy Technology in Perspective

John M. Deutch
Arthur C. Cope Professor of Chemistry
Massachusetts Institute of Technology

This paper discusses certain principles concerning the role of energy technology in meeting future energy needs. Particular attention is given to the relationship of technology to economic considerations.

Several issues are discussed about the role of energy technology. These include (1) objectives for energy technology, (2) the importance of time in energy technology, (3) the relation of energy technology to energy economics, (4) the role of government, and (5) the opportunity for international cooperation in energy technology.

The final section of the paper discusses an application of the principles that emerge from consideration of the role of energy technology to the specific case of the new synthetic fuels program of the United States. It is argued that such an effort is needed and that the present program design places the government in an appropriate role.

INTRODUCTION

The purpose of the International Energy Symposia Series is to address the complex and interrelated worldwide issues of energy supply and demand. The theme of the Symposia Series, quite appropriately, is "increasing energy productivity and production." This is the correct central issue concerning the world's energy future (when combined with a concern for distributional and equity effects that arise from energy developments), since energy should not act as an unreasonable constraint on improving the

95

quality of life for all nations. For if energy acts as such a constraint, the world must be reconciled *either* to devoting progressively greater amounts of real resources to meeting energy end-use needs *or* to enduring regional and perhaps global strife arising from progressively scarcer and more expensive energy.

*The energy problem is particularly complicated by the heterogeneous geographical distribution of our society's current chief energy source—oil—*and by the externalities that flow from one consuming nation's action to others. With respect to distribution, it is evident that known oil reserves and production capacity are located in areas of the world that are politically unstable, vulnerable to Soviet political/military influence, and not entirely friendly to the United States or the free world. With respect to externalities, it is clear that if the United States and other major consuming countries do not moderate their historically increasing demands for energy, the marketplace and the price mechanism will assure, unless compensating measures are adopted, that smaller and less developed nations will bear sharply higher real economic and environmental costs.

In sum the energy problem is inexorably tied to the problem of world security. The economic consequences that result from both the vulnerability of and the dependence on imported energy (chiefly oil but also other energy resources) quickly create political and diplomatic problems that have been dominating the world stage since 1973. A failure to manage collectively the problems in an acceptable way has the potential to lead not only to disagreeable relations among nations but also to war.

Symposium I focuses on defining the nature and extent of this problem. This paper has been prepared for the session entitled "Improving World Energy Productivity and Production: the Role of Technology." From the preceding discussions, it should come as no surprise that this author does not believe that technology can or will provide an instant solution to the energy problems. Technology may well play an important role in the world's muddling through the energy problem. But the role of technology is unlikely to be either dominant or decisive. *Dealing properly with the energy problem requires that technological considerations be blended with economic and political considerations in adopting policies and implementing programs.*

This essay deals with selected aspects of the role of energy technology in relation to broader energy policy. In the next section several issues are discussed: (1) objectives for energy technology; (2) the importance of time in energy technology; (3) the relation of energy technology to energy economics; (4) the role of the government in energy technology; and (5) international perspectives. This discussion attempts to set some principles for thinking about the role of energy technology. In order to make these principles and the issues they raise more concrete, the final section discusses the appropriate strategy to be pursued in synthetic fuels development. The discussion is intended as a rough illustration of application of the principles to a particular energy supply technology of importance.

THE ROLE OF TECHNOLOGY

In this section the role of technology in energy development is discussed. This discussion is not intended to be a handbook for the energy planner, but rather it is meant to highlight several important concepts that should be borne in mind when discussing how technology should be viewed in relation to energy. The subjects that will be discussed are:

- objectives for energy technology,
- the importance of time in energy technology,
- the relation of energy technology to energy economics,
- the role of government and its relation to the private sector, and
- the need for increased international cooperation.

Objectives of Energy Technology

Most simply, the American dream is that American ingenuity and technology will bring forth a cheap and inexhaustible new energy source that will cause all these messy energy problems to go away. Indeed, one hears persistent refrains of "technology breakthroughs," "Manhattan project," and "putting a man on the moon," all in an effort to discover or at least believe in a modern version of Ponce de León's Fountain of Youth. Specific technologies attract advocates and seek congressional support with such rhetoric. While, of course, it is impossible to rule out such a breakthrough, one is most unlikely to emerge, and it certainly would be a serious mistake to count on such an occurrence. Moreover, it is well to bear in mind that nuclear power, the last technology viewed in this manner, unfortunately has not worked out as expected.

In practice, technology has a more modest and complex role. Basically *the technology effort should be directed toward providing more candidate alternatives for achieving desired objectives.*

Which particular technology or set of technologies is selected depends, of course, on normative political judgments on the net balance of benefits and costs in meeting the specified objectives. With regard to energy technology the correct objective is directed *not* toward developing new energy *supply* technologies but rather toward more fundamental societal objectives, in particular:

- economic health, including prosperity, growth, and equity;
- national security, including concerns for world stability; and
- environmental protection.

From the perspective of these objectives it is clear that *energy technology should be as equally concerned with improving energy productivity and mitigating environmental insult as with increasing energy supply*. Moreover, energy technology efforts must necessarily be evaluated continually in terms of economic costs and benefits. And these costs and benefits must be counted both in the real political world, where prices are often only remotely related to marginal resource costs, and in the notional world of the energy policy planner, where the choice among alternatives is governed by the analytic rule of marginal costs equal marginal benefits.

Of course, calculations that relate energy technology efforts to the Olympian objectives listed above cannot be carried out with any precision. There is a simpler set of objectives that can be viewed as governing energy technology efforts. These are:

- increasing energy supply,
- increasing energy productivity, and
- decreasing the environmental costs associated with energy production.

There are many critics who argue that the present research and development (R&D) efforts of the United States and of its allies are too sharply weighted toward increasing supply at the expense of increasing productivity or reducing environmental costs. In some measure this is a consequence of the American dream mentioned earlier. It is also a consequence of the structure of American industry and the modern industrial marketplace. The private sector knows how to deliver (and get paid for) energy goods and services but has not learned how to deliver profitably conservation or environmental protection. On the other hand, the public sector, with a few notable exceptions, has proven at best clumsy and at worst incapable of substituting regulation for the profit motive as an effective incentive for improving energy productivity and environmental protection.

It should also be stressed that the joint technology objectives of supply, productivity, and environmental protection should be achieved by *the combination of measures that involve the lowest resource cost*. There is little purpose to deploying a technology that will provide energy services at greater cost than available alternatives. To do so would mean that resources were being wasted, and in the energy field the scale of such waste has the potential of being very large.

The best means of avoiding the dangers of resource waste is *to focus on formulating an energy strategy, not on the technology but upon the end-use need*. If this is done, there is a greater chance that important tradeoffs will be uncovered and seriously examined. For example, instead of focusing on liquid fuel supply, one should examine transportation needs. It is only in the latter approach that the appropriate balance will be struck between fuel supply and vehicle efficiency.

The end-use perspective is most important in assessing policy decisions that encourage the *deployment* of technology because it is in the deployment phase that the greatest amounts of resources are committed. But the end-use perspective is also necessary in guiding *technology development*. For, as will be discussed below, one of the most important issues in energy R&D planning is the *timing of when a new technology is needed*. In order to allocate available R&D resources properly, it is necessary to make a judgment on when the new technology can make a cost-effective contribution to a particular end use.

In contrast to deployment of technology, *early technology development is almost always a good thing*. The reason is that the technology development provides the information about a particular alternative required to reach the necessary judgment on its cost effectiveness relative to alternative courses of action. The earlier that an R&D program can provide such information, the better it will be for the energy policy planner and the private sector investor who is continually making large investments that are not expected to be economical for many years. *Every energy R&D program should be undertaken with the principal purpose of narrowing the uncertainty in key characteristics of the technology*. The three key characteristics are (1) the performance of the technology in meeting a specified objective; (2) the economic costs of meeting those objectives (in terms of both investment and life-cycle operating costs); and (3) the environmental, including health and safety, consequences of employing the technology. This type of information is vital for a private or public sector decisionmaker in selecting the appropriate course of action to follow.

The Importance of Time in Energy Technology

A major underlying reality of energy is *the long time necessary to introduce change.* The principal reason why this is so is the vastness of the energy enterprise—consider as an example the US transportation problem, where over 100 million vehicles are fueled by approximately 8 million barrels of oil per day. The energy sector is dependent upon an enormous capital stock that requires years and in some cases decades to replace, even with an aggressive investment program. *The consequences of this long time scale, set by the magnitude of the capital required to make a quantitative impact on energy production and consumption patterns, are far-reaching.*

It is useful to separate the future into three time frames. In the *near term,* the next five years, there is little hope that new energy technology can make any quantitative impact on the means of energy production or on the patterns of energy consumption. The basic reason for this is that the economy is stuck with the existing capital stock that limits the efficiency of production and of end use. Essentially, technology is limited to improving present practices and introducing technical fixes to existing plants and equipment. While this is by no means a minor possibility (for example, residential conservation retrofit could make a big difference within five years), it does severely limit both the quantity and type of energy supply possible as well as the level of end-use energy productivity that can be realized.

In the *long term,* say after the year 2000, the possibilities for innovation in both production and efficient consumption are much greater; some would say limitless. The reason is that there is time not only to develop new technology but to undertake the vast amount of investment needed.

The critical period from the point of view of energy technologies is the *mid term,* extending from, say, 1985 to 2000. For it is in this time period that the transition should be made from the unsatisfactory circumstances of the present to the future. During this mid-term period, technologies must be developed and demonstrated if they are to be deployed in quantity in the post-2000 time period. Of course, if no action is begun today, the three time frames simply are displaced into the future without resolution of the underlying problem of greater depletion of progressively more costly oil and gas production.

The separation of the time scale into near, mid, and long term is, of course, not meant to be entirely rigid. In all cases the transition from newly developed technology to deployment will in fact occur gradually, and in many instances it will prove difficult to classify precisely candidate actions into the three time periods. However, the separation into these time periods is particularly important for understanding several fundamental energy policy issues.

First, consider the problem of what energy goals should be adopted. The preceding discussion makes it clear that in the near term little can be done about *dependence* on OPEC oil imports. There is neither the time nor the capital to change appreciably the pattern of energy use. This means that in the near term the important policy objective must be to reduce *vulnerability* to supply interruption. This must be accomplished for the sake of world stability and western security. It is essential that programs be put into place today designed to protect against supply interruptions in the near term. These programs include, but are not limited to, prompt filling of the Strategic Petroleum Reserve; establishing a standby gasoline tax with rebate (or, less desirably, standby rationing) to be employed in event of a major shortfall; and energy conservation/curtailment measures.

Second, *it must be recognized that supply and demand response to price changes will be much less in the near term than in the long term.* The reason is simply that in the near term the capital stock is fixed and there is limited ability to accept energy substitutes or introduce more productive energy use technology. In the long run, such innovation and readjustment should be expected to occur, moderating the demand for high-cost energy.

Third, in energy R&D planning, it is important to bear in mind the time period in which a new technology is *needed* (in contrast to when the R&D must be performed). For example, there is little purpose to deploying a technology in the near or mid term that will provide energy services at greater cost than available alternatives. A particular weakness of the US technical community is to favor new technologies that are exceedingly sophisticated (and attractive to the technical fraternity primarily for that reason) over simpler solutions that may not be as technically glamorous but, in fact, involve lower resource cost. This weakness is most frequently exhibited in arguments over the timing of a new technology's introduction. In short, there is little reason to urge early *deployment* of new technologies such as fusion, electric vehicles, and synthetic fuels as long as lower cost technology alternatives remain available—for example, light-water reactors, more fuel-efficient spark ignition cars, and enhanced oil recovery. In the United States the energy R&D debate is much confused—for example, in the breeder controversy—over the difference between the timing of deployment and the timing of development of new energy technologies.

Finally, it is important to acknowledge the long lead times required to develop and demonstrate a new technology. It is an unfortunate fact that almost all significant energy technology development efforts require between five and ten years to be carried from the laboratory to a point of demonstration where adequate data are available on cost, performance, and environmental consequences to permit private sector investment decisions. These long development times are constrained not only by funding but also by the requirements for adequate system development and field testing. In pursuing energy R&D programs it is essential that realistic schedules be established which reflect honestly the time and resource requirements, including technical personnel, needed to complete successfully a technology development effort.

In sum, time is critical to every aspect of energy policy, from establishment of realizable policy goals to understanding the interplay of economic forces and needed technology development efforts.

The Relationship of Energy Technology to Energy Economics

In general, technology development and technical innovation must be viewed as subordinate to economic considerations. It does little good to develop a new idea if it will not survive the test of the marketplace and perform a desired service at a lower cost than available alternatives. Energy technology is no exception to this principle. Accordingly, those interested in energy technology must essentially be concerned with both micro and macro energy economics.

A strict economic viewpoint might be that energy differs little from other commodities in international trade—for example, chrome, tungsten, and other strategic materials (where the United States has both vulnerability and dependence) or grain and other agricultural products (where major trade balances exist). But energy is perceived to occupy a special position. There are several reasons why this is so. Most importantly, in the short run, with a fixed capital stock, it is difficult to substitute for energy. This results in politically unacceptable vulnerability to supply interruption. Second, energy is a major cause of domestic inflation and an adverse trade balance of an unacceptable magnitude. Third, the impact of high energy costs and the indirect effects of inflation and the adverse balance of trade are widely believed to be a potential cause of lower long-term economic growth. The extent to which energy growth will be correlated with economic growth in the future remains one of the most important outstanding questions in energy economics.

Finally, and most importantly, energy is correctly perceived to have national security implications. The fact that energy substitution is so inelastic in the short run means that energy is usable for political leverage. Nations will pay large, and possibly premature, development or deployment costs to avoid this "security external-ity."

If economics is to be an accurate guide for technology development, it is important that the prices of energy properly reflect the relative marginal cost of resources. This most emphatically is not the case today where there are price controls on domestic crude oil production and regulation of natural gas. One cannot expect to see the private sector invest resources for new technology development if artificially low average prices are maintained. For example, if the average price of natural gas in California is $2.50 per thousand cubic feet (MCF), there is little reason for industry to think that consumers will replace gas with solar hot water heating systems. On the other hand, if the price of gas to some category reflects the higher cost, say $5.50/MCF, of the marginal source of supply—for example, liquified natural gas (LNG) or Canadian gas imports—the solar hot water industry almost surely will have a market.

In a regulated energy market one does not have equilibrium between relative prices and marginal resource costs, and prevailing prices cannot be used by the technology planner in deciding on resource allocation among alternative energy R&D opportunities.

The United States is presently on a path that will lead to oil and gas price deregulation. And this is an important step. On the fuel supply side there is some reason for optimism that the US system is moving in the direction where prices more accurately reflect marginal cost. The same is not true on the demand side. Particularly in the utility sector (both electric and gas) there are major institutional problems in assuring that prices reflect marginal cost. Any move in this direction leads to higher costs for the consumer and quite probably higher profits for producers—both politically unpopular.

The major consequence of this discrepancy between the average price to the consumer and the higher marginal cost for producing the energy has led to systematic underexpenditures on energy conservation and energy productivity measures. It also, of course, has led to underinvestment by the private sector in new technology for these purposes.

In sum, efforts at rational energy technology planning are frequently complicated by imperfections in energy markets. If, in the long run, energy prices do not reflect marginal cost, one runs the danger of developing new technologies that will not be adopted in the marketplace.

Another serious problem arising from economics that confronts energy technology concerns uncertainty, a not uncommon circumstance. If one seeks to develop a technology that takes five to ten years to bring to the marketplace, the uncertainties concerning the future in terms of price, demand, alternative source of supply, and the regulatory environment are enormous. These uncertainties in each of these areas are so major that one is tempted to hazard that they are even greater than in other areas of the economy. Nor is there any reason for a company to believe, after the recent history of petroleum regulation and environmental constraints, that it will be possible to capture greater return for running larger risks associated with these uncertainties. As a result, available private capital, as exemplified by the recent actions of many oil companies, is diverted into nonenergy development ventures.

The federal government can, of course, choose to take actions that will remove or at least compensate for imperfections in the energy market. Steps can also be taken to narrow uncertainty, particularly where it results from government regulation. But these actions are politically difficult to accomplish, and the evidence of the recent past indicates that this nation will do so only very slowly.

Nowhere is uncertainty more evident than in long-range forecasts of future energy consumption by fuel source and by end-use sector. If technology planners are looking for guidance to such long-range forecasts, they will be disappointed, especially if comparisons are made to estimates of the recent past. Not only have the estimates of overall energy consumption fallen dramatically, but also the composition has shifted significantly from nuclear to coal.

It is tempting to conclude from these falling projections of total primary energy consumption that increases in primary energy supply are not needed for the United States. Indeed, it has become fashionable to ridicule past estimates of US primary energy requirements in comparison with today's much lower estimates. The difference between today's conventional wisdom estimate of approximately 100–110 quadrillion Btu's (quads) and the estimates of over 160 quads made in the early 1970s is frequently cited as evidence that this country can and will make do with less energy. And indeed this is so, and it is welcome. *However, attention is less frequently drawn to the fact that the lower estimates of year 2000 energy needs are in large part a result of lower estimates of real economic growth for the next two decades*. Nor is it always pointed out that the favorable declining trend in the ratio (percent energy growth/percent economic growth) from 1 to 0.5 or lower has been achieved mainly as a result of sharply higher relative energy prices that have meant that energy is much more expensive for the consumer. However, in the short run one should expect a sharp decline in the relationship of energy growth to economic growth as the economy adjusts to the greatly increased real cost of energy relative to other inputs. In the long run, once this adjustment (including capital modifications) to higher real energy prices has occurred, one should anticipate reestablishment of the historically close relationship between energy and economic growth. Attention is still justified on technology that will increase energy supply as well as productivity. Moreover, in the unlikely (but most welcome) event that the economy takes a sharp, sustained upward turn, and for the more immediate purpose of reducing the real resource burden of meeting energy needs, further effort on energy supply technology is merited.

The Role of Government

In the United States and many other nations, it is widely (but not universally) accepted in all particulars that *energy should be left to the free operation of the market*. The reason for this is the conviction (supported by ample historical evidence) that more efficient allocation of resources is accomplished by the market than by government regulation.

There are several important consequences of this basic assumption. The most important is *the decontrol of energy prices so that the market and investment decision can be made on the basis of prices* that more accurately reflect marginal resource costs. For example, the decontrol of domestic crude oil prices in the United States and the deregulation of natural gas prices can be expected to moderate energy demand, stimulate the production of additional conventional domestic oil and gas, improve market allocation, and—most important from the technology viewpoint—improve the competitive position of new energy technologies.

The federal government role is appropriately limited to:

- removing market imperfections that impede competition,
- eliminating institutional barriers to efficient activity,
- undertaking programs that compensate for inequitable or undesirable income effects resulting from free market operation,
- supporting R&D that the private sector cannot be expected to undertake, and
- dealing with national security externalities.

From this list, it is clear that the government plays an important but subordinate role to operation of the private marketplace. Here is the place not to discuss all these elements of government responsibility but rather to draw particular attention to those aspects of the federal role that bear upon technology.

First, consider the adoption of quantitative energy goals as an element of governmental energy policy. The call for energy goals is an understandable reaction to those who want to embark upon a firm course of action to resolve some of the most pressing outstanding energy problems and who view with concern the apparent vicissitudes of US policy. *But the adoption of such goals, whether they be for solar energy in the year 2000, oil imports, synthetic fuel production, or nuclear power, leads in the wrong direction.* The specification of a quantitative goal implies a degree of certainty about energy developments and pertinent economic factors such as price and availability of alternatives that simply does not exist twenty years into the future. Moreover, it is possible that goals will be adopted because of their political attractiveness and that indeed a set of quantitative goals will be inconsistent or at least not adequately reflect interactions of different energy sectors. The place for goals in energy policy is limited, and they should be invoked with great caution.

The second important implication of the assumption of a relatively free energy marketplace concerns the federal government role in research, development, and demonstration (RD&D) of new energy technologies. At the outset it must be recognized that governmental RD&D efforts are ultimately intended to lead to private sector innovation and investment. This is in sharp contrast to the more familiar role of government R&D in defense and space (and to a large degree health) where the government is also the buyer and end user of the new technology. This means that governmental energy technology development efforts must be market oriented and concerned with commercialization. As a result, government planners and managers of energy R&D must grapple with many thorny issues in deciding at what point to begin and end subsidization and how most effectively to induce industrial participation. The types of issues that are encountered in this area range from patent policy to rules that should govern joint government/ industry ventures (such as large demonstration plants).

Perhaps the most difficult and persistent of these issues concerns governmental intervention through direct purchase of early production of a new technology. Examples include purchase of photovoltaic devices, windmills, and solar heating/cooling units. Here the argument is made that the technology will not be cost effective until unit costs are driven down by a market of sufficient size, and that the private sector is unwilling to take the risk of the initial investment without governmental assistance. The argument has merit in some circumstances, but it evidently also invites open-ended government subsidies (through market guarantees and by direct purchase) that not only divert scarce resources from technology development but may also delay true private sector activity. The argument over the desirability of direct procurement support is an example that shows that the process of commercialization in energy R&D is imperfectly understood. There is little agreement on the bounds of needed or desirable governmental involvement in the commercialization process.

It is useful to review the circumstances under which governmental RD&D is justified. *In the first instance, governmental support is justified for basic and applied research on generic technology* and scientific subjects as well as for the exploration of advanced concepts that cannot be expected to pay off, except possibly in the very long term. Examples in this category include materials science, computational science, combustion, catalysis, engineering unit operations, superconducting transmission lines and motors, and many others. Also included in this category are such topics as acquisition of fundamental physical and chemical data and instrument development that are necessary underpinnings of any technology effort.

In all these cases one cannot expect the private sector to undertake sufficiently large and diverse activities because it is difficult to predict the ultimate utility of the research. Furthermore, it is not possible for a private firm that undertakes investment in this research to be confident of capturing the potential long-term benefits.

The second circumstance concerns R&D that supports an activity which for one reason or another is directly undertaken by the government. The reasons why the activity is in the public sector may be national security (naval reactors), perception of important health and safety consequences (nuclear waste management), or a history of diverse involvement (nuclear fuel enrichment and hydropower). It is clear that R&D is required to support such activities and that the R&D programs should be managed by the federal government, although the costs of the R&D may be legitimately charged to the end user. The case of enrichment provides an interesting example of how direct federal R&D can be successful—both the centrifuge and the advanced isotope separation technologies presently under development provide major improvements over the present diffusion process.

The third instance where federal R&D support is justified is in areas that involve social costs (or benefits) that for one reason or another are not adequately internalized by the private sector. This occurs in the environment, health, and safety areas—for example, in air and water quality and understanding the hazards of toxic chemicals. The classic area where there are external benefits for the common good is *national security*. There may be instances where national security externalities can be captured by technology development, in which case federal support for such efforts is justified. But one should be cautious about invoking a national security justification since it can easily be misused by too broad application.

These three circumstances are widely accepted as justifying governmental R&D support. The final circumstance is much more controversial. It involves federal RD&D activities that *accelerate* private sector commercialization of energy technology. The private sector makes decisions on deploying new technology based on risk/return judgments. Here it is argued that the private sector will not move sufficiently rapidly without direct government support in the latter stages of technology development and demonstration. In many instances—for example, coal liquefaction and gasification—the initial investment is so large (quite possibly over $2 billion) that the risk is simply too great for any individual firm to bear. Fundamentally this argument is based on the assumption that either because of market imperfections or because of differing estimates of the future trends in demand, supply, and price, the government is in a better position than industry and the private market to make judgments about net economic benefits. Examples of government programs that to one degree or another are justified on this basis include the liquid-metal fast breeder, solar heating and cooling demonstration plants, the US Department of Energy's (DOE's) industrial energy efficiency program—all expensive ventures. A judgment on the validity of this argument depends upon the detailed context of the particular case and should demand great clarity about the reasons why a subsidy is needed as an additional incentive of private industry to proceed more rapidly than they would on their own accord.

In this regard, it is important to remember that *the federal government has three different mechanisms available to encourage the development and deployment of new energy technologies*. These are:

- direct R&D support,
- financial incentives (e.g., tax credits), and
- regulatory requirements.

It should not be assumed that direct R&D support is the most efficient or quickest means of achieving desired acceleration in technology development or deployment. There will be important instances where sufficient technical, cost, and environmental information is available but the private sector is not prepared to proceed at a pace judged necessary by the government. The government policymaker then can make the choice about whether it is better to accelerate the technology by an indirect incentive rather than direct RD&D support. A recent example where the US administration urged this course of action is for surface retorting of oil shale; a $3 billion tax credit has been proposed in lieu of direct governmental participation in construction of demonstration modules.

The government role in R&D must accept the primacy of the marketplace and always coldly address the question of why federal subsidies are justified. If governmental assistance for technology development is judged desirable, it should be provided in a manner that is temporary and distorts the private market as little as possible. The basic objective should be to provide information on technical performance, cost, and environmental impacts that will permit the private firm that is considering financing an investment to judge the risk and return that is being undertaken.

Addendum on Prospects for Conservation

The intended purpose of this paper is to discuss the role of energy technology with emphasis on energy supply. Discussion at the conference indicates that it might be useful to summarize the author's views on energy productivity improvements (many of which are scattered throughout the paper).

(1) Energy productivity improvements should be evaluated on a comparable basis with energy supply options in the short, mid, and long terms. For the purpose of such evaluation an end-use perspective is particularly important. Selection among these alternatives should be based on market economics with federal government action directed toward removing market barriers and assuming the RD&D support is based upon end-use cost-effectiveness criteria.

(2) The record, at least in the Untied States, of adopting energy productivity measures in all sectors (transportation, industry, residential-commercial) is not strong due in part to market imperfections (e.g., lack of marginal cost pricing), instiutional barriers (e.g., lack of financing), and in large measure to the absence of an active industry delivery system. For the residential-commercial sector, many potential conservation measures are best taken on a highly individualized basis—as suggested by the term "appropriate technology"—for which large-scale federal programs are not well suited.

(3) In the near and mid term, as a consequence of the sharp increase in the relative price of energy, many productivity measures (including new and reconfigured capital investments) are likely to be the most cost-effective means to lower dependence on foreign oil and to minimize the real resource cost of producing desired goods and services. However, once these measures have been undertaken and a new equilibrium established appropriate to higher energy prices, further increases in economic output will require proportional increases in energy inputs, in the absence of energy productivity technology changes. This latter point indicates the need to undertake energy conservation R&D which, until recently, has been relatively neglected.

(4) There are those who appear to argue that all energy needs for the foreseeable future can be met through energy productivity improvements. In this author's view, such a strategy would be misguided in both the near and long term. There are supply options, including both large-scale power plants and small-scale renewable technology, that make sense in economic terms relative to conservation measures for both industrialized and less developed nations.

Energy Technology and International Cooperation

In an International Energy Symposium, held as part of a

World's Fair, it is particularly pertinent to address the subject of international cooperation. *In fact, despite long and sustained rhetoric, there has been far less concrete action on international energy technology cooperation than there should be.* Too much attention has been given to popular communiqués at bilateral or multilateral meetings calling for energy R&D cooperation and to symbolic projects announced by political leaders on visits to foreign countries. Serious programs, supported by significant resources, sponsored either by international organizations (e.g., the International Energy Agency and the United Nations) or by bilateral and multilateral arrangements, remain inadequate.

There are several reasons for this lack of progress. *First, the energy problems, resources, and economic conditions in nations differ widely.* Even among the developed nations the energy circumstances of the United States, United Kingdom, Japan, and France differ markedly. The problems facing less developed countries also display wide variation depending, for example, on whether the country has indigenous energy resources or not. Moreover, less developed nations face more constraints on capital and technically trained personnel than do the industrialized countries. These differences in energy circumstances and rational energy objectives are in large measure responsible for the lack of progress in this important area.

A second reason concerns the problem of commercial interests. Since energy production and utilization remains largely in the private sector and frequently in the hands of big industrial companies, there is always considerable concern that commercial advantage will be gained by the industry of one nation at the expense of the industry of other nations participating in the cooperative venture. An example of the type of difficulty that can arise concerns patents and proprietary data, say in a cooperative project involving the US and FRG governments and industry. The practice in the Federal Republic of Germany between industry and government is entirely different from that in the United States, where both data and patents are usually more accessible in a cooperative venture with the government.

Third, in energy technology cooperation, as in other forms of international cooperation, overarching political relationships between the countries make negotiation, planning, and program execution more difficult.

Finally, it should be remembered that international cooperation is frequently slowed by the mismatch in the procedures and bureaucratic attitudes of the national ministries that participate in the international cooperation. Not only are such important matters as budget cycles and management practices entirely different, but the participating national agencies may be quite unwilling to agree to a particular arrangement because of implications to their interdepartmental bureaucratic relationships at home.

Despite all these difficulties, the case for increased international cooperation is strong. There are several reasons that point in this direction, the most important of which are the existence of common problems, the economies that can be realized by cooperation, and the opportunity to encourage technical innovation generally. If these reasons are to receive the weight they deserve, nations must begin by recognizing that progress on energy problems in one country benefits all. For example, conversion of utilities in the United States from oil to coal reduces aggregate demand for world oil, thus at least moderating upward pressure on price, which benefits everyone. The wide deployment of small-scale renewable energy systems in nations that presently are dependent on oil or seek to expand their energy use without oil serves the same purpose.

Many examples can be given of the common energy technology problems that are faced by countries and where expanded international cooperation would be desirable. Environmental problems are particularly attractive since many of these problems (e.g., carbon dioxide, acid rain, and radioactive waste disposal) have important potential consequences on a supranational level. Basic research programs that produce needed technical or economic data are also worthy of expanded effort, especially since this type of activity does not have a direct impact on the commercial interests mentioned above.

A strong impetus should also be given to international cooperation because it is a means of sharing the very large costs involved in energy technology development and demonstration projects. As yet, the developed nations have moved only very tentatively in this direction. The International Energy Agency has a fairly modest joint R&D effort including, for example, a $20 million solar central receiver project in Spain, and it was only at the 1980 Venice Summit that attention was first given to expanded cooperation on major billion-dollar demonstration projects. To date the singular example of cooperation on such a large scale is the SRC-II joint venture [solvent-refined coal project] between the United States, the Federal Republic of Germany, and Japan.

The magnitude of these large-scale development efforts is truly enormous, involving a sequence of billion-dollar projects. Examples of major technologies that could benefit through sharing of development or demonstration costs from expanded international cooperation include (1) coal conversion and utilization, (2) magnetic fusion, and (3) nuclear technologies (safety and radioactive waste management now, breeder programs perhaps in the future).

Magnetic fusion perhaps presents the easiest starting point for such an expanded effort. Fusion is a long-term technology option that has the potential to provide essentially inexhaustible energy. The expense required to establish the scientific feasibility and engineering practicality are truly immense—at least $20 billion over a twenty-year period. With such an uncertain and long-term payout, sharing the risk and costs among nations makes particularly good sense.

Fortunately, a rather good record of accomplishments has already been compiled on magnetic fusion cooperation. Four nations are supplying the large superconducting magnetic costs for an experiment held in the United States in Tennessee; the United States and Japan have entered into a major joint project on the Doublet-III fusion device; and there is the Joint European Torus (JET) under construction in England. In addition there are numerous other important and productive smaller collaborations, so that the workers in the field have established and are familiar with cooperative programs.

There are also important opportunities for expanded cooperation in small-scale renewable energy technologies. Here the industrial countries have much both to offer and to learn from less developed nations. The US record in supporting such efforts, is not as strong as it should be. It was only in the fiscal year 1981 (FY 81) budget that the US administration proposed and Congress enacted separate funds for international solar energy projects.

While there are many mechanisms that can be envisioned for encouraging increased cooperation in this important area, the approach that could be most productive is *the establishment of a major international technical center, sponsored and managed internationally but financed by the industrialized nations, dedicated to the development, testing, and demonstration of simple, small-scale renewable technologies*, for example, biomass, wind, and passive and active solar. This idea has been suggested in the past, but exploration of alternative mechanisms has not been done nor has serious program planning for such an international technical center taken place. Such an enterprise should not be undertaken unless significant multiyear resource commitments are made—on a scale of $100 million per year.

In sum, energy technology presents important opportunities for international cooperation. To date there has been much talk but little concrete action toward international cooperation on a scale that appears desirable.

THE UNITED STATES STRATEGY FOR SYNTHETIC FUELS DEVELOPMENT

Recently the United States has embarked upon an ambitious synthetic fuels program. This program grew out of President Carter's July, 1979, energy message and resulted in the passage of an omnibus energy bill by the Congress that included the establishment of a quasi-public *Synthetic Fuels Corporation* (SFC). The purpose of the SFC is to accelerate the deployment of various synthetic fuel technologies, including primarily shale, synthetic gas and liquid from coal, and heavy crude/tar sands. The government cost of the program in the first phase could reach $20 billion with a possible second phase that could reach a total of $80 billion.

The program has both strong supporters and critics in the United States. The supporters claim that private industry, in the face of regulatory price controls and technical uncertainty, will not undertake investments in these billion-dollar synthetic fuel projects without governmental support. The establishment of a synthetic fuels industry is regarded as an important way (but not the only way) to define the costs and technical practicality of an important alternative to dependence on foreign oil. Supporters correctly argue that establishing a US synthetic fuel industry would have a major impact on the expectation of oil producers and lessen the likelihood that oil could be and would be used as a political weapon.

Critics of the program argue that there are more cost-effective measures to reduce dependence on foreign oil, for example, conservation subsidies or expanded non-OPEC oil and conventional gas development. In addition, some critics view government intervention in synthetic fuels production as likely to be inefficient as well as improper interference in energy matters.

The subject of synthetic fuels development and the present US governmental strategy for this development is both important and timely. In this section, the principles discussed in the previous section concerning the role of energy technology are applied to the subject of synthetic fuels. It will be argued that *the present approach being followed by the United States is essentially correct* but that unfortunate rhetoric has led to a good deal of confusion about the purpose of and justification for this effort.

Related Governmental Activities

Before proceeding to discussion of the SFC and the accelerated synthetic fuels production effort it addresses, it is essential to note that *there are other government technology activities with respect to synthetic fuels that are clearly justified and probably deserve much greater support and attention.* These activities include:

- *Research both basic and applied* that bears on the improved utilization of coal and shale. Areas of importance include combustion research, catalysis, materials science, process engineering, acquisition of basic physical and chemical data, and development of instrumentation and controls. These types of activities continue to require governmental support—as well as the accompanying education programs that provide trained scientists and engineers.
- *The development of new processes to the pilot plant level* that explore new and improved concepts and methods for converting and utilizing coal and shale.
- *Environmental consequences of synthetic fuel utilization.* Here the array of important subjects is quite broad, ranging from research on the carcinogenic and other toxic properties of synthetic fuels to the more global issue of carbon dioxide's impact on climate.

In short, even if one does not believe that the government should subsidize synthetic fuel demonstration or production plants, there remains a substantial body of synthetic fuel technology that requires governmental support. The justification for this support is that the private sector has insufficient incentive to undertake these types of activities at a level that is socially desirable.

Objectives for the Synthetic Fuels Program

Simply put, the objective of the US synthetic fuels program is to accelerate the deployment of technology for producing synthetic liquid and gaseous fuels that displace high-priced imported oil. As such the purpose of the synthetic fuels program serves only one of the energy technology objectives discussed in the prior section—increasing energy supply. Accordingly, the synthetic fuels program should not be viewed as the only or even the most important action that is undertaken to meet future energy needs.

Synthetic fuel development is one measure that must be taken in combination with others—for example, utility oil back-out, residential conservation, and improved automobile fuel efficiency—to lessen long-term dependence on oil.

Those who support the present US synthetic fuels program, as this author does, believe it *unrealistic to expect the private sector to make investments at the levels required*. There are three reasons for this. First, the private sector makes investments on the basis of projected world oil prices that are highly uncertain. One cannot expect industry to risk capital at a scale of $1 billion to $2 billion per project as a hedge to "higher than expected" oil prices; this is particularly true during a time of high interest rates. Yet, the evidence is compelling that both the private and the public sectors have badly underestimated the pace and magnitude of increasing foreign oil prices. Today DOE estimates that oil will cost about $40 billion (in FY 80 dollars) by 1985; last year the estimate for 1985 was $25 billion. National policymakers must be concerned, in a way which the private sector cannot be expected to be concerned, with hedging against the possibility of significantly higher than expected price levels for oil.

Moreover, the cost of an imported barrel of oil to the nation can exceed the price because of indirect effects on both the economy and national security. The economic effects include, for example, the impact of balance of payments deficit increases and the *possible* impact that importing additional quantities of oil at the margin has on the price of the average barrel. The national policymaker can legitimately think in terms of a "security premium" that should be added to the price of an imported barrel to reflect properly its social cost. While, of course, the means of calculating this premium is open to debate, it clearly is an important reason why the private sector can be viewed as not investing in synthetic fuels at the level needed.

Estimates of the magnitude of this security premium differ widely according to whether one is considering short-run supply interruption scenarios (to which the synthetic fuel effort can contribute at most only indirectly) or long-run import reduction possibilities. For the latter case, responsible experts have estimated premiums ranging from $5 billion to $70 billion. If one adopts a security premium in the range of $10–20 billion, with the present price of imported crude and conservative assumptions about synfuels costs, *a US synthetic fuels program almost certainly is a cost-effective (although not necessarily the most cost-effective) import reduction measure.*

Second, the private sector, particularly in the area of energy, has faced *unbelievable regulatory uncertainty*. This regulatory uncertainty includes both environmental and price regulation. Environmental regulation includes air and water quality, waste disposal, surface mining, toxic chemical control, and worker health and safety. The 1970s saw an enormous amount of new regulation, all of it desirable, but all of it also contributing to higher private costs and, most importantly, unanticipated change in requirements. There is also concern about the regulations and costs that may flow from legitimate local and regional concern about the impact of synthetic fuels projects. Petroleum and natural gas price regulation and even coal rail transportation rates are notorious for their changing character. Under such regulatory circumstances, it is not surprising that the private sector views synthetic fuels projects as being especially risky.

Third, there is *important demonstration value* to the entire economy from the deployment of pioneer plants in a variety of synthetic fuel technologies. The benefits of early deployment are that it demonstrates technical and economic feasibility to other firms and it puts the nation in a much-favored position to expand production rapidly if necessary. If rapid deployment becomes necessary, more plants can be built at a lower cost as a result of the learning made possible by the SFC-sponsored initial production facilities. An individual firm that takes the risk of a pioneer plant can, at best, capture only part of this benefit.

The demonstration value of synthetic fuels as a "backstop technology" for imported petroleum of course depends strongly on the cost of producing these oil substitutes. Once synthetic fuel production costs are demonstrated, there will be a reduced incentive for OPEC to keep oil in the ground or to raise prices beyond synfuel production cost. It follows that keeping the costs of synfuels low should be a principal objective of the US synfuels program.

In sum, the objective of the synthetic fuels program is to provide incentives for the private sector to accelerate synthetic fuels production more rapidly. The reasons for doing so are (1) that private sector investments will not be forthcoming at the level desired, (2) regulatory uncertainty, and (3) the demonstration and preparedness value of early production plants.

There are many important counterarguments that can be raised to the line of argument just presented, and they deserve serious consideration. The counterargument that synthetic fuel plants are not the most cost-effective way to reduce dependence on foreign oil has already been mentioned. And there are many who would prefer to see a direct resolution of the regulatory issue—a reconsideration, so to speak, of the balance that has been reached in this nation between environmental protection and energy production.

But this author believes that all these arguments and counterarguments are dominated by the simple idea that the United States must assure that experience with technology to produce needed liquid fuels is available. Then, if circumstances dictate, production of synthetics can be expanded in large quantities at lower cost. The history since 1973 is clear: the private sector has not been willing to undertake the massive initial investments needed, so federal assistance is necessary.

The Government Role in the Synthetic Fuels Program

There are four particularly evident roles for the government in the synthetic fuels program. These are enumerated below.

1. Targets for synthetic fuels production. Unfortunately, a good deal of confusion has resulted from the initial US administration proposal and subsequent congressional debate on *targets for synthetic fuel production*. Targets suggested by responsible parties for the quantity of fuel to be produced from coal and shale have ranged from 1.25 million barrels of oil equivalent per day (BOE/D) to 2 million BOE/D in 1990, and even larger quantities have been suggested in some quarters. In the case of some of the targets, it appears that insufficient attention has been given to the time required for deployment, so that the target is simply unreasonable.

Such targets are perhaps helpful in gaining momentary political support, but as discussed in the previous section, targets do not make much sense from the point of view of rational energy policy. At present there is little certainty about the cost for producing synthetic fuels, or, as importantly, the cost and effectiveness of alternative approaches to reducing dependence on foreign oil. Moreover, one can have only little confidence in the future demand for and price of imported oil. If principal reliance is to be placed on the private sector and the marketplace for production and distribution of energy, it is not helpful and may indeed be counterproductive to establish goals that are based on highly uncertain assumptions and that may prove desirable to be changed.

In fact, the legislation establishing the SFC has adopted a more sensible approach. The SFC is to proceed in two phases. In the *first phase* up to $17.7 billion will be available to the Corporation to undertake approximately ten large-scale projects that will have technical diversity. Examples of the types of technologies that require relatively less R&D and shorter time to bring to technical readiness are (1) surface retorting of shale, (2) medium-Btu gas from coal, (3) high-Btu gas from coal, (4) indirect Fischer-Tropsch liquefaction of coal, and (5) methanol from coal. The coal technologies could employ either eastern (caking) or western (noncaking) coals. The emphasis on technical diversity is consistent with the government role of providing information on candidate technologies. Continuation into the second phase, in which an additional $60 billion might be expended, would be dependent upon the experience gained in the first phase, for example, environmental consequences of shale development and prevailing energy conditions. Thus, major expansion of the synthetic fuels effort would not take place until sufficient information about the technology and likely costs was in hand. Indeed, the first phase of the proposal should better be characterized as a program of demonstrating technical and commercial feasibility in a serious manner rather than a crash program for synthetic fuels production. This is entirely appropriate—the nation should assure an early capability to produce synthetic fuels and be in a position to expand rapidly the production of these fuels if necessary.

2. The mechanism to encourage synthetic fuels production. In a major departure from past practice, the administration proposed that *the synthetic fuels program be carried out by a new quasi-public corporation* rather than by DOE. This decision was based in part on a realistic appraisal of the practical ability of DOE (or indeed any federal government bureaucracy) in today's climate, with today's regulations, to do the job. But more importantly the decision was based on the recognition that the basic purpose of the synthetic fuels effort, to accelerate private industry activity, required that the venture be undertaken in a manner that resembled, as closely as possible, normal private sector practices. This led to the proposed formation of the SFC, as a quasi-public corporation, governed by rules quite different from those which apply to a federal agency and its personnel.

Further, the objectives of the synfuels program suggest that the incentive mechanisms that the SFC should employ should be indirect financial incentives that are designed to assist the project sponsors to obtain the capital required. The financial incentives include loans, loan guarantees, price guarantees, and product purchase agreements. Note that these indirect incentives differ markedly from the normal governmental practice of providing government-owned, contractor/operator facilities or of providing direct project financial support as is done by DOE in joint industry/government demonstration projects. The SFC presumably will rely upon the indirect financial mechanisms and not become directly involved in project management, thus avoiding one of the principal causes of inefficiency in governmental R&D programs.

Once the decision has been made for government support of synthetic fuel production, the SFC model is an appropriate means, in light of the principles presented in the last section on the governmental role, for carrying out the policy.

3. Relation of the SFC to the DOE demonstration plant program. This paper has argued that there is justification for governmental sponsorship of large-scale demonstration projects and for the SFC approach to encouraging coal and shale commercialization. But why both?

In practice, the answer to the question is partially found in politics. Constituent groups successfully develop congressional sponsorship for a plant to be built by DOE and industry. There is little enthusiasm for shifting the sponsorship of these projects from DOE, which enjoys annual congressional authorization and appropriation oversight, to a quite independent SFC.

However, there also is the important question of technical readiness. *The focus of the SFC is intended to be on those technologies where no major technical uncertainties exist.* This means either that a full-scale process plant employing the technology is in operation somewhere in the world or that all the major subsystems are or will soon be in operation. *The demonstration program sponsored by DOE is intended to focus on processes that have been proven on the pilot plant scale but require scaling up* for both key components and the overall systems before the technology may be regarded as proven. Examples of DOE demonstration projects include direct liquefaction, modified on-site shale extraction, and second-generation medium- and high-Btu gasification. These fossil demonstration plant projects are extremely costly, and it is doubtful that DOE, even with its impressive laboratory system, has the technical and management resources to run many such efforts.

Present US administration policy is that both programs, namely, the SFC- and DOE-sponsored fossil demonstration projects, should be pursued. There remains an important outstanding question: How much of each program? This question cannot be assured definitely. However, while there will always be instances where the DOE-sponsored demonstration plants can be justified, there probably should be relatively greater emphasis placed on the SFC to carry out demonstration projects with the use of indirect financial mechanisms, for example, project completion guarantees, to compensate for the technical risk that may be present.

International Cooperation on Synthetic Fuels

Several nations possess major coal and/or shale resources, including Canada, the United States, Australia, and Morocco. One should expect that over time these resources, along with heavy crude, tar sands, and unconventional gas, will be developed and produced. The process for developing technology for this purpose is evidently quite expensive, so that one might anticipate a good deal of international cooperative ventures to share these costs. Yet, today the single example of such large-scale cooperation is the SRC-II direct liquefaction project being done jointly by the United States, Japan, and the Federal Republic of Germany.

As discussed above, a great deal more could be done to share the risks and costs of such development. There are several synthetic fuel projects that would be reasonable candidates for bilateral or multilateral cooperation. The examples of Venezuelan heavy crude development, Canadian tar sands, Australian coal and shale, and natural gas conversion to methanol in New Zealand and elsewhere are often cited. Additional examples, primarily dealing with coal, were presented to the 1980 Venice Summit by the International Energy Technology Group.

There are three elements that are required to undertake these types of projects. These elements are the natural resources, the capital, and the technology. In order to achieve a sensible basis for international cooperation, there must be a recognition that each of these factors is an appropriate basis for participation. And there must be recognition of equitable sharing of the benefits of the project in terms of product guarantees and technology ownership as well as return on investment.

The nations of the world that do not possess oil and gas will need to make a transition to a nonpetroleum world in the coming decade. It is likely that during this transition synthetic fuels, particularly from coal and shale, will be needed on both economic and security grounds. It is in the interest of all countries to cooperate in order to assure that such technology is practically available, with as little uncertainty as possible about operating characteristics, costs, and environmental effects. Additional efforts should be undertaken now to establish international cooperative ventures in synthetic fuel development and production.

Acknowledgments

I thank Professors Paul Joskow and Joseph Nye for commenting on an early draft of this paper and my former colleagues William Lewis and James Harlan for helpful remarks.

CHAPTER SIX

If I Had A Hammer

Amory B. and L. Hunter Lovins
Friends of the Earth, Inc.

Abraham Maslow once remarked that if the only tool you have is a hammer, it is remarkable how everything starts to look like a nail.

As a former high technologist aware of the seductions of various hammers, and a sociologist and political scientist aware of the dangers of nailing everything in sight with one's favorite hammer, we appreciate this opportunity to advance a seemingly paradoxical thesis. We shall argue on the one hand that the role of technology in improving world energy productivity and production is far greater than is dreamed of by the previous speakers most enamored of technological solutions. The energy problem does not arise from a lack of adequate technologies. On the contrary, technologies exist which, in technical terms, can meet virtually all long-term global needs for energy services, and can do so faster, cheaper, more surely, more safely, more cleanly, and more easily than would be possible with the frontier oil and gas, coal, or nuclear technologies emphasized by many previous speakers. Yet at the same time we shall suggest that the most interesting, difficult, and important choices in energy policy are not technological but rather political and ethical. Discussing technical options is one of several prerequisites for knowing what is possible, but such

The conclusions in this paper rely on extensive published technical analysis and supporting documentation. The citations are contained, directly and indirectly, in two survey papers: (1) A. B. Lovins, "Economically Efficient Energy Futures," in W. Bach et al., eds., *Energy/Climate Interactions* (Dordrecht: Reidel, 1980); and (2) A. B. & L. H. Lovins and L. Ross, "Nuclear Power and Nuclear Bombs," *Foreign Affairs* 58 (Summer 1980), pp. 1137–77 (especially sections III-V). The latter paper is rehydrated, with much fuller documentation, in A.B. & L.H. Lovins, *Energy/War: Breaking the Nuclear Link* (San Francisco: Friends of the Earth, December 1980; and New York: Harper & Row, May 1981).

a discussion cannot reveal what is wise or how to achieve it. Indeed, purely technical analysis mistakes the fundamental nature of the energy problem, encourages us to ask the wrong questions, and may well obscure solutions or create new and less tractable problems. Technical and economic logic, therefore, is not the end of energy policy but one of its many beginnings. Thus while we shall suggest that, in one sense, technology is the answer, we shall also ask—with a persistence perhaps uncomfortable for technologists—what was the question?

It is ironic that those analysts who have placed the greatest emphasis on technological progress have taken the least account of it in their studies of further energy needs. That progress, mainly outside the official programs, has been extraordinarily rapid and effective. We shall start by summarizing some of its main elements, in order to show that the global energy supply problems that have preoccupied most previous speakers are an artifact of economically inefficient energy policies. The state of the art which we shall describe is based entirely on *empirical* cost and performance data from devices and techniques already in or entering commercial service. Though the technologies are being further improved very rapidly, and though some more speculative innovations (such as cheap photovoltaics or photolysis) would, if successful, make the energy supply problem almost trivial, we do not count on any future developments but only on what is available now. The further refinements that would be desirable offer interesting problems of detail but are not of a fundamental character: they concern rather the best way to do something that can with assurance be done well even now. Some may consider our examples "extreme." We consider them rather a fair representation of what can be practically done by carefully shopping around for the best present art, and we make no apologies for seeking out good engineering.

One further caveat is in order. If you have preconceptions about what sort of future society we are assuming, please put them out of your mind. Today, as in all our published analyses, we shall *assume* traditional, rapid, heavy-industrially based economic growth in all countries. Our five-cars-and-a-boat-and-a-helicopter-in-every-garage scenario may be "spherically senseless"— it may make no sense no matter which way around one looks at it—but we shall assume it anyway, in order to show that if your goal is to Los Angelize the planet, you will be able to meet the resulting energy needs most cheaply and effectively with a soft energy path. If you think such a goal is unworthy, or if you consider today's values or institutions to be imperfect, then you are welcome to assume some mixture of technical *and social* change which would make the future we describe easier to realize. We, however, have not done that. We have assumed a "pure technical fix": that is, a policy relying only on technical measures which are presently available, are presently cost effective (at least at the margin), and have *no* significant effects on lifestyles. We assume only noncoercive deployment measures. To repeat, we assume no significant changes in where people live, how they live, or how they run their society—just a "more of the same" economic extrapolation. If there are to be, as Dr. Franssen suggested [chap. 3], "massive changes in consumer habits," those changes will come from the assumed economic growth, not from our energy policy. We do, however, take seriously *as the foundation of our analysis the orthodox economic criterion of supplying each desired energy service in the cheapest way*, that is, at lowest private internal cost. We leave out of our analysis, as a conservatism, all externalities, even though they are probably more important than internal costs as a basis for public policy.

With those essential provisos, let us quickly survey the state of the art in "technical fixes" for raising energy productivity in some main applications.

HOW MUCH ENERGY DO WE NEED?

Space heating and cooling is the largest single energy use in most industrialized countries. Table 6-1 shows that, with best present art, it is cost effective to reduce the space-conditioning energy requirements of new and most old buildings to approximately zero. In cold climates, this entails superinsulation, perhaps insulating night shutters or shades on windows, and tight construction—but with excellent ventilation through a simple air-to-air heat exchanger. (Analogous measures can reduce domestic water-heating needs by about half, with a payback time of a few years.) In hot climates, the technique is to prevent heat gain—by efficient lighting, insulation, heat exchangers, window overhangs or coatings or blinds, trees, and so forth—and to use passive solar cooling techniques if necessary. In short, whether in a subarctic or a tropical climate, any new building that needs significant amounts of energy to maintain comfort, or virtually any existing building that after retrofit still requires more than a small fraction of its present energy needs, is simply badly designed or underinvested in efficiency improvements.

Table 6-1. "Technical Fix" Energy Savings in Buildings[a]

North American new houses, examples	Approx. annual space-heating load per degree-day[b]		Extra capital cost (1979–80 US $)
	Btu/ft²DD(°F)	kJ/m² DD(°C)	
US 1976 stock, average	15	313	(reference case)
USHRAE[c] 90–75 standard	8	163	~0
Proposed BEPS[d] standard (subarctic climate)	3.2	67	payback in a few years
Leger (Massachusetts)	1.3	28	0
Phelps (Illinois)	1.2	25	≤0
Pasqua (Saskatchewan)	0.85	17	$2,000–3,000
Balcomb (passive solar, New Mexico)	0.46	9.4	payback in a few years
Saskatchewan Conservation House (not passive solar design)	0.22	4.5	≤$3,000
Best passive solar designs	~0	~0	~$0–3,000

Canadian office buildings, examples	Total annual energy consumption for comfort and lighting[e]	
	MBtu/ft²	GJ/m²
Present stock	~0.18-0.26	~2-3
Hydro Place (Toronto, best art, ~1975)	0.05	0.6
Gulf Canada Square (Calgary, best art, 1976)	0.015	0.17
	(payback in a few years)	

Retrofitted old buildings
Detailed analyses based on empirical costs show payback times of a few years, against 1977–78 fuel prices, for retrofits saving at least two-thirds of space heating in the US, British, and Danish housing stocks. European experience with exterior retrofit insulation suggests that savings from 80% to about 100% pay back in fewer than ten years when compared with synfuels ($40 per barrel retail). Well-designed basic retrofit programs in North American buildings consistently cost about $6–7 per barrel saved (about 0.4¢ per kilowatt-hour saved), or about $100 per square meter of commercial floor space.

[a]Compiled from sources cited in A. B. and L. H. Lovins, *Energy/War: Breaking the Nuclear Link* (Friends of the Earth, 1980), esp. footnotes 209–11.
[b]In British thermal units per square foot, divided by the number of degree-days in degrees Fahrenheit. Equivalently, can be expressed as kilojoules per square meter, divided by the number of degree-days in degrees Celsius.
[c]American Society of Heating, Refrigerating, and Air Conditioning Engineers.
[d]Building Efficiency Performance Standards.
[e]In millions of British thermal units per square foot. Equivalently, can be expressed in billions of joules per square meter.

Table 6-2 likewise summarizes the state of the art in making cars efficient. Without sacrificing comfort or performance, car efficiency can now be improved by nearly an order of magnitude from the present North American norm.

Table 6-2. **The 1980 Art in Efficient Cars**[a]

Examples	*EPA composite*[b]		Status
	mi/US gal	*km/1*	
1980 US fleet	16	6.8	
1980 W. European/ Japanese fleet	20-25	8.5-10.6	
Av. import sold in US in 1979	32	13.6	commercial
VW diesel Rabbit (Golf), with 10% less interior volume than av. 1978-model-year US-manufactured car	42	18	commercial
Turbo-charged version, same performance	60-65	26-28	marketable early 1980s
VW advanced diesel, bigger than Rabbit	70-80	30-34	phototype tested; marketable 1984–90
VW diesel-electric series hybrid retrofit in 350-lb (1590-kg) car using off-the-shelf components	83	35	prototype tested
Same, optimized drive/low rolling resistance	>100	>43	theoretical based on VW & BL[c] prototype tests
Same, crashworthy but very light body design	150-200	64-85	theoretical
SERI/EPA[d] "acceptable performance" compact	70-100	30-43	no new technology needed

[a]Compiled from sources cited in A. B. and L. H. Lovins, *Energy/War: Breaking the Nuclear Link* (Friends of the Earth, 1980), esp. footnotes 202–08.
[b]In miles per US gallon. Can be expressed equivalently as kilometers per liter.
[c]Volkswagen and British Leyland.
[d]Solar Energy Research Institute/US Environmental Protection Agency.

Table 6-3 summarizes the main cost-effective savings practically achievable in the use of electricity in typical Organisation for Economic Co-operation and Development (OECD) countries—savings sufficient to eliminate the need for *any* fossil or nuclear

Table 6-3. **Some Major Opportunities for Saving Electricity**[a]

Application	Saving in that use (%)	Typical payback (in years)[b]
Industrial motors: improve sizing, coupling, controls, clutches (UK, FRG, US data)	≳50	3-4
Household appliances: redesign (Danish analysis)	70	4
Lights: comfortable levels, task lighting, daylighting, efficient bulbs and fixtures (US data)	65-80	1-5
Alumina smelters: switch to best processes (European data)	40	5-10
With efficiency improvements & passive solar, replace the electricity now used for low-temp. heating & cooling (1/3-1/2 of all electricity demand in many OECD countries)	~100	1-5

[a]Compiled from sources cited in A. B. and L. H. Lovins, *Energy/War: Breaking the Nuclear Link* (Friends of the Earth, 1980), esp. footnotes 116 and 210.
[b]Against marginal delivered price of about 8¢/kWh, 1980 $.

power stations, old or new, to power today's economy in such countries as the United States, Canada, France, Switzerland, Sweden, Japan, and New Zealand (assuming only present hydroelectric capacity, readily available small-scale hydro, and a modest amount of windpower in the United States, and committed 1985 geothermal capacity in Japan).

What if such state-of-the-art, cost-effective improvements are added up in hundreds of sectors throughout an industrial economy? Very careful and detailed analyses of this question in several diverse countries, including the United States, Britain, the Federal Republic of Germany, and Denmark, have shown a total efficiency improvement of three to sixfold. The most detailed study, done by our colleague David Olivier and funded by the UK Atomic Energy Authority, has shown more than a sixfold efficiency improvement over about fifty years, both for primary energy and for electricity—using "technical fixes" that are all cheaper, and usually several times cheaper, than building synfuel plants or power plants to do the same tasks. Thus a growth in real Gross National Product (GNP) to 2.9 times and of industrial production to 2.3 times the 1976 level could be supplied by total primary energy use 0.42 times the present British level, and by total electricity use under 200 watts per capita (about one-sixth of the present US level), simply by using the energy in a way that saves money.

Developing countries should be able to achieve the same ultimate technical efficiencies faster and cheaper, since, lacking infrastructures, they can build them efficiently the first time rather than having to retrofit existing stocks as do the industrialized countries. If national case studies are used as an existence proof, checked by global and regional scoping calculations, then a world with a population doubled to 8 billion, with complete heavy industrialization of every country (assuming that to be possible and desirable on other grounds), and with everyone enjoying the present Western European standard of living, would probably use somewhat less total energy than the present 8–9 terawatts (TW)— and several times less than the economically inefficient levels shown by the scenarios of the International Institute for Applied Systems Analysis (IIASA).

There are two main reasons why most official analyses show much smaller energy savings than are actually possible. First, these analyses are aggregated and therefore omit large numbers of collectively significant, if individually small, opportunities for raising energy productivity. Second, they seldom assume the present best art even compared with present fuel prices, let alone with the marginal costs implied by their assumed supply systems.

SUSTAINABLE ENERGY SOURCES

Similar methodological errors lead many analysts to conclude that appropriate renewable energy sources are inadequate or unattractive. Only a fine-grained, locally oriented analysis—of the kind Amulya Reddy's paper [chap. 14] describes—can identify unique local resources. Only a careful matching of supply technologies to each task in scale, energy quality, and complexity can capture essential synergisms to optimize performance at least cost. In general, the supply curve representing the cost, difficulty, or nastiness of supply as a function of demand rises steeply and discontinuously after the shallowly sloping lower portions. Those lower portions represent the economically optimal balance between investment in energy productivity and supply. At that level, say around 8 TW for the world, virtually all long-term energy needs can be met cost effectively by presently available renewable sources. In particular, a plausible structure for 8 TW of demand would be about 1 TW or less of electricity—which can be met by present large-scale hydro and by readily available small-scale hydro modestly supplemented with wind—plus about 1 to 1.5 TW of portable liquid fuels for transport (available from farming and forestry wastes consistent with maintaining soil fertility). The rest is heat, which can be collected at temperatures sufficient for most industrial processes even on a cloudy winter day in Leningrad. This surprising result—that present "soft technologies" suffice, with careful selection and efficient energy use, to run expanded industrial economies in such countries as Britain, the Federal Republic of Germany, France, Denmark, Japan, and the United States—has been confirmed by many regional, national, and local analyses and appears to be robust. Though the mix of sources varies widely from one country to another and between adjacent parts of the same country, every country so far studied—more than fifteen, including the most difficult—appears to have enough. Such countries as France and Japan, though poor in fuels, are singularly rich in energy if only they seek a least cost solution.

Soft technologies—by which we mean the diverse renewable sources that are technically sophisticated but understandable to the user, and that supply energy at the scale and quality that will minimize the costs of energy distribution and conversion respectively for each task—are not cheap. But they are cheaper than not having them. Properly designed, they are consistently cheaper in capital cost, and several times cheaper in delivered energy price, than are the synfuel plants and power stations which one would otherwise have to build to do the same tasks. Further, they have such short lead times and fast paybacks that their velocity of cash flow is nearly an order of magnitude higher than that of competing hard technologies. Therefore, however difficult they will be to finance, they will be vastly easier to finance than Professor Häfele's sort of future, which increases the quantity of supply severalfold and its unit capital intensity typically tenfold or more above levels which most countries, and especially developing countries, cannot afford today.

The same short lead times, the wide markets, and the diversity and independence of constraints on the deployment of the immense variety of soft technologies likewise permit them to be deployed far faster than hard technologies, which have long lead times, narrow markets, and generic constraints. That is, soft technologies give more energy, money, and jobs back sooner per dollar invested than hard technologies. They also have many external advantages such as avoidance of nuclear proliferation and climatic change—advantages which this paper ignores.

The same relative quickness is also an advantage of efficiency improvements, reinforced by their large cost advantage and ready accessibility to many actors. For example, during 1973–78 in the nine European Economic Community (EEC) countries, about 95 percent of effective new energy "supply" came from energy savings; only 5 percent came from actual supply expansions. In the United States in 1979, about 97 percent of the real GNP growth (which was 2.3 percent) was fueled by energy savings; only 3 percent by energy supply expansions. During 1973–78, the United States *got* twice as much energy-"supplying" capacity twice as fast from efficiency improvements as synthetic fuel advocates say they can provide at five or ten times the cost. In short, the experiment John Foster urged [chap. 3] has actually been going on since 1973. Its result is that the centrally planned supply programs are already being outpaced approximately 20 or 30:1 by millions of individual actions in the marketplace. This is indeed remarkable in view of the many market imperfections (institutional barriers) still in place which prevent people from using energy in a way that saves money. Enormous tax and price subsidies—over $100 billion per year in the United States, and proportionately even larger in Japan—conceal true marginal costs from consumers. Some individual examples of the speed of efficiency improvements are striking: for example, last autumn, by a program of door-to-door citizen action, the people in the US municipality of Fitchburg, Massachusetts, weatherized at least 15 percent of their housing stock in ten weeks.

Though we have not been able in this short overview to do justice to a very rich technical background, we have tried to convey the message that technologies exist which, wisely and promptly used within the framework of conventional economic decisions, can effectively "solve" the energy problem. But the right technologies will not be used at the right time by the right people without much clearer thinking about their *purpose* and about the *criteria* for technological choice. We therefore turn now to the areas that many technologists do not like to think about.

WHAT KINDS OF ENERGY DO WE NEED?

So far we have considered only *how much* energy will be needed. But an equally important and seldom-asked question is *what kinds* of energy will be needed. Most analysts still treat demand essentially as homogeneous. They seek to get more energy, of any kind, from any source, at any price—even though there are in fact many different forms of energy whose different prices and qualities suit them to different applications. We instead take seriously a principle to which several previous speakers have paid lip service: that one should seek exactly the *amount, type and source of energy that will provide each desired energy service at lowest cost.*

This criterion, if used as a basis for policy, immediately casts the most serious doubt on the economics of any sort of central electric system at the margin (and thus makes it unnecessary to answer such questions as the technological or social characteristics of fast breeder reactors). This is because electricity is a very special, high-quality, and expensive form of energy. Based purely on economics—not on thermodynamic ideology—electricity in most countries has no marginal market. In today's dollars, electricity delivered from a new central station ordered today in most OECD countries will bear a price in the vicinity of 8¢ per kilowatt-hour. This is equivalent to buying the heat content of oil priced at $130 per barrel, or four times today's OPEC oil price. The special, premium uses which can justify this high marginal cost are extremely limited. As shown in Table 6–4, motors, lights, electronics, electrolysis, and other "electricity-specific" uses are typically only 7–8 percent of all delivered energy needs in the industrialized countries, and usually an even smaller share in the developing countries. The remaining energy needs—typically 92 percent or more of the total—are for heat (mainly at low temperatures) and for portable liquid fuels for vehicles. Both of these are applications for which marginal electricity is grossly uneconomic *in principle*, even if one assumes implausibly efficient heat pumps and electric cars. Worse still, present electrical supply is much larger than the electricity-specific needs (even before efficiency improvements), so a substantial fraction of the OECD countries' electricity is already being used in the way in which still more could *only* be used: namely, for low-temperature heating and cooling—rather like cutting butter with a chainsaw.

This implies that debating what kind of new power station to build is rather like shopping for the best buy in brandy to put in your car, or the best buy in antique furniture to burn in your stove.

Table 6-4. **Percentage of Total Delivered Energy (Heat-Supplied Basis) Required in Various Forms in Selected Industrial Countries, circa 1975[a]**

Form required	USA	Japan	Sweden	UK	France	FRG	W. Eur.[b]
Total heat	**58**	**68**	**71**	**66**	**61**	**75**	**71**
<100°C	35	22	48	55	36	50	45
100-600°C	15	31	14	6	14	12	13
>600°C	8	15	9	5	11	13	13
Portable liquids	34	20	19	26	29	18	22
Electricity-specific	8	12	10	8	10	7	7
ind. motors	5	7	6	4	6	4	4
other elec.[c]	3	5	4	4	4	3	3
(Supplied as elec.)	(13)	(16)	(18)	(14)	(12)	(13)	(11)

[a]Complied from Japanese data calculated by the authors from national statistics. Other data from A. B. Lovins Report ECE (XXXIII)/2/ADD. 3/Part II/1.G (Geneva: UN Economic Commission for Europe, January 1978). Reprinted in slightly revised form in *Energy Policy* 7 (September 1979), pp. 178–98.
[b]Average for Western Europe.
[c]Lights, electronics and telecommunications, electrochemistry, electrometallurgy, household nonthermal appliances, electric rail, arc-welding, etc.

It is the wrong question. It does not matter in the least whether one kind of new power station will be able to send out cheaper electricity than another, because *no* kind of new power station can come close to competing with the *real* competitors—the cheapest way to supply the same unsaturated end-use services. Those real competitors are familiar measures like weather stripping, insulation, heat exchangers, window overhangs and coatings and shades, greenhouses, pyrolysis of logging wastes—measures which, intelligently done, are cheaper than the running costs *alone* for even a new nuclear power station. Thus, if you have just built such a station, you will save your country money by writing it off and never operating it. Why? Because its additional electricity can only be used for low-temperature heating and cooling (the premium markets being already saturated), but it is not worth paying any more for that heating and cooling than what it costs to do it in the cheapest way: through efficiency improvements and passive solar. Those measures cost only about 0.3–0.4¢ per kilowatt-hour, whereas running a new reactor will cost about 1–2¢ per kilowatt-hour even if building it were free, so you are better off not running it. (Under US tax laws, the additional savings from not having to pay the reactor's future profits and tax subsidies will probably suffice to recover the sunk capital cost as well!) This same argument applies all the more to plants that are partly built, partly amortized, or fossil fueled. It applies right down to the level of capacity needed to meet electricity-specific end-use demand at a level of technical efficiency that is cost effective against new power plants. That level of electricity demand is generally at least fourfold below present levels to provide today's economic output in industrial countries with no changes in lifestyle (see Table 6-3 above).

WHAT IS THE ENERGY PROBLEM?

To emphasize the importance of this end-use-oriented view of the energy problem—what's the job? what's the best tool for the job?—consider this sad little story from a major industrialized country, which shall be nameless. The story concerns a "spaghetti chart," as shown in Figure 6-1, representing the flow of primary fuels via conversion processes to final uses.

A few years ago, the energy conservation planners in our anonymous country started (wisely) on the right-hand side of the spaghetti chart. They observed that their single largest energy use was for heating buildings, and found on analysis that the most uneconomic way to heat buildings, even using heat pumps, was with electricity. They fought and won a battle with their nationalized utility, as a result of which their government agreed (at least nominally) to discourage, or even try to phase out, electric heating because it is so wasteful of money and fuel. But, meanwhile, down the street, the far more numerous and influential energy supply planners in the same government were starting on the left-hand side of the spaghetti chart. Alarmed at the nasty imported oil, they mused, "Oil is energy; we must need some other source of energy. Aha! Reactors give us energy; we'll build reactors all over the place." But they paid little attention to what would happen after that. The two sides of the national energy establishment thus proceeded with their respective solutions to two different, and indeed contradictory, national energy problems: more energy of any kind, versus the cheapest kind for each task. And so they went on—until last year they collided in the middle: the Ministry of Industry suddenly realized that the only way to sell that extra nuclear electricity would be for electric heating, which they had just agreed not to do.

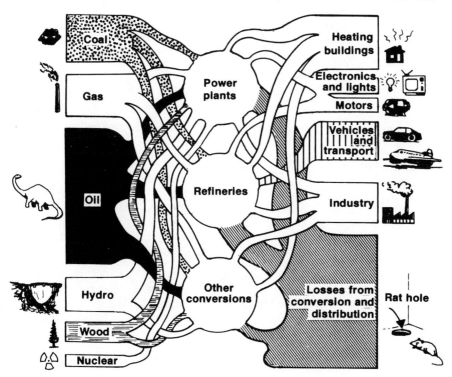

Source: A. B. Lovins, "Soft Energy Paths: How to Enjoy the Inevitable," *The Great Ideas Today 1980,* Encyclopedia Britannica, Inc., Chicago.

Figure 6-1. **A Schematic, Stylized "Spaghetti Chart"**

Every industrial country is now in the same awkward position as this anonymous country, and some developing countries are getting into it as fast as they can. What the example shows vividly is that which end of the spaghetti chart one starts on—what one thinks the energy problem is—is not an academic abstraction; it determines what one buys. People starting on the left side are led by hermetically sealed logic (based, admittedly, on false premises) to the view that the only major choice for energy policy is whether to build coal or nuclear power plants. People starting on the right, and taking their economics seriously, find at once that power plants are the *last* thing they would consider building.

If we did want more electricity, we should surely seek to get it from the cheapest sources first. In most countries these are, in order of increasing approximate price:

1. Eliminating pure waste of electricity (like lighting empty offices at headache level).

2. Displacing with efficiency improvements and passive solar measures (and in some cases even with active solar measures) the electricity now used for low-temperature heating and cooling.

3. Making appliances, motors, lights, smelters, and so forth cost effectively efficient.

4. Adopting measures such as industrial cogeneration, combined heat-and-power stations, low-temperature heat engines run off industrial waste heat or solar ponds, filling empty turbine bays in existing large dams, microhydro in good sites, modern wind machines in good sites, and possibly—though our analysis does not assume these—some new developments in photovoltaics with cheap optical concentrators.

5. Building new central power stations: the last resort, worth considering only after the four cheaper tiers of opportunities have been demonstrably exhausted, for it is the slowest and costliest known way to get more electricity—or to save oil.

HOW TO SAVE OIL

If we wish to save a lot of oil in a hurry, there are, at least for most developed countries, only two important methods. The prescription is distressingly simple: stop living in sieves and stop driving Petropigs. Illustrative numbers for the United States, applicable by analogy elsewhere, are instructive.

From the work of the Princeton group, the August 1980 Santa Cruz Summer Study, and others, we know in detail from empirical data how a basic weatherization program for US buildings can save at least 2.5 million barrels of oil per day over the next ten years, at a price typically around $7 per barrel or less. That one saving would be about two-fifths of the 1980 average rate of net oil imports. The rest of the imports, and more, can be saved by a single measure: turning over the private road-vehicle fleet faster to replace it with a fleet as efficient as the best already on the market.

The trade-in value of North American gas-guzzlers is so low that they have been filtering down to the poor people who can least afford to run them or replace them. The turnover is thus slowing down. But the incentive to accelerate it is enormous. A car fleet averaging only 60 miles per US gallon, or 26 kilometers per liter—less than the efficiency of existing turbo-charged diesel Rabbits (including one prototype safe in a 40-mile-per-hour or 65-kilometer-per-hour head-on crash)—would save nearly 4 million barrels of oil per day. This is more than the US imports from the Persian Gulf, two and one-half Alaskan North Slopes, eighty big synfuel plants, and more than seven times the pre-1979 imports from Iran. By combining this savings with a similar one in light trucks and with the weatherization program, the United States could eliminate the entire 1979 net rate of oil imports (7.8 million barrels per day) by about 1990—before a synfuel plant or power plant ordered today could deliver any energy whatsoever, and at about one-tenth of its cost. How, then, can car efficiency be raised so quickly?

Rather than building synfuel plants, if would be cheaper and faster to save oil by using the same money to pay between half and all of the cost of giving people *free* diesel Rabbits or Honda Civics or an equivalent American car if Detroit would make one—provided they would scrap their Brontomobiles to get them off the road. Alternatively, giving people a $200 cash grant for every mile per gallon by which their new car improves on their scrapped Petropig would give an average five-year payback against synfuels retailing at $40 per barrel. There should also be a bounty, based on inefficiency and residual lifetime, for gas-guzzlers scrapped and not replaced.

Can Detroit, which at the start of World War II shifted to completely different products in less than a year, leapfrog straight to cars averaging 50 miles per gallon (22 kilometers per liter)? Detroit is planning to spend about $50 billion in the 1980s on tooling for two new generations of cars. If transforming the cars incurred an extra retooling and retraining cost as implausibly high as $100 billion, that cost, spread over a complete new fleet of 100 million new cars and 30 million new light trucks, would average $769 per vehicle. The payback time against the 1980 US price of gasoline would be *under two years*. (Those of you concerned with the depressed car, steel, and coal industries, or with the inelegance of mandatory efficiency standards where a market-pull strategy would do better, may also care to contemplate the political advantages of redirecting John Sawhill's synfuels budget in this way.)

Our valued friend Dr. Sawhill is not unaware of these and similar opportunities. Indeed, he did our nation the great service last year of commissioning the Solar Energy Research Institute and its consultants to analyze a least cost energy strategy for the United States. Their draft report, being released for review around now, assumes a *two-thirds increase in the real GNP of the United*

States by the year 2000. Yet just by grasping cost-effective opportunities for efficiency improvements and renewable supply, taking conservative account of plausible deployment rates (not even assuming accelerated scrappage of gas-guzzlers), the United States could simultaneously *reduce its total primary energy use by a quarter and its use of nonrenewable fuels by nearly half*, thus rendering virtually all the federal supply programs unnecessary. Total electrical demand would also decline: indeed, the analysts' biggest problem was that after backing out all the oil, gas, and nuclear capacity, just the remaining hydro and coal capacity, without cogeneration, was still producing more electricity than could be used to economic advantage.

From the failure so far of the US government, or any other we know of, to implement such a least cost strategy, we can only infer that force of habit is stronger in official circles than conservative economic principles. This inference is strengthened by the observation that not one of the fifteen or so countries in which we work has ever based its choice of energy technologies on a fair and symmetrical comparison of all marginal investment opportunities for meeting each end-use need. It is instead normal to compare with each other in cost the systems that governments like to build—coal versus nuclear power plants versus synfuel plants—and then to compare less desired alternatives, like efficiency improvements and renewables, not with those competing hard technologies, but instead with the historically cheap (and often heavily subsidized) fossil fuels which the world is running out of and which all these investments are meant to replace. The hard technologies would fail this same test by a far wider margin. Thus, the US government dismisses many renewable sources in the $15–25 per barrel (bbl) range because they are said to be "uneconomic"—compared with old natural gas temporarily priced at about $10/bbl. But at the same time, to replace $30/bbl imported oil, the same officials encourage vast commitments to synfuels retailing at over $40/bbl and to electricity at over $100/bbl. That's just nuts (or, more formally, that leads to a misallocation). If instead we compare all marginal investments in each end-use category *with each other*, not some with each other and some with the old formerly inexpensive fuels, then we find, as the Harvard Business School energy study found last year, that the best buys are the efficiency improvements, *then* the soft technologies, *then* synfuels, and *last* of all the power stations. As a nation we have been taking those investments in reverse order, worst buys first. Virtually every other nation has been doing the same.

THE LEAST COST ENERGY PATH

The technologies forming the backbone of the IIASA energy future could never survive in a competitive market. Trying to finance and build them would require global equivalents of the Energy Security Corporation, to evade market forces, and of the Energy Mobilization Board, to evade democratic forces. The hard energy path requires massive central planning and intervention which the market-oriented soft path does not require and indeed probably cannot tolerate. Likewise, the hard path seriously compromises national security because its central electric and gas distribution grids are extremely vulnerable to disruption, whether by accident or by malice. The soft supply system, conversely, is diverse, dispersed, renewable, redundant, and highly resilient. The hard path allocates energy and its side effects to different people at opposite ends of the distribution system, creating inequities and tensions that are already causing more than sixty "energy wars" in the United States and that have brought parts of the Rocky Mountain states close to civil war. Soft technologies, instead, automatically give their costs and benefits to the same people at the same time so they can decide for themselves how much is enough.

It is this convergence of low political costs with relatively low economic costs that makes a soft path inevitable and permits a soft path to appeal to many constituencies for many reasons. If you are an economic traditionalist, most concerned with what's cheapest for you, you can put up your greenhouse because it's cheaper than not doing it. If you're a worker, you might want to build it because it gives you more and better jobs than building power stations. If you're an environmentalist, you can build it because it's benign; if you're a social transformationalist, you can build it because it's autonomous. But it's still the same greenhouse. You don't have to agree, before or after, about why you built it.

Nearly all countries today enjoy a virtual consensus that efficiency improvements and the benign renewable sources are desirable, but there is no consensus on any other energy options, and probably never will be. Perhaps, then, we should be pushing hard on the things we mainly agree about, so they'll be enough—and then we'll be able to forget the things we don't agree about, because they'll be superfluous. Nations have never tried to construct an energy policy around an existing, broadly based consensus, but it seems time they started.

The problems of doing that are messy and localized: purging institutional barriers and subsidies, and moving gradually and fairly toward marginal-cost pricing (or equivalent allocation mechanisms which can largely evade the efficiency/equity problem of energy pricing). But there is no energy future free of difficult problems. One can only choose which kinds of problems one would rather have. There is no free lunch, but some lunches are cheaper than others. We happen to believe that it is politically easier, especially in a democracy, to achieve insulated houses than a plutonium economy, but this is not a technical question and cannot be settled on technical merits.

SURPRISES

One thing we can all agree about, however, is the desirability of resilience in the face of surprises. In 1974, one of us (Amory B. Lovins) made up, for the Massachusetts Institute of Technology's Workshop of Alternative Energy Strategies, a list of the twenty most likely surprises in energy policy over the next decade or so. Near the top of the list were a major reactor accident and a revolution in Iran. The last item on the list—of which no examples could be given—was "surprises we haven't thought of yet." Many energy policy specialists who spend their professional lives coping with the effects of a singular event in 1973 cheerfully go on to assume surprise-free futures. It will not be like that at all. And some of the worst surprises may come from misguided efforts to "solve the energy problem" by making it into someone else's problem—a proliferation or climate or equity or land or water problem. Nations that do not take a holistic view will rediscover the hard way that the cause of problems is solutions. Indeed, we suspect that by the end of the 1980s, energy will be the least of our problems as the linkages between the problems of water, soil fertility, energy, and the sustainability of systems in general converge as an integrated resources crisis of enormously greater scale and intractability.

It is perhaps fortunate that the energy crisis came first, because it can provide a valuable metaphor for its successors. We are, for example, making all the same mistakes with water as with energy:

- being supply-oriented rather than demand-oriented;
- seeing water needs as homogeneous and ignoring the fact that different qualities are appropriate for different uses (the analogy to heating houses with electricity is flushing toilets with drinking water);
- pursuing illusory economies of large scale in water and sewage

systems while ignoring potentially larger diseconomies (which the US Environmental Protection Agency has lately come to appreciate); and
• pricing water at a tenth or hundredth of its marginal cost.

But those attempting to follow what Professor Sueishi of Kyoto calls a "soft water path" will find the going much rougher than in a soft energy path: if the analogue of soft solar technologies is dispersed wells, over a third of those wells in the United States are already poisoned.

The United States has long held itself out as an agricultural model for the developing nations. But we are now discovering that industrialized American farmers are unsustainably overcapitalized and overcommitted to energy-intensive, water-intensive soil mines. The United States is losing to wind and water erosion some 20–50 tons of soil per hectare per year; the maximum replacement rate is about 12. Iowa, one of the world's rich farming areas, has already lost a third of its topsoil, to say nothing of the soil that is compacted, depleted, or sterilized. From erosion alone, a dump-truck load of topsoil is passing New Orleans in the Mississippi River every *second*. The loss today is faster than in the Dust Bowl years, temporarily hidden from view by massive energy-intensive chemotherapy. In 1945 Illinois farms averaged 50 bushels of corn per acre; twenty years later, 95 bushels per acre. But over the same period the chemical fertilizer applied to those fields rose from 10,000 to 400,000 tons per year. The near doubling in production required a fortyfold increase in the application of energy-intensive fertilizer.

Forestry is likewise being run as a mining operation. High-yield intensive forestry, wisely outlawed in the Federal Republic of Germany, is merely an ingenious way to strip the soil of its organic content—the finely pulverized "young coal." Some analysts worry that the Pacific Northwest may be deforested in fifty years. As David Rose mentioned [chap. 3] , the United States is indeed overcutting its timber resources. The villain is, however, the high discount rates and biological illiteracy of some lumber and paper companies, not woodlot harvesting for firewood. (There is abundant firewood in Maine, but if there were not, the remedy would not be selling the stove to someone else, as Professor Rose suggests. Instead, the remedy would be insulating the house—just as in Nepal the solutions to the very real firewood shortage there might include efficient stoves even before biogas. No kind of biomass fuels program makes ecological or economic sense without greatly enhanced end-use efficiency first.)

These energy, water, and land problems feed on each other. Eighty percent of the water used in the United States goes to agriculture. In California in 1972, to produce a dollar's worth of alfalfa used 9,000 gallons of water, or 34,000 liters; of cotton, 4,000 gallons, or 15,000 liters (plus massive amounts of pesticides); of grapes, 2,000 gallons, or 7,500 liters. These three crops accounted for one-third of the entire use of land area and water to produce the more than two hundred commercial crops in California that year. To support such unsustainable farming operations, we have in many areas to mine fossil ground water. For example, the amount of water pumped in the four dry months of each year from the Ogallala Aquifer, a giant formation underlying the High Plains states in the United States, exceeds the full annual flow of the Colorado River through the Grand Canyon. The aquifer is being drawn down 1 to 3 meters per year and recharged less than 1 centimeter per year. Half of it is already gone.

The expensive, energy-intensive chemicals used in modern agriculture do not stay where they are put. They migrate, usually into the water supply, imposing massive water treatment costs. In midwestern US farm communities one now hears of rental arrangements that allow a livestock grower to back out of the deal if the water on the land turns out to be so nitrate rich that the livestock are threatened with illness or death. Runoff from feedlots and fertilized fields has poisoned aquifers and streams, causing miscarriages in farm families, headaches, blindness in children, and spontaneous abortions of pig litters, calves, and foals. Cleanup costs can run as high as $100,000 to install nitrate removal equipment and an additional annual cost of $20,000 to operate it.

Now add in energy. In recent years the feedstock for fertilizer has amounted to well over a fifth of the interruptible supply of US natural gas. As the water retreats towards the People's Republic of China, the energy costs of pumping are rising exponentially. Synthetic fuels are extremely water intensive and can outbid farmers for water. The same is true of the power plants needed to run the synfuel plants and coal mines. We recently were invited by farmers in Garden City, Kansas, to help them fight a proposed coal-fired power station to provide pumping power to mine Ogallala water faster—and the plant itself was to be cooled with mined ground water. The state of Washington has recently built a project called the Second Bacon Siphon to divert water from rainy western to arid eastern Washington. The pumping power needed to lift the water over the mountains plus the lost hydroelectric capacity add up to about 500 megawatts. This turns out, however, to be the same as

Washington's share of the proposed Colstrip Unit 3 and 4 power stations in Montana. If there were some mechanism (which there is not) for considering the net effect of the two projects together, it would become obvious that they substitute coal for hydro and export embodied water from eastern Montana (which is more arid still) to western Washington. In return, the Montanans do not even get cheap electricity from old Washington dams, but only deserts and smog.

A badly done biomass fuels program—one which is not used as a vehicle for reforms to make farming and forestry sustainable— could make large deserts quickly. A corn-based ethanol/gasohol program such as is now being proposed will lose at least 2 bushels of topsoil for every bushel of corn grown. Corn is also water intensive. Forty percent of US feedlot cattle are fed on corn and other grains grown with Ogallala water. To grow enough corn to add enough weight on a steer to put an extra pound of meat on the table consumes about 100 pounds of lost, eroded topsoil, and over 8,000 pounds of mined, unrecharged ground water. Enjoy your hamburger.

WHO HAS THE ANSWERS?

Solutions in this wider area of making our biotic systems sustainable may be much slower in coming than in the relatively narrow area of energy policy. But problems of this complexity and inter-linkedness are likely only to be solved, as the energy problem is starting to be solved, from the bottom up, not from the top down. Washington will be the last to know. This is because the energy problem is made of billions of little pieces scattered throughout a very diverse society. In that sort of problem, central management tends to be more part of the problem than part of the solution. As we deliberate here, individuals and communities across the country and around the world are getting on with the job of solving their energy problems, thus quietly aggregating the solution sought by so many in this room.

A few examples will give the flavor of what is happening in one pluralistic society—the United States. We have well over 200,000 solar buildings, of which about a half are passive and half of those are retrofits. About 15 percent of all housing contractors now use passive solar designs. In our more solar-conscious areas, anywhere from a quarter to all of the new housing starts are passive solar and solar heating already has 5 percent penetration in the whole housing stock. Over one hundred fifty New England factories have switched from oil to wood, on private initiative, and over half the households in the northern areas did the same. A handful of stove foundries blossomed into more than four hundred. Private firewood use grew sixfold, and the United States now gets about twice as much delivered energy from firewood as from nuclear power. The two biggest 1979 commercial commitments by the more than forty main wind-machine manufacturers totaled almost one-quarter billion dollars. Most states have fuel alcohol programs. Small-scale hydro reconstruction is flourishing and may be boosted further by cheap turbines starting to be imported from the People's Republic of China.

Why is all this happening? Sometimes, as in the Fitchburg example, it's because people are scared about oil. Sometimes, as in some midwestern towns we know, it's because the local utility has been asked to buy the next piece of a coal plant with a capital cost of over $2,000 for every woman, man, and child in town. That sort of number concentrates the mind wonderfully. It makes people say, "Wait a minute! That's an awful lot of money. For that much money we could fix up every building in town so it would never need heating or cooling again, and we'd have money left over. Doesn't that make more sense?"

But the best way to get to a soft path, truly to harness technologies to meet people's energy needs, may be to realize the consequences of not doing it. Perhaps the best example of this process is Franklin County, the poorest county in the state of Massachusetts. It has some decaying old mill towns, rocky and tenuous farms, many people out of work. It is cold and cloudy and depends on imported oil. Because they were worried about the oil, about fifty people from many walks of life got together with a $30,000 grant from the US Department of Energy a few years ago to study their energy problem. After a year they had an informal sort of town meeting to discuss the results.

The first thing they found was that every year, the average household in the county was sending out more than $1,300 to pay for energy. Someone held up a bucket with a hole in it, to symbolize the drain of money out of the county: $23 million per year was going from Franklin County to OPEC. They never saw it again. That $23 million turned out to be the same as the total payroll of the ten biggest employers in the county! That made people pay very careful attention. If, in the year 2000, they were lucky enough to achieve the *lowest* official forecasts of energy needs and prices, things would be four times worse: the average household would send out $5,300 per year to pay for energy, not counting inflation; and to keep that leaky bucket full, the biggest single employer in the county today would have to duplicate itself every couple of years for the rest of the century.

At this point the utility and Chamber of Commerce people turned white. They said, "That's absolutely impossible. We can't do that!" But people had worked out what to do instead. They could stuff up the holes totaling a square meter in each house, insulate, use passive and active solar heat, run their cars on methanol from the sustained yield of some unallocated public woodlots, and meet their electrical needs with wind or six times over with small-scale hydro within the county. They figured that the local machine shops with no work to do could make all the equipment, and the total cost would be about the same $23 million per year that they were already paying. But the difference is that they'd have plugged up the hole in the bucket. The money, the jobs, the economic multiplier effects would stay in Franklin County. Before the 1973 embargo, a dollar would circulate twenty-six times in the county economy before it went out of the county to buy something else. Today it goes around fewer than ten times. They're bleeding money. It was the Economic Development Commissioners who first understood that a soft energy path, stopping that hemorrhage, was the only hope for the economic regeneration of their county. So they and the County Commissioners and other constituencies got behind it, and as a result, it's no longer just a paper study; it's a project. With various fits and starts, they're doing it.

Another, final example. The San Luis Valley in southern Colorado is a cold, high-latitude plateau the size of Delaware, containing several of the poorest counties in the United States. Its traditional Hispanic people lost their source of firewood when a corporate landowner fenced their commons-land and began shooting at woodcutters. The people were too poor to buy wood, let alone any commercial fuel or electricity. But a few people in the valley knew how to build cheap (averaging $200 or less) solar greenhouses for heating and food growing. They started giving hands-on workshops. Word got around. In the past few years they have gone from a documented four to over four hundred greenhouse retrofits. They suspect there are at least twice that many. The doubling time is under one year. There are solar trailers, a solar post office, even a solar mortuary. The local Baskin & Robbins ice cream store has installed a high-technology solar systems. The greenhouses have spawned many other renewable projects: farmers are building a geothermally heated alcohol still using cull potatoes and barley washings. Wind machines are springing up. The valley is starting to think seriously about energy self-reliance with soft technologies—simply because, as in many pioneering projects in America's cities, the people were just too poor to afford anything *but* solar energy.

A similar process of community energy planning and action is occurring in at least hundreds and probably thousands of cities, towns, and counties around the country. Some, such as the San Luis Valley, are poor areas; some, such as Davis, California, are affluent. The common thread seems to be that in this explosively growing grassroots movement, the energy problem is being examined at such a detailed and concrete level that for the first time people can see it as *their* problem—something they can start to address with their own resources in ways they already know a lot about. It's no longer meaningless, abstract statistics that bureaucrats in the capital worry about; it's the cracks around my window.

We have abundant evidence coming in now, from around the country and around the world, that most people are pretty smart. Given incentive and opportunity they can go far to solve their own energy problems, just as we have always solved problems. There's nothing mysterious about energy. It isn't too complex or too technical for ordinary people to understand—though it may be too simple and too political for some technical people to understand. Technology can very largely be the answer—if only we remember to ask the common-sense questions.

In short, we are all relearning in energy policy what Lao-tse said some 2,500 years ago:

> Leaders are best when people scarcely know they exist,
> not so good when people obey and acclaim them,
> worst when people despise them.
> Fail to honor people, they fail to honor you.
> But of good leaders who talk little,
> when their work is done, their aim fulfilled,
> The people will all say: "We did this ourselves."

CHAPTER SEVEN

Selected Comments

NAIM AFGAN

I would like to express my point of view regarding some statements made in the Deutch paper. In order to clarify my point of view, I will use the same time scale as was used in the paper. As I have pointed out, the energy problem—the short-term energy problem—is a political problem, and I agree that there is no technical solution to this issue, but regarding the medium-range problem, Professor Deutch has failed to mention the possibility of transfer of presently available technology to less developed countries, which seems to me might be one of the ways to increase energy productivity in these countries. As we are all aware, transfer of technology, especially in the energy field, requires a time period of more than five years in order to get benefits from it.

Regarding long-term energy problems, I would like to support the idea expressed here that we need more international cooperation, particularly between developed and developing countries. If we assume that in a period of twenty years or more from now, the present developing countries will be developed by the present standards, this means that there should be room even now for them to be included in the long-term energy development programs. If we do not follow this program of international cooperation in the long term, we will again be leading our society to the point where the energy problem must be solved by the political means.

To be specific, I would like to mention two examples. The first example: currently, oil shale technology development is oriented to oil-rich shale, but in the developing countries, the great reserves of oil shale are low grade, and technology is needed to enable their exploration. If the present program of oil shale technology development was extended to also include research on low-grade oil shale, where possible using people from the developing countries, it would be of great benefit to all of us, because the future worldwide demand for energy would be less. The second example to which I would like to draw your attention is the problem of coal use.

There are many interesting development programs that are devoted to the problem of gasification and liquefaction of coal. As well as with oil shale, there is a great reserve of low-caloric coal in the developing countries, but use of this reserve is limited by technological development. Obviously, there will come a time when these countries will need new technologies for the better use of their reserves. It thus would be wise for the benefit of all to have more joint programs between the developed and developing countries.

Finally I would like to comment on the presentation of the Lovinses. I would like to support their positive attitude toward saving energy, but only to the extent that we do not forget that we really need new energy sources. It reminded me of a poster I have seen where an electric plug is shown—a big electric plug—with the inscription, "Why do we need new energy power when we have electric plugs?" I don't want to end up in that situation.

JOHN M. DEUTCH

Professor Afgan, let me quickly say that I hope I did not give the impression that I don't believe that there is room for technology transfer in the short run on energy conservation measures or other energy productivity steps. I would certainly strongly subscribe to that, and you would find me a complete supporter. I just did not manage to cover every aspect possible and every sensible thing that should be done. With respect to the oil shale, I would again quite agree that here is an example of a technology where international cooperation makes sense. As you quite correctly point out, there are very important oil shale deposits in other countries of the world than the United States—Morocco certainly comes to mind, and it would be precisely a venture of that sort—a cooperative venture of that sort—that I think is so urgent and important for the international community. Finally, let me again say that I don't want to leave you or anyone else with the impression that I do not think that more efficient use of coal is an urgent and important problem to deal with in the near and mid term. I would point to efforts on coal exports, coal trade—we do already have rather successful and welcome international cooperation through the International Energy Agency, Organization for Economic Cooperation and Development (IEA, OECD) on fluidized bed combustion, and all of the points that you mentioned would be ones that I would certainly subscribe to, and I regret if I didn't stress them adequately in my presentation.

HERMAN FRANSSEN

A short comment on so-called facts. When one takes a careful look at recent national data and pronouncements on conservation, one gets the impression that the entire reduction in energy demand in recent years has been due to more efficient use of energy. In fact, energy demand reduction or declines in energy growth rates are due to a number of factors, but we are still unable to separate actual conservation (using energy more efficiently) from the other factors. What has been happening on energy demand in recent years is partly the result of much higher energy prices. In addition to actual conservation there has been a great deal of deprivation, and there have been changes in the composition of the Gross National Product that led to a decline in energy-intensive industries. So we have to disaggregate very carefully our energy demand data to see what has actually happened and not to label as "conservation" every bit of reduction in energy demand.

On the issue of the soft versus the hard path, I find that I end up somewhere in the middle and that I don't see them as mutually exclusive but as actually complementary, because neither one is likely to prove correct in its entirety when we move into the future. Due to numerous uncertainties built in all scenarios, it is very difficult for policymakers to completely endorse either the soft or the hard path. In the real world, I think we will go ahead and try both paths: from the bottom, as the soft paths would suggest, and at the central government level, as implied in the hard path.

On the side of energy demand, the impact of both higher prices and government policy on reducing energy consumption may turn out to be greater than we currently estimate. (In fact, it has been the practice of the past decade to underestimate the impact of price increases on demand.) On the other hand, we cannot entirely rely on the soft path because of all the inherent uncertainties that exist in projecting the numerous uncertainties related to such drastic societal changes. Therefore, we have to go ahead, I believe, with the kinds of programs John Deutch discussed, to see which one of the various new technologies are likely to pay off in the future. And if it turns out that numerous demand-reducing actions taken at the individual and company level will in fact substantially reduce demand well beyond our current assessments, the better it will be. But at this point, I think we cannot really say that either one of the two paths is likely to lead us back to the promised land of stable energy prices and a high level of sustained economic growth.

AMORY B. LOVINS

Let me first pick up a comment of Professor Afgan. The technology transfer that I think ought to occur runs in both directions, and those of us in the highly industrialized countries need to learn some humility. I think, for example, of the item in *Peking Review* last December reporting that the People's Republic of China since 1972 had installed about 9 million biogas plants and that most of the rural electricity is from micro hydro. I think of the Brazilian ethanol program, of some of the things going on, say, in Papua-New Guinea, and it's quite clear that many developing countries are technical and institutional leaders in developing technologies of worldwide interest.

We weren't forgetting about energy supply; we were only saying that from an economic point of view, upwards of 90 percent of one's analytic effort and investment ought to be going into efficiency improvements, just because they are the cheapest and quickest thing to do. The transition to essentially complete dependence on renewables would take forty or fifty years and would require the intelligent transitional use of fossil fuels; we didn't have time to go into how that should be done. It might be a little faster in countries with special social advantages like Japan, but our point was that this transition should begin *now*, because if one does not choose the cheapest and fastest solutions at the margin, then the money spent on other things like, for example, synfuels or power stations actually *retards* oil displacement because of the opportunity costs: the same money and other resources cannot be spent on faster and cheaper measures to replace oil.

Dr. Franssen is absolutely right about efficiency improvements versus curtailments. People like Lee Schipper in this country have looked in some detail at the improvements in technical coefficients in sectors and subsectors, and they do find very considerable gains in technical efficiency quite aside from changes in composition and in intensity. In the Federal Republic of Germany, the projected energy savings in industry over the next fifty years or so are the same from cost-effective efficiency improvements as from the changes in composition that the Economics Ministry was expecting anyway but that hadn't gone into the energy projections.

We didn't have time in our talk even to define hard and soft energy paths, let alone to go into why we think they are exclusive. I think there may be some confusion here between *paths*—which we define partly by their political characteristics—and mere collections of *technologies*, which are in no sense technically incompatible, and we've never said they were. But we cannot afford, and we do not need, to do everything at once; and some options exclude others. I never have understood why over three-quarters of the energy investments and energy research funding in the Common Market countries goes for central electric systems (I think that's still roughly the right number), when that is a form of energy that appears to have no economic market at the margin, and when the institutions that can buy the technologies are facing very serious long-term financial problems. We see this problem of investment now with the incipient bankruptcy of much of the US utility sector. It's perfectly clear that because of the long lead times involved, one can overinvest in inappropriate or uneconomic technologies and then not have the financial flexibility left to do what one should have been doing instead. So in an era of real resource constraints, it is really very difficult to try everything at once. If you go into a food shop, you don't get caviar just for the sake of having something off each shelf; you try to get a balanced diet within your money budget. I think we ought to do the same sort of careful shopping with energy.

VACLAV SMIL

I have six things, and I again will be very quick and crisp. If the Lovinses like to quote Lao-tse, I would like to remind them—since I read Chinese philosophy just about every evening—that in Lao-tse's poems they would also find one which says "hard and easy are complementary, long and short are relative, high and low are comparative"—and that might interest them. And they would also find another line which says "Get rid of the experts!" So quoting Lao-tse is like quoting Mao or Lenin—you can find almost anything you want to find.

That was number one. Number two: I am one of the, to quote

the Lovinses, "millions of individual actions in the marketplace." I tried to insulate my twelve-year-old house as much as I could, and I sunk some $1,500 into it, and I carefully monitored my savings. During the past four years I cut my natural gas bill by about 40 percent, and I will get my money back in six to nine years. This personal experience makes me question any statement that in North America at the 50th parallel one can insulate an older house to attain a zero energy level and have the cost of insulation paid back in a few years.

Number three: a factual correction. The Lovinses state that a savings of nearly 4 million barrels of oil per day resulting from a highly efficient US car fleet would equal "several pre-1979 Irans." Not true. In 1978 Iran was producing 5.24 million barrels of oil per day, and so the savings would be less dramatic—just less than one Iran.

Number four: another factual correction. China currently has 7 million biogas digesters, not 9 million, and a good number of the existing units are not producing any biogas; they are just anaerobically fermenting manure.

Number five. Soil erosion losses in the United States are really tremendous, but there is a quiet revolution under way called reduced tillage or minimum or zero tillage. By the year 2000 as much as two-thirds of seven major annual US crops might be grown by the no-tillage system, which greatly reduces and often nearly eliminates erosion.

Number six. We heard a statement, and I quote, "It has to be this way." I thought that only God Almighty when he was creating this earth in six days, or supreme leaders of the extreme left, would say that. When we are dealing with human affairs, to say "it has to be this way" is not only rigidly dogmatic but, I think, intellectually very impoverished. We have more choices than one.

AMORY B. LOVINS

I think the economics of retrofit are about as we have described them here, using, in fact, Canadian data. It is quite possible to do worse than this by not doing measures in the correct order or by not using the most cost-effective techniques. One can also find, for example, that the cost per square meter of exterior retrofit insulation added can be two or three times in the Federal Republic of Germany what it is in Sweden for exactly the same technique, simply because it is not a familiar technique yet to the German contractors as it is to the Swedish ones.

Briefly on finance: suppose we require electric and gas utilities (in this country, for example, where they are regulated) to loan out their money at their own cost of money, repayable at or below the borrower's rate of return, for any efficiency improvement that is cheaper than a new power plant. That loan removes the capital burden on the consumer, which is otherwise a very serious impediment, as Professor Smil says; but it also keeps the utilities solvent by increasing the velocity of cash flow roughly tenfold, avoiding the high marginal costs of outside capital, saving capital, and keeping them from going broke by building more power plants than they can pay for. Analogous mechanisms, I think, would work quite well in most countries. This method has the further advantage that if you directly compare the cost of all options for providing end-use services at the stage where one is deciding whether to build a new power plant, then you are comparing, for example, efficiency improvements and soft technologies, not with the old cheap natural gas, but instead with the *marginal* cost represented by the proposed new plant. In this way you could thus allocate most of the capital going into the national energy system *as if* energy were priced at the margin, without first having to achieve those unpalatably high prices. By a rather simple measure of this sort, which California and Idaho have already taken at the state level, you could largely do an end run around the whole energy pricing problem.

L. HUNTER LOVINS

Professor Smil is quite right in stating that there may be technical fixes to many of the problems that I posed. Some of the leading research stations in this country are the New Alchemy Institute in Woods Hole, Massachusetts, and The Land Institute in Salina, Kansas, which is working on a radically new version of agriculture that would involve absolutely no plowing. That institute's philosophy is that the plowshare may well have been as destructive to humankind as the sword. However, most of the low-till and no-till agriculture being practiced now substitutes chemotherapy—heavy inputs of chemicals and pesticides and even greater amounts of fertilizer—for previous tillage practices. You can't just substitute one problem for another. We do, however, urgently need to work toward solutions, and I think they exist; we simply need to start finding and implementing them. I am less sanguine about the integrated resource problems beyond energy than I am about the energy problem, but I also feel that it is perhaps even more urgent that we begin to look for solutions to these problems, because they will surely be far more disastrous in their impact on humankind than will the energy problem.

JOHN M. DEUTCH

Hearing discussion about alcohol leads me to wish to raise some questions here. The first is that we should try to keep distinctions among different uses of biomass energy. I notice that the remark was made that wood-burning in houses and stoves produced more energy now than nuclear power in the United States. That's not been my experience in my part of the country. Perhaps it's correct. I do worry about the widespread use of wood; there are, of course, serious environmental consequences associated with wood-burning. I doubt that one will ever see this country produce the same amount of kilowatt-hours from wood as from nuclear energy. But another aspect of biomass is of equal concern to me—and I'm not quite sure where my friends and colleagues Amory and Hunter stand on the question—and that has to do with the use of corn to produce alcohol. I think that it's very important to be sure that we all clearly understand, first of all, the enormous subsidies in this nation—upwards to $60 per barrel subsidies for gasohol production, which is handsome compared with even some of our favorite high-technology items—and second, to note that the trend of using valuable food and chemical feedstocks—sugar and corn—strikes me as being a direction which one should not be eager, worldwide, to push, in contrast to the use of cellulose, which will take some longer period to develop. But I do think that it's important, because I was not clear, Amory, on where you stood on the use of corn in the United States to produce ethanol for cars.

L. HUNTER LOVINS

As Amory indicated to you, we think a massive corn-based ethanol-gasohol program is spherically senseless. It makes absolutely no sense, economically or ecologically. The problem of subsidies to the alcohol fuels program is a very real one, partly because these programs are to a large extent being done wrong. They are emphasizing outdated distillation techniques; they are emphasizing collection of feedstocks from a large area to be brought to a centralized plant, for conversion and then redistribution to dispersed users. This will probably not be a suitable model for an alcohol fuels program. We need a wide range of conversion technologies, from small mobile pyrolyzers that run around on the back of a pickup truck to local conversion plants about the size of a milk-bottling plant collecting feedstocks from a several-county region. And I suspect that the use of any kind of special crop for biomass will prove to be economically and ecologically unstable. We specifically recommend farming and forestry *wastes* only. If you have a very efficient vehicle fleet, the total amount of liquids that you need is quite low: in the United States it would be in the range of 5 to 6 quads per year to provide the sort of transportation system that we are running now.

We quite agree with you that methanol is, especially in the long term, probably the product of choice—far more important than ethanol, not only because there are more cellulosic feedstocks, but because it is often a more attractive technology.

The food-to-energy problem can be a very real one. In the United States I think it is a red herring, because if we are seriously concerned about that, then we will take our cattle, pigs, and chickens off the world's largest welfare program and put that grain into human food. In some other nations, I think it could be a real problem. But it is an unnecessary problem if we do a biomass program right. I think it is imperative that we do so, converting biomass *wastes* with efficient *processes* to run efficient *vehicles*, and using the programs as a vehicle for structural reforms that can make our farming and forestry *sustainable*.

PAUL DANELS

As, I think, the sole representative at this table from a community-based organization, I would like to direct my question towards what I can see is a community-based concern, but one which we share with the concerns expressed by Professor Sadli yesterday—those concerns of the less developed countries. Professor Sadli characterized the feelings of less developed countries in the energy crisis as being hapless victims of a big market struggle. In many cases, our National Urban League constituents—the poor, disadvantaged minorities in the United States, primarily living in cities—feel very much the same way and share that feeling. So I was very interested in the political and social comments, especially the comments of Professor Deutch on social equity in the US energy program. It seemed that Professor Deutch was endorsing subsidies for low-income people, such as those that have been-proposed earmarking 25 percent of the windfall profits tax for

low-income energy assistance. Professor Sadli's discussion yesterday of less developed countries spoke of the need to balance sub-
sidy programs with programs that promote self-reliance for those
countries, and I think an analogy can be drawn to this country and
the needs of the poor people in this country. And my question is, to
the speakers this morning: What are your views on the appropriate
roles of governments in achieving a balance between government
subsidies for energy costs and promotion of self-reliance through
economic development and employment strategies, and what
strategies would you suggest that we should take?

JOHN M. DEUTCH

Mr. Danels, let me just say that I believe that the single most
important aspect, from the point of view of the disadvantaged and
of social equity within our system, concerns the relationship of
energy to jobs, and it is most important to assure that energy
problems do not reduce job opportunities. Whether this can best be
through direct community development on energy-related projects or not, I would be open to evidence and discussion. But the
main point about helping the less advantaged in our society is
through the assurance of jobs and of reasonable economic growth.
If that comes through local development, that's fine; if it comes
through some other way, that is also fine. The second most important item is to keep the prices down as much as possible. Prices
must rise—the cost of energy is inexorably going up in Tennessee
or in Boston or anywhere. You must keep prices down for the
consumers as much as possible. But, finally, let me say that it is my
very strong view—and I wish that we had been more successful in
the windfall profits tax negotiations—that substantial funds
should be made available to the elderly, the disadvantaged, the
unemployed in the United States, not only to help protect them
from unreasonable price increases but also to help weatherize their
homes and to take other measures that in my judgment are required by humane energy policies.

L. HUNTER LOVINS

I totally agree with your point of view. The TransCentury Report
released a year or so ago reported that one in four Americans is at or
below the poverty level. Developing nations tend to think of the

United States as a very wealthy nation, but many of our citizens wonder where that wealth is going to. Merely subsidizing fuel purchases locks individuals—or nations—into welfare programs, which not only destroys their dignity but offers no ultimate escape from this sort of dependence. In a very short-run sense subsidization may be necessary for individuals or for nations, and certainly when you have steep fuel price increases, something has to buffer the impact. But wherever possible, the instruments should be geared to relieving that dependence, either through elimination of the need—through efficiency improvements—or through least cost new supply that is appropriate to that particular community or culture. An example along this line comes from the black community in San Bernardino, California, where Valerie Pope and some other welfare mothers banded together to bootstrap a very small Comprehensive Employment and Training Act (CETA) grant into the training of their own youngsters in installing insulation; they have now built that up into a several-million-dollar-per-year industry, not only installing insulation but building and installing solar collectors. They simply were tired of being on welfare. Those sorts of government programs, I think, can be exceedingly effective. However, in the Fitchburgh example, one of the problems that they had to overcome was that the CETA regulations prohibited non-CETA people from installing weatherization materials which had been sitting locked in a warehouse for the better part of a year. It took some people up from Washington, DC, from the federal agency ACTION, to bust that particular barrier and then have the good sense to leave the community so that the Fitchburgh people felt that it was their program. But that was government action of a very sophisticated and sensitive type. I think analogies of that need to be applied, not only in this country, but throughout the world.

JOSÉ GOLDEMBERG

I have a short question for Professor Deutch. It's not usual for a scientist, mainly a scientist of the Third World, to be against more support for science and technology. But I think that Professor Deutch's suggestion, which is in his paper, to establish a major international technical center dedicated to simple, small-scale renewable technologies, is completely misdirected. Most of the needed technologies exist, and in addition to that, there are many technological centers all over the world and in the United States doing that. What's really needed are not more studies but more practical actions, such as long-term loans and technical assistance that is not directed to promote commercial interests.

JANE CARTER

First of all, may I congratulate the organizers of the Symposium for what is a very well-structured program and the speakers both yesterday and today for their excellent presentations. We started off with a very able description of where we stand at the moment—a depressing starting point for our discussions. We heard where we might get to in the years of the next century—again, very interesting. This morning we came to Professor Deutch's very able presentation, but as head of the Energy Conservation Division in the UK Department of Energy I was a little distressed that so far conservation had been rather the poor relation of the discussion— the Cinderella. However, the balance was very quickly redressed by Amory and Hunter Lovins—indeed, the seesaw, if I may use the analogy, went right down on the other side.

I do agree very much with some of the points made by Amory and Hunter Lovins. I think, first and foremost, that conservation is a resource in its own right and that this should color our thinking and our approach to the problem. Conservation is within our own reach with existing technology; it is within our own control with the benefits remaining within our communities; and it is the only resource, I would suggest, that can make a really major contribution in this decade to reduce the dangerous pressures on the oil market that we have faced and that, I think, we must continue to expect for the rest of the decade and, indeed, to the end of the century. So from that standpoint, I think that there is a genuine need for a continuing cool look at the balance between supply investment and conservation investment. I wouldn't agree that one can pose them as alternatives, but bearing in mind the fragmented nature of decisionmaking on energy conservation, the work that we have done in the United Kingdom does show that some conservation measures emerge as much more cost effective than investments in increments of supply. However, as Dr. Franssen indicated, because of uncertainty we can't just back one horse in this race, and the uncertainty attaches, I would suggest, not only to demand forecasts but also to conservation itself. We have many of the technologies within our grasp now. We see prospects of others coming forward in the rest of the century. But we cannot be certain that all these techniques will be used to the utmost limits and produce the results we know could stem from them. And it is because of this inherent uncertainty that I think we need to look much more closely at the barriers and impediments to conservation investment.

Now I know Amory won't mind if I quote what someone said about him at another conference I attended. He said, "He's a brilliant speaker, but he's always got a 'for instance,' and by the time you examine the relevance of the 'for instance,' he is on to another, and you never catch up with him." Well, the "for instances" I was interested in were the ones that were quoted as examples of how conservation can work, how it's moving forward at the community level; and the "for instances" that came up were all poor and deprived communities, where necessity was the spur to action. Elsewhere in the economy I would suggest that conservation is not getting the priority it demands. We know that in industry, for example, companies look for shorter payback periods for conservation investment than for other investments. We know that more affluent consumers also are not giving full priority to conservation. An example was quoted to me this morning in a discussion of the five-car family. The family is accustomed to that standard of comfort and independence of transportation. The decision to have fewer cars or more efficient cars may not have the same priority for that family that it ought to have from a strictly conservation point of view.

So, after Minister Altmann's very telling reminder of the fact that without efficient conservation measures and more efficient use of energy we will be beggaring not only ourselves but our neighbors who can afford it much less than we can [see chap. 3], I would like to raise the question of what avenues of research we should be undertaking at the moment into motivations and barriers to conservation, and also—a point which I think is very interesting and was raised by Dr. Franssen—the question of the problems encountered in measuring conservation. These might perhaps be subjects for the following Symposium to take up in due course.

JOHN M. DEUTCH

Madame Undersecretary, I am delighted to hear, from the United Kingdom Department of Energy, this strong interest in conservation. I might say that when I was an Undersecretary of the United

States Department of Energy, I seemed to hear the UK Department of Energy talking a great deal more about breeder cooperation. But I welcome your interest in conservation.

Let me proceed to say that I did not mean—and was not able to address—all aspects of conservation issues before this country and others in my remarks; indeed, I understood the organizers of the conference wished for me to direct my presentation towards the supply side. I am extremely well aware of two facts about which nobody should kid themselves. At the present time, conservation is the most cost-effective way for the United States to proceed, and I hope that there is nothing in my presentation that suggested that conservation at the present moment, in energy productivity improvements, are the most cost-effective way to proceed— certainly more cost-effective than synfuels. I am also well aware of the deplorable record that we have in this country in this regard. I am aware of the poor record we have in automobile efficiency, the appalling record we have in weatherization—in building energy performance standards generally. Nevertheless, the issue before us at this session is not: Is an appropriate balance necessary, or how do you strike an appropriate balance? That is perhaps a bit too bland. We would all certainly agree that an appropriate balance is proper. Rather, I believe that the underlying hypothesis of this session is: How long can one go on without making large use of large-scale technologies? How long can one delay the development of synthetic fuels, responsible utilization of nuclear power, and matters of that sort? And it is on this point that I tried to dwell in my remarks. I do not wish to slight small-scale technology. I do not wish to slight the importance of cost-effective and rapid conservation measures. They are of priority concern. The point I'm trying to make is that we cannot, over the long haul, hope to proceed without paying some attention to the development and the deployment of large-scale energy supply technologies as well.

DAVID J. ROSE

Some very short comments regarding Professor Deutch's paper. The idea of internationalization of the entire debate and of trying to match the scale of the problem to the scale of the response is very important, and also, I welcome the comment that the synfuels program should be to keep option space open and not to try to imagine that we are going to satisfy some preordained yet uncertain goals. Regarding the Lovinses' paper, I welcome very much the comment that many big problems are not just those of energy but of food, resources, and other things. Energy, coming upon us now, is in some sense a paradigm of the others to follow. But in your paper this morning you said that this is noncoercive, which I find difficult to rationalize with what you say in the article you submitted to the program committee, which appeared in the summer, 1980, issue of *Foreign Affairs* [A. B. & L. H. Lovins and L. Ross, "Nuclear Power and Nuclear Bombs," *Foreign Affairs* 58, pp. 1137–77]. In that article, you state that a necessary and sufficient condition for nuclear stability is to outlaw all nuclear power or make it commercially not available. Those were your exact words.

AMORY B. LOVINS

I think that's a misunderstanding. We tried to point out in some detail in the article that nuclear power is not commercially viable and does not meet the test of the marketplace. We did not suggest outlawing it; we suggested accepting the verdict of the marketplace. In fact, we said toward the end of the article that "our thesis is certain to be misrepresented as 'trying to stop proliferation by outlawing reactors.' We have not said that. We have presented three main elements, and many sub-elements, of a coherent market-oriented program ...." and so on. [Ibid., p. 1174.]

DAVID J. ROSE

Just to cut that short, may I just read one sentence of yours. "The foregoing reasoning implies that eliminating nuclear power is a necessary condition for nonproliferation." [Ibid., p. 1147.] That implies coercion.

AMORY B. LOVINS

I really think that's a misreading of the article. We can take this up later, David, but the whole thrust of the article is *not* to suggest passing a law saying nobody can use nuclear power, but merely to suggest refraining from the heroic measures needed to bail it out from its incurable attack of market forces.

May I take this opportunity just to reply very briefly to Jane Carter. I've lived in Britain for the past fourteen years or so, and we've been in this together for some time. All we were pleading for in our paper was a symmetrical comparison in costs, rates, risks, difficulties, and so on among all opportunities at the margin. So far as I am aware, there is no Department of Energy in any country that has written down all the options for each end-use category on one piece of paper, with their costs and so on listed, so that they can be compared in this way. I am sure we'd all agree that we have fallen far short of doing what is cost effective: indeed, we are still fighting in Britain for the level of roof insulation that Sweden had in 1940. I think that what we most need to get a handle on in conservation is end-use-oriented data systems and some idea of what is happening locally, because institutional barriers are very often at the equivalent (to use US terms) of a county or municipal level, not necessarily at a state or national level. The United States has three thousand obsolete building codes, and so on. Purely national solutions can't even identify those barriers, let alone clear them.

JOHN H. GIBBONS

I have two questions and a brief comment. The first question is primarily directed toward Professor Deutch, and it is as follows: We've spent a good deal of time this morning—and I think appropriately so—balancing the issues of production (supply) and productivity (conservation). Your paper, by design, focused mostly on the production side, but at the same time, your remarks seemed to indicate a feeling that somehow the productivity options are mostly short term, mostly near-term-focused activities. I can appreciate that approach, because those, in fact, are the main actions in the energy marketplace today and contribute in no small portion to the recent reductions in US oil imports. At the same time, it seems to me one must be very careful about focusing on the near term because analysis indicates that the truly major options available from increased productivity are not those achievable in the short term but rather those that can be achieved in the long term, as new capital stock replaces old stock. Therefore, my question concerns the relative lack of emphasis—even, Amory, with your paper this morning—on options driven by the higher price of energy for new kinds of technology on the side of use efficiency. Amory, you stressed this morning that you were simply assuming things we already know about. It seems to me that it is rather a sad statement to imply that we don't expect many new discoveries about how to use energy more productively.

My second question has to do with Hunter and Amory Lovins.

This morning the energy/food interaction via the biomass option came forth; it is an argument that is common for all nations— industrial or Third World. Because of this commonality of interest it would seem that we should think together about the mutual needs and opportunities for cooperation in improved plant species development; in growth and cultivation methods that enable people to treat the land more carefully so that it can yield in a more long-term, sustained way. That same cooperation is needed in developing improved conversion technologies of biomass to readily utilizable fuels while at the same time using residuals of biomass production to maintain soil productivity. Therefore, in terms of government involvement in energy from biomass, should we not think about using government involvement in biomass energy as a mechanism to revolutionize the way we have been generally treating our land and water, our agriculture, our forestry? For instance, could we not use, as critieria for agriculture and forestry loans or development assistance, the extent to which biomass energy development also adds to and complements the way we use the soil? In other words, doesn't it make sense to use the leverage of energy from biomass to encourage a general improvement in agriculture and forestry management and conservation practice?

Finally, one brief comment. It appears to me that we seem inexorably drawn to a supply-side focus whenever we think about energy—toward making more. It strikes me, at least for the United States, that this reflects a very deep-seated way we think about the world and about how we solve problems and maintain human progress. We have thought in the past about moving west or producing more as our main hope for solving problems. It strikes me now that we have some new ways of thinking about how to solve problems: not so much by moving west or making more, but by using our brains a little bit more elegantly and applying this strategy, through technology, to the delivery of these goods and services while consuming fewer resources along the way.

JOHN M. DEUTCH

I regret that my use of the phrase "short run" has prompted some misunderstanding of my view on prospects for energy productivity improvement. One can expect a declining trend in the ratio of energy growth to economic growth from the expected 1:1 relationship as the economy adjusts to the sharp change in actual and expected relative real prices between energy and other inputs. This process takes place both in operations and retrofit and most importantly, as John Gibbons points out, through changes in the capital stock. Once this adjustment has taken place, *if there is no change in technology*, one should expect a return to the traditional 1:1 relationship in growth although, of course, from a lower absolute base in energy per unit output. Perhaps unwisely I refer to this entire adjustment period as the "short run"; that is, I include the adjustment in the capital stock. I certainly do not wish to imply that these savings are small. Both John Gibbons and I would agree, I believe, that the "savings" could amount to hundreds of quads over the next two decades. In the long term (in the sense I use it) the question is how well we can introduce new energy end-use technology to offset the anticipated need for new energy supply if economic growth is to be achieved, on a relatively economical basis.

AMORY B. LOVINS

With regard to John Gibbons's comment, I couldn't agree more that there is enormous scope for new technologies, especially biotechnologies. We are very concerned, however, that things are now going in just the opposite direction. The seed banks are being bought up by, for the most part, oil and chemical companies, and they are even trying to get US legislation passed making it illegal for other people to have access to the original, unhybridized germ plasm. That's a very scary development—narrowing our genetic base.

The point of the Land Institute's work that Hunter mentioned is to come up with a perennial polyculture agriculture instead of an annual monoculture—imitating, for example, the tallgrass prairie, because if there were a more efficient way to use the sunlight, it would have been there already. The New Alchemy Institute is another center of excellence developing integrated shelter-energy-food systems that are within everybody's economic and managerial reach. They can design a house, for example, that supplies your shelter, warmth, other energy, food, and cash crops. In principle, I think we agree about the direction of your question on biotechnology, but it is not clear that we know how to do that. Our understanding of biological systems is really in its infancy.

Finally, I think a propitious beginning on the last point John Gibbons made would be to use accurate terminology. When John

Deutch talks about "producing" oil or coal, he means mining it. I wish I knew how to "develop" my bank account by pulling money out of it. If we stop using these euphemisms and talk about the real depletion of resources after which there is no second crop, we will be a lot further along in understanding what we're doing.

JAN K. BLACK

Professor Deutch and the Lovinses are in agreement on one crucial point: that the central questions surrounding energy supply and distribution are not technological but rather political. From that point, however, they diverge sharply, and I find Professor Deutch overly generous in his assessment of the performance of the US government. I am more inclined to agree with the Lovinses that, at least in the case of US energy policy, the cause of our problems is solutions. I might add that one of the solutions commonly suggested—sacrifice—is a solution only to those who do not experience it. For those who do, sacrifice is not the solution, it is the problem.

Once again we are being told that national security calls for personal sacrifice. The crisis mentality that has been generated has made it possible for those who have the greatest interests at stake to stampede Congress into going along with extreme measures without adequate consideration. The first casualty of the energy crisis—the "moral equivalent of war"—has been the public interest.

The US government's acceptance of the thesis that national security requires freedom from dependence on foreign oil is a classic case of using the inherent nebulousness of the concept of national security to advance sectoral interests in a way seemingly contrary to any common-sense apprehension of that concept. This national security problem is now to be remedied by the creation of a national strategic petroleum reserve; one is thus presented with the extraordinary spectacle of a government that encourages more rapid pumping of oil out of its own soil while it pays inflated prices to purchase foreign oil which it then pumps under adjacent ground to store as a "strategic reserve."

The beneficiaries of the curious formulation of the national interest as requiring more rapid depletion of national reserves are, in the first instance, petroleum companies having substantial holdings in the United States but not abroad. Petroleum is inherently a highly profitable product, and further profit has been derived by them from the various forms of favorable special treatment that the political power of the oil industry has caused to be written into the tax laws.

In addition, it should not be overlooked that market manipulation has extended to the demand as well as the supply side of the equation. There exist fully documented cases in which oil companies have successfully opposed the development of mass transit systems, which would use less energy than the generalized use of the private automobile. The companies have also purchased controlling interests in less profitable alternative energy sources, not only to be in a position to profit by them should their use become necessary, but also, at least in some cases, in order to retard their development as competitors to oil.

The attempt by President Carter to develop and implement a concerted national energy policy is commendable and overdue. It is clear, however, that his attempt has encountered the fate typically meted out to other attempts to impose a vision of policy genuinely in the public interest. His proposed energy policy has been subjected to the push and pull of the private interests so well represented by those who have the ears of members of the Congress.

However, despite the fact that President Carter has leveled some harsh criticisms at the private oil industry, in general the pronouncements of government spokespeople, coupled with the media blitz of institutional advertisements by the industry, have left the public with the vague and confused notion that the enemy is OPEC, or "shortage" itself, or perhaps their own habits of wasteful consumption. In fact, however, these shortages have, to a large extent, been contrived and are not real: instead, what we are confronted with now is a balance of payments problem and fluctuating short-term shortages based not on absolute resource availability but on deliberate market manipulation, political (or economic) blackmail, or technical breakdowns.

The new quasi-public Synthetic Fuels Corporation, in which Professor Deutch places great hope, is almost a good idea—"almost" because even though it represents a threat to the environment and a waste of resources such as water, more precious by far than the energy resources to be extracted, and an extravagance unwarranted by the actual energy supply situation, it also represents recognition that government has a role to play in the development of energy resources. Unfortunately, it is, in this case, the wrong role. It is yet another means of diverting the taxpayers' money into the coffers of private corporations. There remains an urgent need for a genuinely public corporation to compete with the oligopolistic private ones, to assert control over the untapped resources beneath public lands, and to serve as a purchasing agency for foreign oil. Such a corporation might at least develop an information system independent of the private companies and a corps of technicians and managers skilled in energy production who are servants of the public first and foremost.

The energy crisis, as it affects North Americans, is, in the first instance, a domestic political crisis; no technological breakthrough and no degree of shrewdness or enlightenment in our dealings with foreign governments will compensate for a failure of political will and intelligence at home. But the US government, having arrived very late at a recognition of its responsibility to guarantee that the energy resources of the national domain will not be squandered and that the energy needs of the nation will be met, stands to learn a great deal from other governments that have been dealing with these issues and problems for several decades. I sincerely hope that symposia such as this one will help to further that learning process.

BOGUMIL STANISZEWSKI

With respect to Professor Deutch's presentation, I would like to comment that international cooperation in the field of energy is necessary and very urgent. However, this cooperation requires communication and discussion (which will happen during one of the future International Energy Symposia).

In addition, I would like to note that the program of synfuels development is very important, not only for the United States but for other countries as well. There are several reasons for this: (1) world coal resources are much larger than the resources of oil and gas; (2) it is very likely that some yet-undiscovered coal deposits exist, some of them in less developed countries; (3) successful development of synfuels creates the possibility of improving coal-mining techniques even at the present level of consumption; and (4) synfuels enable coal mining on a big scale and extend the period of mined fossil fuels as energy sources. This can give more time for developing new technologies such as fusion.

Finally, with respect to the Lovinses' presentation, I would like to mention that the introduction of "soft technologies" and efforts in energy conservation deserve support. However, in forecasting the expected results one should be very careful, realistic, and even a little conservative.

PIERRE JONON

I don't have a hammer to offer to the Lovinses, but only the result of numerous studies and applications performed with important means. So, I would pass on to them three remarks about their presentation:

1. How can electricity be considered any longer as an expensive energy? At the beginning of the century, the cost of 1 kilowatt-hour was equivalent to one hour of work for a laborer. It is now equivalent to one minute of work. Over the coming years in France, because of the development of nuclear energy, the cost of electricity will become less, in the off-peak periods, than the cost of heavy oil used in industry. If you consider that electricity is much more efficient for users than oil is, it seems to me that it would be a very attractive mode for the future.

2. Your presentation of the energy-saving problem is largely unrealistic and wrong. As an example, the French program for saving energy, which we consider as drastic, foresees "only" a savings of 3 million tons of oil every year in the near future. That savings will be obtained in different ways, among which will be the use of electrical processes in industry in place of oil-steam processes, thereby leading to an important savings of primary energy—50 percent on an average.

3. We work intensively on the development of what you call "soft energies," but at this time our conclusions are quite different from yours: "soft energies" are essentially inefficient and expensive. So, their contribution to our energy supply will be very small in the future; perhaps 1 or 2 percent by the end of the century, and that mainly issuing from agricultural sources and high- and low-temperature geothermics. Now, if you know of other cheap devices using "soft energies," tell me, please, where to go to purchase such devices. We will rapidly develop them.

JOHANNES PIETER VAN RIJ

The point was raised by one of the precedent speakers that in the European Community most research and development (R&D) was directed towards the production of electricity and would lead to overproduction of electric power. It is not easy to say how European Community research funds (throughout the complicated systems of direct, indirect, and cooperative actions) are distributed over different categories. A very considerable part is destined to energy conservation, which is an inevitable short-term objective.

Another part is directed towards the development of alternative forms of energy, such as solar and geothermal. And a third large share goes to nuclear technology and safety. However, it would be erroneous to consider this third share as wasted on an uneconomic source of energy. *First of all*, most present electricity generated in the European Community is based on imported sources, and replacing these by cheaper homemade nuclear energy through advanced reactor techniques will contribute to the European Community's decreasing import dependency and consequently to more international stability. *Second*, additional advanced nuclear techniques possibly can also supply considerable contributions to economic and environmentally safe uses of heat for industrial purposes. *Finally*, I want to point out that much of the European Community's R&D is linked with projects of other countries and organizations through international cooperation agreements, thereby significantly increasing its scope and its efficiency.

ROBERT L. LITTLE[1]

Speaking as a forester turned wood technologist, I feel compelled to comment on the presentation of Amory and Hunter Lovins.

I am in complete agreement with their stand on the fullest use of existing technologies to attain energy efficiency. In fact, to carry their arguments for building construction efficiency a step further, using wood products as both structural and decorative components will save energy in many ways:

1. Less energy is required to convert a standing tree to a ready-to-use building component than would be needed to produce a comparable product from a competing material such as steel, aluminum, or concrete.
2. Wood in and of itself is a much better insulator than are other building products—for example, steel, aluminum, or concrete.
3. Less energy is required to assemble wooden components into a usable building than with other materials.
4. Due to wood's high strength and relatively low weight, energy expended in getting it from the point of manufacture to the point of use is less than for other building materials.
5. Wood is renewable.

Point five brings me to the indictment that the Lovinses made against the wood products industry. I will be the first to admit that forestry is an inexact science and that many indiscretions have occurred in the past. I maintain, however, that sound forestry practices are the best way to achieve full production on our forest lands.

The science of forest management has made quantum leaps in the last few years. Because wood is renewable, it is being looked to for the supply of more building components, furniture, insulation, and fuel. Forest managers have the responsibility for producing wood for an increasingly wide variety of products for more people. This expansion is expected to happen despite urban encroachment on forest lands which takes those lands out of production.

To the untrained eye, practices such as even-aged management, clear cutting, and monoculture may seem to be raping the land. Research has not borne this out. Closer, more factual examination of all aspects (species, topography, hydrologic conditions, markets, and landowner desires) involved in making wise management decisions frequently shows that these practices have merit. Indeed, in some cases they are the only feasible alternatives.

Offsetting these problems are the expanded use of logging slash, use of wood residues, and recycling of wood products. The last ten years has seen a boom in the development of composite wood products made from wood residues, such as medium-density fiberboard, particleboard, paper, and structural flakeboard. Unfortunately, these higher uses mean that wood residues are essentially priced out of the cheap energy market. Wood to be used directly as a fuel or as a raw material for alcohol production can be obtained most economically from logging slash. Currently, large volumes of this material are left in the field due to the nonavailability of a suitable collection system. Estimates on the volume of this material left in forests in the western United States go as high as 227 tons per acre.

The "bottom line" of my comment is that we must rely on good forest management to produce the wood that the world needs in the short run to fill the energy gap. At some point in the future, technological advances may, we hope, erase the energy supply gap, enabling wood to be put to other, more productive uses such as furniture, where its natural beauty can be used to the fullest.

DEE ASHLEY AKERS[2]

Despite its tendency to create new problems, technological development is obviously a necessity in dealing with the world's energy problems. But the more dependent the world becomes on energy technology, the more caution is needed. Technology's propensity for begetting technology almost invariably extends the developmental time frame, leads to higher costs, and jeopardizes the original technological mission. Risks of failure and of adverse impacts seem to vary inversely with the length of the time frame within which the technological development takes place. Thus, the application of technological solutions to the world energy problem under "crisis" conditions carries with it a larger than usual set of risks.

Absent adequate technology assessments prior to development, diversity becomes crucial in the mix of technological solutions—as to both specific technologies and to the context within which they are developed. The need for a variety of energy solutions warns of a need for a degree of technological hedging through diversification of developmental contexts, especially where so much of the technology is large scale. To increase certainty of positive mission results, and to reduce skewing of these results in favor of one set of interests to the detriment of others, it is important that there be a special *public interest* development program separate from and supplemental to projects being attempted with the private sector. Such an approach not only enhances the prospect of success but also provides a standard against which to measure private sector results, to avoid excesses as well as to assure quality results. Governments are justified in taking developmental risks as a means of evaluating new technologies on other than a marketplace test, especially since such energy markets are not generally free to operate according to competitive market principles. Thus, there is a need for vigorous public sector research and development programs to offset the risks and costs associated with complex developments that may have only marginal significance. Such an approach will increase the prospect of timely solutions to the energy problems at affordable prices and decrease the prospect that the developmental process will serve mostly the interests of technology. Low-cost technologies are not likely to be developed by energy companies who do not deal in small-scale business. Government leadership and support are essential to the development of such technologies.

NOTES

1. Robert L. Little, Ph.D., Associate Professor; Department of Forestry, Wildlife, and Fisheries; University of Tennessee, Knoxville, TN.

2. Dee Ashley Akers, Director, Government Law Center (Technological Law); University of Louisville, Louisville, KY.

For identification of the other discussants, please refer to the list of participants in Appendix II.

CHAPTER EIGHT

Summary

Hiroo Tominaga
Professor of Synthetic Chemistry
The University of Tokyo

In this session we had three distinguished guest speakers, John M. Deutch and Amory B. and L. Hunter Lovins. First of all, I would like to express our thanks to them for their presentations on the role of technology in improving world energy productivity and production. They provided many things to think about and plenty of material for discussion. We had also many questions and answers, comments and arguments.

All of these have brought out so many new ideas that I myself cannot at this moment comprehend all of their implications. However, we have identified to some extent the complicated nature of energy, especially in light of technological development. I don't think we had complete consensus on the role of technology in solving energy problems. The truth will emerge, I hope, from continued dialogues, not only between the so-called energy experts of different disciplines but also between the energy experts and laypersons, in the following Symposia II and III.

Now, I would like to summarize our session very briefly. Let me begin with the lecture given by Professor Deutch, who has discussed some principles concerning the role of technology in meeting future energy needs.

Professor Deutch told us that, due to the complex and interrelated nature of the energy issue, technology cannot provide an instant solution to the problem, but it can or should offer the materials for economic and political consideration and planning. He told us also that the effort to develop energy technologies should be directed not only to the increase of supply and productivity but also to more fundamental social objectives. These objectives are economic growth, national security, and environmental protection.

Incidentally, the policy decisions on the development of energy technology should take into account the following:

- The end-use perspective, or the end-use-oriented view mentioned by the Lovinses, is essential.
- It is vital to have timely allocations of the resources needed for research and development for the near term (1980–85), mid term (1985–2000), and long term (2000 and on).
- A long lead time is required to establish a new technology.

In comparison with other commodities, the characteristic nature of energy was described by Professor Deutch as follows: energy is basically difficult to substitute and closely correlated to economic growth and national security. Besides, the incentives for energy research and development are deeply affected by forecasts of the future energy supply, demand, and alternatives and by environmental regulations and so on. Accordingly, the role of government is so large that the following questions should be deliberated:

- Should energy be left to the free market or not?
- Will decontrol of energy prices bring forth a better solution or not?
- To what extent and by what means should government intervene in the energy industries' research and development programs?

The choices may depend on the political-social structures of the respective nation.

However, let me point out the following: first, the primary energy resources, especially oil and gas, are distributed very unevenly over the world; and second, the supply prices of primary energy vary widely from one country to another. Because of these facts, an energy technology could be technically feasible but economically infeasible in some countries. In such a case, subsidization of the technology by the government is desirable if the technology is indispensable to achieve the country's energy policy goal—for example, diversification of the primary energy supply for national security.

Professor Deutch also referred to international cooperation in energy technology. He made the point that there has been far less concrete action than there should have been, for several reasons and because of several difficulties.

If we have many technical problems in common, why don't we share the very large cost and risk of the development of energy technologies? There appears to be growing concern that the deployment of efforts to develop energy technology is sometimes beyond the capabilities of single nations. Several bilateral or multilateral cooperation efforts have been started, but no international cooperation in a global sense has emerged yet.

Professor Deutch proposed his idea of the establishment of a major international technical center dedicated to the research and development of renewable resources, namely, biomass, solar, wind, geothermal, and so on. I am in favor of his proposal, but let me add a few more words. Specifically, this center also should be dedicated to the research and development of technologies for more efficient energy use or for elimination of energy waste, and to education and training for these purposes because energy conservation is, I believe, one of the first things we have to pursue to mitigate the energy problem. Professor Deutch concluded by describing the research and development program of energy technology in the United States.

Now, I would like to sum up the Lovinses' presentation. They have identified the energy issue with which we are faced today to be a problem of how to provide the amount, type, and source of energy to satisfy each desired end use in the cheapest way. This is, I think, basically in line with the statement given by Professor Deutch concerning the policy for research and development of energy technology.

Next, the Lovinses told us that the energy problem does not arise from a lack of adequate technologies. Technologies do exist, but the real problem is the wise choice of the technologies, which is instead a political and ethical issue.

They cited that there are many opportunities and great possibilities for increasing energy efficiency in space heating, car driving, electric power utilization, and so forth. Energy savings can be achieved, they mentioned, by using "technical fixes" in accordance with an end-use-oriented view. Coupled with the above, they maintain that the deployment of an immense variety of soft technologies can supply virtually all long-term energy needs from renewable resources, namely, hydro, wind, biomass (such as farming and forestry wastes), solar, and so on. These technologies are now either available or in progress and will enter commercial application very soon. They are cheaper, cleaner, and more attractive than the conventional hard technologies.

One might hope that all of these messages were perfectly true and believable. Frankly speaking, I cannot follow them without hesitation.

In the long history of mankind, the reality is that we are running out of oil and other fossil fuels. I do share the concerns about the possible global climate change due to combustion of fossil fuels. There seems to be no lack of supply of renewable energy resources. In view of this, fossil fuels should be replaced by renewable resources, as soon as possible and to the best practicable extent.

The day of soft-path energy, however, does not seem to me to be in the immediate future. In fact, the renewable resources are almost infinite in their integrated volumes, but there are limitations on their individual availability for several reasons. Sunlight, as an example, is very dispersed (165 watts per square meter on an average) and fluctuates by day, weather, and season. Accordingly, for the full utilization of solar energy, many technical problems remain to be solved. Hence, the immediate roles of solar energy and other renewable resources may be supplementary to the existing hard-path energy system, and there is a need to explore the compatability of the soft and hard energy paths for the long future.

Incidentally, all of the natural resources cannot be turned into "reserves," the true economic assets for human beings, without wise human action in which science and technology are essential. Petroleum, for example, was good for nothing before we learned how to find, extract, refine, and make use of it. The same applied to uranium. In this respect, the role of technology in solving energy problems is, I should say, essential and fundamental, even if it has limitations.

Over the next ten years or so it may be political/economic factors rather than technical factors that will have major significance. This is due to the specific character of energy issues in the near term. Namely, for the moment, interruptions of oil imports and increases in oil prices are seemingly the greatest concerns of almost all of the developed countries, and of developing countries as well. In the mid and long term, however, technology can and should provide the alternative choices for political and economic actions to prevent or to mitigate the energy crisis.

In view of this, it is obviously essential to learn what the energy technologies are; in other words, what their current status and their possible future developments are. This is prerequisite, I think, to discussing the role of technology in solving energy problems—different technologies' capabilities and limitations, including their impacts on environment, and so on. This has not been done enough today, but I hope it will be taken up in the next Symposium, so let me point out just a few subjects very briefly.

Fossil fuels are finite in their size and exhaustible, so from an historic point of view, the days of fossil fuels may be a temporary phase. However, the transition to the days of nuclear power and/or renewable resources should be gradual. Because of this, the research and development of fossil fuels still remains one of the most important subjects. This research and development should include the following:

- enhanced recovery of oil and gas from known deposits;
- exploitation of new zones—deep offshore, Arctic, and geo-pressurized zones;
- recovery of tar sands and oil shales; and
- liquefaction and gasification of coal.

I don't want to give a detailed description of these subjects because of the limitation of time, so let's move on to the next research subject—nuclear energy. I myself am not well qualified to talk about this, but to my knowledge, social acceptance of the nuclear energy system is the most important problem, and from a technical standpoint, the establishment of fuel cycles and safeguards for the nuclear system are vital.

The last but the most important category of energy research and development is improvement in energy efficiency. This should be given top priority, especially in energy-consuming countries like the United States, the European countries, and Japan. There are many possible fields for increasing energy efficiency in energy conversion processes and in the industrial, transportation, household, and commercial sectors.

I am afraid there is not enough time to get into a detailed discussion of every energy savings possible, but let me show you how the demand for energy in industrial production has changed in Japan since the last oil crisis in 1973. In Figure 8-1, several indices of economic activities are plotted against the fiscal years 1973 to 1978. The Index of Industrial Production (IIP) was 100 in 1973. It fell sharply in the following two years but rose thereafter, attaining 105 in 1978. The energy use for the industrial sector is shown by a broken line. In the two years immediately after 1973 it decreased, and it remained around 95 during the next three years. This means that, in the past three to five years, we have realized an IIP increase of 5 percent with 5 percent less energy consumption. Namely, a 10 percent increase in energy use efficiency has been achieved in our industries. This might have come in part from an improvement in technology and in part from a structural change of industry; a detailed investigation is now under way. We are expecting that another 15 percent or so increase in energy efficiency will be attained in the industrial sector by technological advances.

Now, what are the priorities for implementing the research and development of energy technologies? It depends on geography and climate, and also on the economic, industrial, technical, and cultural structures of the society or community. Different countries or communities may have different priorities for energy technology development. This is, I believe, one of the subjects to be explored in the next Symposium.

In conclusion I would like to repeat what I said in the begin-

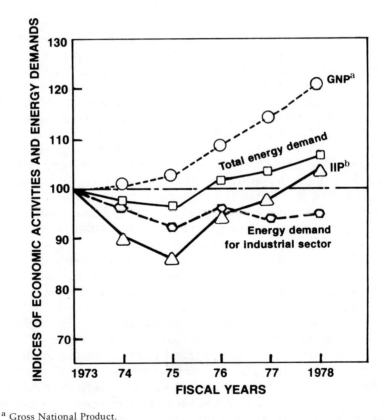

^a Gross National Product.
^b Index of Industrial Production.

Figure 8-1. **Improvement in Energy Efficiency in Japan since 1973.**

ning: we now have some discrepancies in our opinions, but it is only through the repeated exchange of information and opinions that we can reach the true answer to the urgent energy problem we all share in common. In this context, the International Energy Symposia Series should be highly appreciated.

Toward an Efficient Energy Future:

Critical Paths, Conflicts, and Constraints

Introduction

Hans H. Landsberg
Senior Fellow
Resources for the Future
Washington, DC

There is very little I can think of in the field of energy that does not easily fit under the title of this afternoon's session. Certainly in this country, and I believe elsewhere, policy decisions in both the public and the private sectors have had to face the stubborn reality that whatever steps are taken in the energy arena have repercussions for other societal goals.

It used to be said that most good things in life are either illegal, immoral, or fattening; in a variation of this theme I sometimes think that most obvious solutions in the energy field are either economically inefficient, environmentally adverse, socially inequitable, or politically infeasible, or all four of them. I have long maintained, and I do now believe, that much if not most of the time that we seem to be discussing energy matters we are in fact merely using the topic of energy as a means, as a pretext, as a proxy. In reality we are discussing the state of mankind locally, nationally, regionally, globally: who we are; who we wish to be; what is right and what is wrong (mostly the latter).

An energy debate often resembles nothing so much as a gigantic group therapy session. The entrance fee consists of knowing some aspects of the energy problem, and then we settle down to display our entire armory of beliefs, fears, prejudices, commitments, and dreams both good and bad. I say this not to be critical (no doubt I am infected by the same virus), but I believe that looking at the energy debate in this way provides perspective, detachment, distance; and, believe me, every little bit of this helps in the debate. We might not reach consensus more easily; we might not reach it at all, but we will have a more accurate sense of what we are disagreeing about. Again and again I discover that what seems at first blush disagreement over fact stems from the consequences of one's profoundly held beliefs in the insights that analysis of these facts points to. That is to say, very often we are arguing back to rather than forward from what we believe are facts. That, more than anything else, makes the energy debate both fascinating and frustrating.

You may wish to speculate about the meaning of having on this afternoon's program three physicists, excluding myself—I am an economist, if you haven't noticed. They surely have gotten a long ride out of a mere, measly two laws of thermodynamics! The US Department of Energy without even half trying turns out many times that much in any day of the week in terms of new regulations. As I introduce the speakers, you will also note that not only are they all physicists but they are all nuclear physicists. I refrain from drawing any inferences whatsoever from that fact, for either the fortunes of the nuclear power industry or the fortunes of the nuclear physics profession. A friend of mine on a similar occasion referred to this phenomenon of the ubiquitous nuclear physicist as "the wandering minstrels of the late twentieth century." If so, I am delighted that they have taken up temporary residence this afternoon around this hall and will be our speakers.

The first speaker of this session is Professor José Goldemberg from Brazil. I shall make my introduction short to not cut into his time. He is widely studied, widely traveled, widely known, a member of several prestigious organizations (including the Brazilian Society for the Advancement of Science, of which he is president), and a writer of books and articles. He is probably known to most people because of his interest and active participation in exploring and doing something about the potential that ethanol has for becoming a significant member of the transportation fuel family, and that is his subject here.

The next speaker, as I noted, is also a nuclear physicist, but with a very different approach. If Professor Goldemberg's can be called "from the bottom up," I think Dr. Nwosu's method, which he is going to present, is "from the top down." They can both argue with me about whether that is true or not.

Dr. Nwosu is from Nigeria. He presently holds the post of Chief Education Officer for Science in the Federal Ministry of Education. Much of his life he has been a teacher and a writer, largely on nuclear energy, but as I looked over his publications, I detect a slow turn which has now, I think, come full circle, to the much broader subject of the role of science and technology in human affairs, and finally to human affairs as such together, science or no science. That is why I say he approaches topics from the top down.

My final duty is to introduce David Rose. I believe unless somebody has tried very hard not to know him it is almost impossible to imagine that anybody has not met him. He has, or could have, been seen over the last thirty years at the Massachusetts Institute of Technology; since 1960 he has been Professor of Nuclear Engineering, unles he has been away consulting and advising and learning, which has been much of the time. So that may explain your encountering him here for the first time. But he needs no other introduction—he has a very deep commitment to human values, a profound knowledge of technology, and an insatiable curiosity; I know, because I have seen him in action. That is why there is no better way of ending this session than to conclude with David Rose.

CHAPTER NINE

A Centralized "Soft" Energy Path

José Goldemberg
Director of the Institute of Physics
The University of São Paulo

As is well known, Brazil has embarked on a large program of substitution of petroleum derivatives by ethanol produced from sugar cane. The first goal of this program was to add 20 percent alcohol to gasoline in a mixture branded "gasohol" which does not require modification of conventional motors. This goal was rapidly achieved in the period 1977–79 by increasing the yearly production of alcohol from 500 to 3,500 million liters (60,000 barrels per day).

The next step was to convert motors to run on pure (or almost pure) ethanol, a step that required an important decision on the part of car manufacturers in Brazil; this decision was taken in the beginning of 1980 and a significant number of the cars manufactured this year will be produced with modified motors—200,000 cars in a total of 1 million. The modification is basically a change in compression ratio from 1:6 (for gasoline) to 1:12 (for pure alcohol).

As a consequence, vast quantities of ethanol will be needed to supply the automobile fleet, which by 1985 will be almost totally converted to alcohol (approximately 10 million cars). Firm plans have been laid to produce 10.7 billion liters of alcohol in 1985. This will correspond to approximately 50 percent of the projected gasoline consumption in that year had not the alcohol program existed. This amount corresponds to 200,000 barrels per day of alcohol, a sizable amount of liquid fuel by any standards. The total consumption of petroleum in Brazil is 1 million barrels per day. Efforts are being made to introduce ethanol in diesel motors and eventually as a replacement for fuel oil, therefore reducing the amount of imported oil.

The use of ethanol as a substitute for petroleum is an interesting solution for the energy crisis. What we will do here is discuss the potential for, and some constraints of, such solution; namely:

a. Is this intrinsically a centralized or decentralized type of solution?
b. How applicable is this solution throughout the Third World?
c. What are the conflicts with social and environmental goals?

CENTRALIZED VERSUS DECENTRALIZED SOLUTIONS

One way to decide if a given energy source is suited for a decentralized solution is to compare its energy density—in watts per square meter (watts/m²), for example—with the consumption energy density of the people it is supposed to supply.

For example, in most parts of the Third World, the yearly average incident solar energy is at least 100 watts/m²; since the efficiency for capturing this power is not higher than 1 percent, in most cases, one has an available energy density from the sun of roughly 1 watt/m². Table 9-1 gives energy densities for a number of renewable energy sources.

The energy consumption density of urban populations is on the order of 5 watts/m². For example, the city of London with 1,100 inhabitants per square kilometer has an average consumption of 5 kilowatts per capita; the energy consumption density (per square meter) is the product of these two numbers—in other words, 5.5 watts/m². It is therefore obvious that to supply the energy needs of London's population from any of the sources of solar renewable energy listed in Table 9-1 would be very difficult indeed, because of

Table 9-1. **Energy Densities for Renewable Energy Sources**

Source	Watts/m²[a]
Fuelwood (natural forests)	0.03
Fuelwood (planted forests)	0.3
Biogas	0.1
Hydropower	0.1–0.003
Wind (weak)	0.4
Wind (strong)	7.0
Ethanol from sugar cane	0.4
Solar direct (heat)	50.0
Solar direct (electricity)	10.0

[a]Figures are approximate.

Table 9-2. **Urban and Rural Densities**

	Population density (people/km²)		Energy consumption (kcal/capita/day)		Energy consumption density (watts/m²)	
	Urban	Rural	Urban	Rural	Urban	Rural
India	6,000	135	41,600	7,200	12.0	0.04
London	1,100	—	108,000	—	5.7	—
Tokyo	980	—	81,000	—	3.8	—
S. Paulo (Brazil)	1,260	13	50,000	9,600	3.2	0.006

1 kW of installed power corresponds to a consumption of 20,800 kcal/day.

the large areas needed (except perhaps for hot water). The same is true for most urban areas, as can be seen in Table 9-2.

This is not the case in rural areas of the Third World where the population density is much lower and energy consumption is also very low. In the case of India, for example, the energy consumption density is 0.04 watts/m², which can easily be supplied by any of the sources listed in Table 9-1.

To supply the energy needed in urban areas, one therefore must either use fossil fuels or collect solar energy from vast areas of land and transport it to cities. This is precisely what is done in hydroelectric generating stations, where hundreds of square kilometers of land are covered by water in suitably located reservoirs and the energy generated is transported to cities in the form of high-energy-density electric power.

This might seem a cumbersome method for supplying the energy needs of urban populations, but it is the method used since the most remote antiquity to feed people in cities. The density of agricultural products is also very low—less than 1 kilogram of cereal per square meter of land per year—and the only method of solving the problem of food supply in cities is by cultivating large areas of land, collecting the foodstuff, and transporting it to the consuming centers.

Table 9-3 shows typical yearly productivities of some agricultural products in Brazil.

Under Brazilian conditions, it is necessary to use 1 hectare of land (1 ha, or 10,000 square meters) to supply the food necessary to feed an average family (five persons) which consumes approximately 1 ton of cereals per year.

It is very interesting to notice—again under Brazilian conditions—that an average family in the city of São Paulo requires 2 kilowatts of power for transportation (mainly in automobiles).

Table 9-3. **Yearly Agricultural Productivity in Brazil**

Crop	Yield (kg/m²)
Rice (with irrigation)	0.2 – 0.4
Rice (without irrigation)	0.1 – 0.2
Black beans	0.08 – 0.15
Soybeans	0.15 – 0.2
Corn	0.15 – 0.3
Wheat (with irrigation)	0.1 – 0.25
Wheat (without irrigation)	0.08 – 0.2
Coffee	0.15 – 0.2

This amount of power can be obtained from 1 hectare of land if ethanol is produced from sugar cane, as can be seen from Table 9-1.

One concludes, therefore, that by "harvesting" ethanol from large tracts of sugar cane, one can solve the transportation energy needs of urban areas. This, however, requires sophisticated, mechanized systems of planting, harvesting, and industrial processing which are characteristic of "centralized" technologies.

In addition to the agricultural problems involved in growing sugar cane with high yields (by the use of irrigation and fertilizer), ethanol production from sugar cane requires an industrial phase in which the sugar cane juice must be extracted and fermented before the ethanol can be obtained by distillation. A block diagram of these operations is shown in Figure 9-1.

After considerable experience one has found out that the minimum size, from a technical and economical viewpoint, for industrial units is 20,000 liters per day; this corresponds to the harvesting of approximately 1,000 hectares of a sugar cane plantation per year. In general, five or six of these units are installed together with some basic common parts of the machinery in so-called distilleries capable of producing 100,000 liters per day. Much smaller units can be built, but they have low efficiencies and do not benefit from economies of scale; therefore, the price of the ethanol produced is prohibitively high. The technologies needed for competitive small-scale production might, however, still be developed.

With present technology, the production of large quantities of ethanol involves centralized facilities (including storage and distribution), although on a scale smaller than petroleum refineries, which typically process 50,000 barrels per day (8 million liters per day). Typical investments in the industrial phase are $15,000 (US) per barrel of alcohol produced per day; this compares very well

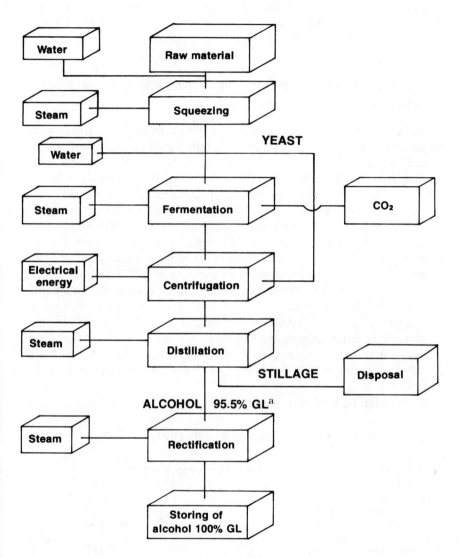

[a] Gay Lussac: scale for measuring purity of alcohol.

Figure 9-1. **Block Diagram of an Alcohol Distillery**

with $40,000 (US) per barrel per day for the fabrication of synfuel from coal, which is being actively considered in the United States.

Ethanol production from sugar cane is typically a renewable

energy source and therefore has many characteristics of being a "soft path" solution, but it is one which requires a "centralized" approach.

POTENTIAL FOR ETHANOL PRODUCTION IN THE THIRD WORLD

In Brazil, ethanol is being produced almost exclusively from sugar cane, which traditionally has been cultivated in that country and requires good land and fertilizer. The average yield of sugar cane is 50 tons per hectare per year. From each ton, one can produce either 100 kilograms of sugar or 70 liters of alcohol; in other words, one can produce 3,500 liters of alcohol per hectare per year.

Currently 1 million hectares are being used for ethanol production (in addition to sugar production, which takes another 1.5 million hectares), displacing other crops that might be more important to feed the population of the country. This is one of the serious problems of the ethanol program in Brazil; we will come back to this point later.

To reach a production of 10.7 billion liters of alcohol in 1985 will require 3.5 million hectares, which is an appreciable fraction of the total currently arable area of the country being used (8 percent of a total of approximately 40 million hectares), as can be seen in Table 9-4.

In order to find out what the worldwide possibilities of this

Table 9-4. **Land Utilization in Brazil, 1977**

	Hectares (x 10^3)
Total area[a]	851,197
Land area[b]	845,651
Arable land and permanent crops[c]	40,720
Permanent meadows and pastures[d]	509,000
Forest and woodland[e]	166,000
Other land[f]	129,931

Source: "Food and Agricultural Organization (FAO) of the U.N.," *1978* (FAO) *Production Yearbook* 32 (Rome: 1979).
[a]Total area refers to the total area of the country, including area under inland water bodies.
[b]Land area refers to total area excluding area under inland water bodies, i.e., lakes and rivers.
[c]Arable land refers to land under temporary crops (double-cropped areas are counted only once); permanent crops include crops that take more than five years, e.g., tea and coffee.
[d]Permanent meadows and pastures refer to land used permanently for herbaceous forage crops, either cultivated or growing wild.
[e]Forest and woodland refers to land under planted and natural forest and bushes.
[f]Other land refers to unused but potentially productive land—built-on areas, waste land, parks, roads, lanes, etc.

Table 9-5. **Sugar Cane Production**

	Area used (x 10³ ha)	Production[a] (x 10³ tons)	Total arable and perma- nent cropland (x 10³ ha)	% of cropland occupied by sugar cane
India	3,220	181,628	169,400	1.9
Brazil	2,413	129,223	40,720	5.9
Cuba	1,246	66,400	3,150	40.0
China	675	47,137	106,500	0.6
Mexico	480	34,500	23,220	1.9
Pakistan	823	30,077	20,300	4.0
Colombia	290	23,100	5,505	5.2
Australia	258	21,500	44,900	0.6
Phillippines	503	20,838	8,100	6.2
S. Africa	250	19,500	14,560	1.7
Thailand	567	19,000	17,650	3.2
Indonesia	180	15,000	17,200	1.0
Argentina	346	14,600	35,000	1.0
Dom. Rep.	174	10,850	1,230	1.4
Bangladesh	150	6,700	9,125	1.6
Total	11,575	640,053	516,560	2.2 (av.)

Source: "Food and Agricultural Organization (FAO) of the U.N.," *1978* FAO *Production Yearbook* 32 (Rome: 1979).
[a]The yield changes from country to country; for this reason, the numbers in the second column are not strictly proportional to the ones in the first column. Yield may change from 50 to 90 tons/ha.

method of producing energy are, one can inquire who the most important world sugar producers are. One could then find out if those producers could follow the path taken by Brazil, in other words, by diverting sugar production to ethanol production in a significant way.

Table 9-5 lists fifteen countries that account for approximately 80 percent of the world production of sugar cane. Brazil produces only 20 percent of the total sugar cane production of the countries listed in this table, covering 6 percent of its arable land with such a crop. The average proportion of arable land used for sugar cane in these countries is 2.2 percent.

Approximately 40 percent of the sugar cane raw material is used for ethanol production in Brazil, as stated before. This corresponds to 60,000 barrels per day (6 percent of the total liquid fuel consumption of the country), produced on 2.5 percent of the total arable land (and permanent cropland) of the country.

If all the other countries listed in Table 9-5 were to follow the example of Brazil (by using 40 percent of their total sugar cane production to produce ethanol), the total production of ethanol would amount to 300,000 barrels per day, which is approximately 5 percent of the total 6 million barrels per day of liquid fuels consumed daily in the non-OPEC countries of the Third World. If the area covered by sugar cane was to increase from the present average of 2.2 percent to 6 percent, the production would increase to 15 percent of present consumption in non-OPEC Third World countries, which is an appreciable amount.

The raw material resource basis for the production of ethanol could, however, be significantly expanded if ethanol could be produced from cellulosic materials. There seems to be no question that the necessary technology is available, although there are not yet industrial installations in the Third World countries following this route. However, the situation will change rapidly in the near future in Brazil because this country is embarking on an ambitious program to produce ethanol from wood to supplement its production from sugar cane, using the experience obtained in this field during World War II in Germany, Switzerland, the Soviet Union, and the United States.

The sustainable forest yield (aboveground annual increment) is generally taken as 8 oven-dry tons (odt) per hectare per year, although higher yields of 20 odt/ha/year are quite common in managed forests. Taking 14 odt/ha/year as an average, one can assume that 2,100–2,800 liters of ethanol can be obtained per year from 1 hectare. (One odt of wood produces 150–200 liters of ethanol, depending on the type of wood.) This quantity of ethanol is lower than the one that can be obtained from 1 hectare of sugar cane, but forests (natural or managed) can use land of lower quality, and the availability of forests is very large in many countries of the Third World.

It is interesting to point out that the fundamental change when moving from sugar cane to cellulose as a raw material for the production of alcohol is to reduce the contribution of the agricultural raw material to the product's final cost. Sugar cane represents approximately 80 percent of the final cost of alcohol. In alcohol production from corn in the United States, the percentage is about the same. Wood, however, represents only 40 percent, the other costs being due to larger expenditures in energy and other products in the industrial phase, as shown in Table 9-6. One therefore can expect technological progress that will lower these expenditures in the case of wood.

Table 9-7 lists the fifteen countries of the Third World with the largest forested areas, representing more than 70 percent of the

Table 9-6. **Percentage Costs in the Production of Ethanol from Different Raw Materials**

	Sugar cane[a]	Wood[b]	Corn[c]
Raw material	79.8	37.7	72.5
Labor	0.7	0.6	4.0
Materials	0.8	23.9	1.5
Energy	0.7	20.4	8
Maintenance & others	2.9	2.5	5.5
Depreciation	15.1	14.9	8.5
Total	100.0	100.0	100.0

[a]Brazilian experience.
[b]Produção de combustíveis líquidos a partir da madeira, Informações básicas, IBDF, Ministério da Agricultura, Brasil (1979).
[c]Report of the Energy Research Advisory Group on GASOHOL, US Department of Energy (April 29, 1980).

total. Assuming that only 5 percent are used and taking the average yield of 14 odt/ha/year, these forests could supply, on a renewable basis, 160–210 billion liters of alcohol per year, which correspond to 2.5 to 3.6 million barrels per day in a total consumption of approximately 6 million barrels per day in all the non-OPEC Third World countries. This is a considerable amount of liquid fuel, although still minor compared with the approximately 60 million barrels per day consumed daily in the world.

CONFLICTS WITH SOCIAL AND ENVIRONMENTAL GOALS

Although rather successful in promoting gasoline substitution in Brazil, the alcohol program has already worsened existing social problems or produced new ones in the following areas:

- production of other foodstuff or subsistence crops,
- changes in the pattern of land utilization and unemployment of the labor force, and
- enhancement of the concentration of wealth in the hands of a small fraction of the population.

From the environmental point of view, two other problems have become rather acute:

- The production of each liter of alcohol leads to approximately 15 liters of stillage—stillage which has been discharged in rivers and meadows without any serious consideration of environmental degradation.
- The use of vast areas of land in a monoculture such as sugar cane might have adverse ecological effects.

Table 9-7. **Forested Areas**

	Forests and woodlands (x 10³ ha)	Land area (x 10³ ha)	Fraction of land area (%)
Brazil	509,000	845,651	60
Indonesia	122,000	181,135	67
China	121,500	930,496	12
Zaire	120,900	226,760	53
Australia	107,000	761,793	14
Sudan	91,500	237,600	38
Colombia	77,190	103,870	75
Peru	73,800	128,000	58
Angola	72,680	124,670	59
Mexico	70,700	192,304	37
India	65,500	297,319	22
Argentina	60,220	273,669	22
Bolivia	56,200	108,547	52
Venezuela	47,970	88,205	54
Burma	45,274	63,888	70
Total	1,641,434	5,214,256	31 (av.)

Source: "Food and Agricultural Organization (FAO) of the U.N.," *1978* FAO *Production Yearbook* 32 (Rome: 1979).

In what follows, we will discuss these points.

Production of Other Foodstuff

Agricultural products in less developed countries (LDCs) serve, in general, two main purposes: meeting hunger as foodstuff (in the form of black beans, corn, or rice) and earning foreign exchange through exports. In the case of Brazil, the growth of the alcohol program has been such that it is using 1 million hectares of land at present; this area will grow to 3 million hectares in 1985. The 2 million hectares of good land which will be used could produce 500,000 tons of black beans, 1.5 million tons of rice, and 1.5 million tons of corn, which represent roughly 20 percent, 17 percent, and 8 percent of the present production of such crops. The production of alcohol is therefore being made at the expense of these basic crops, and one could ask why very-much-needed government loans frequently go into sugar cane production and not into more basic crops.

Another way to express the conflict between the production of fuel and foodstuff is to point out that the existing programs of expansion of agricultural production for liquid fuels, exports, and food will require increasing the present area used at a rate of 7–8 percent per year, while in the period 1968–77 the area increased by only 3.7 percent. There obviously will not be room for all these agricultural production programs except by increasing the arable land area, which is not possible in the more populous regions of the country.

Patterns of Land Utilization

In Brazil, expansion of the sugar cane production has been taking place mainly on the fertile lands of the state of São Paulo, where it has been displacing more traditional crops (coffee, soybeans, and cotton). The argument has been made that such crops are mainly oriented toward exports, that is, the earning of foreign exchange to pay for essential imports, mainly petroleum; therefore, it makes sense to grow sugar cane and produce alcohol, which reduces these petroleum imports. This is a rather controversial question in Brazil.

On the other hand, large sugar cane plantations are being established in regions where many small farms existed; this approach is favored by government policies and the fact that sugar cane production is well suited to mechanized techniques. As a consequence, however, the subsistence crops that existed in the small farms (corn, maize, black beans, etc.) are being eradicated, forcing the importation of food from faraway regions. In addition, this has had the very negative social consequence of forcing the exodus of small farmers and laborers from the fields to small cities where it is difficult to get jobs. They therefore become seasonal laborers for the large plantations, since sugar cane is a six month per year activity.

Concentration of Income

The use of large farms fostered for technical reasons for alcohol production (and the availability of government-subsidized credits) has generated a few very large companies that hold most of the land in many regions of Brazil. This has had a negative effect on income distribution, concentrating resources in the hands of a few privileged entrepreneurs. This is clearly a consequence of the policies followed by the government and could have been otherwise. A system of cooperatives in which individual farmers could grow sugar cane and process it in a collectively owned refinery is possible and is the method used in Australia; this, however, has not been the case in Brazil.

Pollution

The production of large quantities of stillage ("vinhoto") was initially a very serious problem because of the large quantities produced (almost 60 billion liters in 1980). Stillage is very useful as fertilizer, but it has to be transported to the fields and therefore must be concentrated to some extent in order to decrease the amount to be transported. Several small schemes have been developed for this purpose using waste heat from the alcohol refineries. Another scheme is to produce biogas from the stillage through anaerobic digestion and use the final residue as a fertilizer. The production of protein for cattle feeding is also technically possible.

These methods require new investments, and the stillage producers have been reluctant to make them until forced by regulations that forbid stillage dumping in rivers and meadows (a practice which was destroying completely the habitat of fish in many rivers in the state of São Paulo). However, present regulations are becoming quite strict and will probably lead to a solution of the problem.

Ecological Effects of a Monoculture

It is probably too early to notice the consequences of the adoption of an intensive system of land utilization as required by sugar cane, except for the social consequences mentioned above. It is, however, rather impressive to have large areas completely covered by sugar cane plantations, and undoubtedly many species of animals and plants have been eradicated from several of these areas since no havens (such as small sanctuaries or trees) are left intact. The vulnerability of monocultures to disease is becoming an increasingly worrisome factor for the future of the program.

CONCLUSIONS

In light of the above discussion it is possible to expect that in the next ten to twenty years the production of alcohol for use as fuel could increase to approximately 4 million barrels per day in LDCs as a whole. The main candidates for the use of sugar cane as a basic raw material in alcohol production are India, Brazil, Cuba, China, Mexico, and Pakistan. The candidates for the use of wood from managed forests as a raw material are Brazil, Indonesia, China, Zaire, Australia, and Colombia. Cellulosic materials will enter this market slowly, but by 1990 they are expected to make a substantial contribution.

While the environmental constraints seem to be manageable, the use of sugar cane for alcohol production has had some adverse social consequences in Brazil. The use of the wood as a raw material will attenuate the socially adverse consequences, since land used for reforestation does not compete with land used for food production.

The economical units for alcohol production are typically 100,000 liters per day, requiring the use of 10,000 hectares of land; this is not a decentralized way of production. The use of mini (1,000 liters per day) and micro (100 liters per day) distilleries has not yet been proved to be economically competitive. In addition, most of the alcohol produced is trucked to large reservoirs and distributed to large urban centers by the usual methods of distributing gasoline (either pure or in the form of a 20 percent gasohol mixture).

The possible large producers of alcohol (from sugar cane or wood) are more evenly distributed in the Southern Hemisphere than are the OPEC countries, which are largely concentrated in the Persian Gulf. It seems possible, therefore, for many Third World countries to develop their own base for the production of liquid fuels and free themselves from oil imports, using biomass as a resource.

Such alcohol production programs are not necessarily oriented to solve social problems and actually can worsen some of them. Just to give one example, one could point out that in the Brazilian state of São Paulo, automobile owners (who are the main beneficiaries of the alcohol program) are a small minority; 22 percent of the families own 86 percent of the automobiles. One could therefore consider the alcohol program as a program favoring small elites and the car manufacturers (which are all foreign owned). On the other hand, the alcohol program generates jobs and enhances the self-reliance of developing countries, reducing imports and the need for foreign exchange.

CHAPTER TEN

Towards an Efficient Energy Future: Critical Paths, Conflicts, and Constraints

B. C. E. Nwosu
Chief Education Officer (Science)
Federal Ministry of Education, Nigeria

I accepted the invitation to this Symposium with humility, realizing my inadequacies. However, I believe that the main reason I was invited is to provide some perspectives from a scientist coming from a less developed country. This occasion reminds me of a similar situation in which I found myself in 1971 at a consultation held in Nemi, Italy, under the auspices of the World Council of Churches' (WCC's) Department of Church and Society. On that occasion, I was invited to make a statement on the theme of that consultation, "The Future of Science and the Social Responsibility of the Scientist."[1] The theme of this Symposium and that consultation has emphasis on the *future*. The point of view expected of me then, as indeed now, was the point of view of a scientist from a Third World country—a poor country, but one which is euphemistically referred to as a less developed country (LDC).

There are many other parallels that can be drawn between that 1971 consultation and this Symposium. But for now, it is sufficient to mention only one similarity. The Nemi consultation was held close to the time of the publication of *The Limits to Growth* by the Club of Rome. That publication shook the highly developed countries of the world and aroused a great deal of controversy. It identified the major issues arising from the use of nonrenewable resources and the global ecological constraints that the indiscriminate use of these resources are likely to impose on the future of mankind and the earth itself. Similarly, this Symposium is being held as a result of the so-called energy crisis. There is a great deal of controversy on the subject of the energy crisis. The ongoing debate on energy concerns the use of nonrenewable sources of energy as well as the use of other sources of energy such as nuclear energy whose attendant ecological problems have grave consequences for the future of mankind and the earth.

From the above, the concerns of the *future* result from the crisis identification by scientists, intellectuals, and opinion-molders of the highly developed countries (HDCs). They have not originated from within the LDCs. What, then, are the perspectives of a scientist such as myself from an LDC? One reaction is to retort by asking the question, in the same manner as I did in 1971, "Is there a future for science in Africa?" And then ask in 1980, "Is there an energy future for Africa?" The first conflict, therefore, is that of perspectives. For whom are we seeking an efficient energy future?

To understand my line of thought, let us examine a few facts on Nigeria. Nigeria's production of crude oil stands at 2.1 million barrels per day. Out of this, Nigeria's crude exports to the United States totaled approximately 400 million barrels in 1979. Estimates of crude oil reserves, including offshore reserves, are put at 4,036 million tons. Reserves of natural gas are estimated at 3,162 million tons of oil equivalent (TOE). Known reserves of coal and lignite resources are put at 141 million TOE, while inferred reserves are estimated at several hundred million tons. Nigeria is also well endowed with hydroelectric power. Calculations of economically feasible hydraulic resources have recently been put at 7 million TOE annually. Solar energy is abundant, even though actual work in this area has not risen above isolated studies at universities and research institutes. The technical potential for solar energy is estimated[2] at 60 million TOE. Recently, encouraging finds of uranium have been made in two states of Nigeria, Bauchi and Gongola. Exploration has begun as a joint Franco-Nigerian effort.

Thus, from the foregoing, it is reasonable to say that Nigeria is rich in energy resources. The Gross Domestic Product (GDP) figures of Nigeria are shown in Table 10-1. From the figures quoted in the *World Bank Atlas 1979*, the 1977 per capita Gross National Product (GNP) for Nigeria is $510. Compare this with the per capita GNP for the United States ($8,750), Japan ($6,510), and the United Kingdom ($4,540). These figures speak for themselves.

From the above simple analysis, it is apparent that although Nigeria may be considered energy resource rich, it is really a poor country. When one considers some other African countries whose resources are much scarcer than those of Nigeria, the relative positions of LDCs can best be appreciated.

Table 10-1. **Nigerian Gross Domestic Product at 1973-74 Factor Cost** (naira[a] x 10⁶)

Activity sector	1973–74	1974–75	1975–76	1976–77	1977–78	1978–79	1979–80
Agriculture	2,183.3	2,203.8	2,143.1	2,251.9	2,336.6	2,406.0	2,486.6
Livestock	488.8	491.2	393.9	399.6	408.9	422.0	440.6
Forestry	215.0	302.7	328.8	355.1	383.1	412.9	443.2
Fishing	465.0	567.6	573.8	607.1	658.7	698.9	743.6
Crude petroleum	2,771.6	2,797.6	2,345.3	2,676.8	2,015.7	2,480.6	2,866.5
Other mining and quarrying	198.8	247.8	310.5	372.6	436.0	492.7	544.4
Manufacturing	611.0	601.4	729.7	854.4	943.0	1,040.6	1,154.0
Utilities	45.2	51.8	59.7	74.4	95.2	117.6	136.6
Construction	884.1	1,108.4	1,411.4	1,693.6	1,981.8	2,239.5	2,474.7
Transport	429.6	403.1	468.2	636.8	764.1	878.0	966.9
Communication	33.2	38.9	47.7	54.9	60.3	65.8	71.7
Wholesale and retail trade	2,268.1	2,295.1	2,491.5	2,788.5	3,043.9	3,245.2	3,492.2
Hotels and restaurants	32.4	35.6	39.1	43.0	47.5	52.0	57.2
Finance and insurance	140.5	155.0	170.4	187.6	206.4	226.7	249.4
Real estate and business services	61.1	67.3	74.0	81.4	89.5	98.5	108.3
Housing	625.9	688.2	756.6	832.4	915.6	1,006.2	1,107.7
Producer of government services	664.4	743.4	1,049.1	1,082.4	1,208.5	1,299.3	1,399.8
Total	12,118.0	12,798.9	13,392.8	14,992.5	16,285.2	17,182.3	18,740.4

Source: "Guidelines for the Fourth National Development Plan 1981–85" (Lagos, Nigeria: Federal Ministry of National Planning, 1980).
[a]The basic monetary unit of Nigeria.

BASIC CONSIDERATIONS

Why is the situation so bad in the LDCs? At the risk of taking a well-worn line of blaming everything on the exploitation and colonialism of the past, I am afraid that we cannot meaningfully talk about the future without examining the past. It is the view of this author that the mastery and application of wind energy (sailing boats), the effective conversion of chemical energy (gun powder), and the development of associated technologies made it possible for Western Europe to dominate and exploit the resources of Africa, Asia (including the Middle East), and the Americas. With industrialization, Asia and Africa became sources of raw materials and markets for finished products, while North America shook off its colonial yoke about two hundred years ago. With years of independence, nothing significant has changed in the pattern of these relationships, which has now existed for approximately three hundred years.

To some highly placed persons of the highly developed world, there is nothing wrong in this kind of relationship. For example, in the case of crude oil, they argue that Arabs and Africans would probably be still sitting on their oil without knowing that it was there. Therefore, they argue that if someone uses his technology and capital to recover this oil, process it, and make it available for all to use, there is nothing wrong provided he pays a *reasonable price* for the crude oil. But what is a reasonable price for an energy resource such as oil? And who decides what is a fair price? The view of the LDCs, including those who have oil and even those who do not have oil, is that the pricing of oil and the finished products is manifestly unjust. We shall examine this point later in more detail. It is sufficient to record that part of the critical issues in this dialogue on the future is the reconciliation of differing perspectives between HDCs and LDCs.

Next, there is a second major assumption of this paper. The energy problem has not only economic but also political and military parameters that must be considered along critical paths in order to obtain real solutions. Governments in the HDCs, both in the political democracies and in centralized economies, are politically constrained to sustain and even increase their citizens' standards of living. Provision of adequate and reliable sources of energy is cardinal to the attainment of this all-prevailing goal. In fact, I recall that as an undergraduate at the University of Ibadan in the 1950s, when my country was still under British colonial rule, we had a common view on this matter. We held the view that there was no difference in the colonial policy of the Conservative and

Labor parties. We observed that what determined the fortunes of the parties at the election polls was the price of butter and margarine. Therefore, the colonial policy of any British government was predetermined in such a way that whatever happened to the colonies should not affect the price of butter and margarine!

On the other hand, the importance of energy in the present world of science-based technology has only been recently appreciated by the LDCs. The LDCs that are rich in energy resources see possibilities of political control of their resources as a prerequisite of economic independence. However, these same countries are now awakened to the threat of the depletion of their vital nonrenewable resources. Hence their keen interest on the subject of energy for the future.

The LDCs that are poor in energy resources share some sympathy with LDCs who are fighting for fairer prices for their energy resources. They have had common colonial and neocolonial experience. But at the same time, they are seriously distressed by the repercussions on their respective nations of the energy price war between the "poor haves" and the "rich haves." For this class of nations, *energy for the present* is all-important.

Another component that has recently figured prominently in the energy debate is the military factor. But, as indicated earlier, military power has been a major tool of HDCs for the acquisition and maintenance of the world's resources. In an earlier paper in which I discussed the issue of Africa's scientific development,[3] I asserted that "the greatest tragedy in this era (pre-colonial era) was the slave trade. Lack of writing is insignificant when compared with lack of gun powder—that symbol of a superior military technology." In order to understand the military factor, one only needs to read the newspapers and news magazines about Soviet moves in Afghanistan and the Horn of Africa or about the "Carter Doctrine" on the need for a "US strike force to defend the West's oil sources" to feel the gravity of the military factor in considering energy for the future. It is evident that the HDCs need military might to secure energy, and they need energy to maintain their military might.

The LDCs, although weak militarily, are not left out in this military-energy equation. Have we not heard of the oil weapon? The Arabs used it in the Middle East to secure more advantages than they did on the battlefield. Right now, the pressure is still on. Nigeria used it effectively to aid Zimbabwe's freedom. Those countries that are not in the "Big Energy League" or the "Little Energy League" are trying a few small tricks of their own. Hence, we have India's "peaceful atomic explosion," the call for a Pakistani "Islamic atomic bomb," and the Israel-South Africa "outcast" nuclear bomb.

We can deduce from the foregoing paragraphs some major initial conclusions, namely:

- All the nations of the earth are now interested in the energy question.
- The perspectives of their economic, political, and military interests appear, at first sight, not to coincide.
- Indeed, present developments point to confrontation and possible disaster.
- However, there are many international groups and movements, such as this Symposium, that see in the global energy anxiety—which cuts across both HDCs and LDCs—an opportunity to explore models of cooperation for human development on a scale never before contemplated.

JUST, PARTICIPATORY, SUSTAINABLE SOCIETY

On the basis that cooperation is to be preferred to confrontation, let us examine the concept of the Just, Participatory, Sustainable Society (JPSS). The JPSS concept arose from discussions over the last ten years in the World Council of Churches' circles. These discussions were sponsored by the WCC Department of Church and Society. The report of the last major conference on familiar issues of the future has just been published.[4] For those not conversant with WCC literature, let me review the main aspects of JPSS.

Limits to growth exponents argue[5] that the earth is finite; therefore, physical growth cannot continue indefinitely. They argue that due to constraints arising from the depletion of nonrenewable resources and the burden on the environment from population growth, waste accumulation, and so on, the earth will come to a halt by about the year 2060, give or take forty years. A young physicist from the Massachusetts Institute of Technology (MIT) presented at Nemi some interesting graphs.[6] Some of these are shown in Figures 10-1 and 10-2. The figures are self-explanatory.

The *Limits to Growth* document came under severe attack by traditional economists in the HDCs. Their arguments are not covered in this paper. However, it was evident to those of us from the LDCs that the MIT model did not even take into consideration the concerns of the poor nations. In fact, when the young MIT physicist was challenged, he argued, rather naively, that if growth

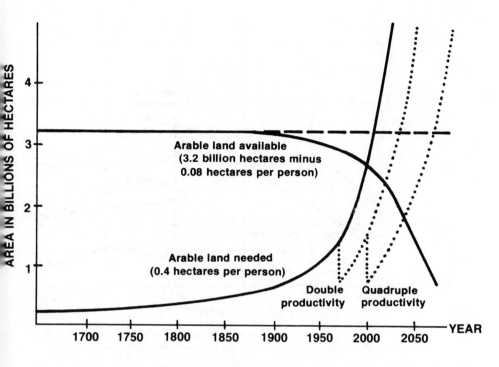

Source: J. Randers, "The Carrying Capacity of Our Global Environment—A Look At the Ethical Alternative," Anticipation No. 8, World Council of Churches Press (1971).

Figure 10-1. **Available and Needed Land Area**

was limited, it would ensure that the gap between the rich and poor countries would remain constant—a better alternative than the present situation, where the gap between HDCs and LDCs was widening every year. Needless to say, this simplistic view could not be entertained in such an ecumenical body as the WCC.

Over the years the concept of limits has been revised by its adherents. But no matter on which side of the debate one was, it was clear to all who cared to study the worldwide situation that, both in the HDCs and in sections of the LDCs, excessive consumption of nonrenewable resources could not be sustained forever. Nor could mounting data on pollution of the environment arising from ever-increasing industrial activities be accepted with equanimity.

Thus, at the 1974 WCC Bucharest Conference the concept of a sustainable and just society emerged. Because of the forum in

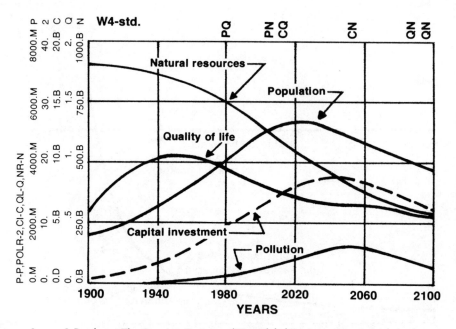

Source: J. Randers, "The Carrying Capacity of Our Global Environment—A Look At the Ethical Alternative," *Anticipation* No. 8, World Council of Churches Press (1971).

Figure 10-2. **Basic World Behavior Model Showing the Mode in Which Industrialization and Population Are Suppressed by Falling Natural Resources**

which this concept emerged, it was evident that the ethical dimension of justice had to be included. The report of the conference summarizes the main points in the concept.[7] However, it is necessary to raise those issues which concern our present dialogue. These are:

- An unjust society may be sustainable at least in the short run. But is such a society worthy of support?
- Can a society be sustainable without the full and effective participation of the members of such a society?
- Assuming that sustainability can be achieved without participation, is such a society just?

Answers to the above questions gave rise to the contention that only a just, participatory, and sustainable society provides a viable option for the future. Science as we know it today is no longer exact. The study of ethics is even less so. Therefore, without going into further details, let us examine some of the aspects of future energy in the context of JPSS.

First, we wish to assert that for the sake of the world's poor, justice demands a redistribution of energy and other resources, both within nations and between nations. Table 10-2 presents the per capita electricity consumption in some selected countries.

Per capita electricity consumption is a good indicator for the comparison of total end-use energy consumption. Also, statistics of electricity consumption in LDCs are relatively easier to obtain than other energy resources data. Most of the developing countries have state-owned electric power companies. It is therefore reasonable to assume that the data in Table 10-2 reflect the orders of magnitude in terms of comparison ratios of total end-use energy consumption patterns. Now look at the table closely. The ratio of electricity consumption between the United States and Nigeria is 180 to 1; between the United Kingdom and Nigeria, 80 to 1.

Table 10-2. **Per Capita Electricity Consumption in Selected Countries of the World**

Country	Period	kWh/capita
United States	1977	10,840
Switzerland	1977	5,780
Fed. Rep. of Germany	1977	5,560
United Kingdom	1977	5,070
Japan	1977	4,630
Liberia	1976	520
Gabon	1976	435
Ghana	1976	400
Cameroon	1976	205
Ivory Coast	1976	190
Guinea	1976	110
Senegal	1976	90
Sierra Leone	1976	70
Nigeria	1978/79	65[a]
Togo	1976	60
Gambia	1976	50
Benin	1976	20
Chad	1976	15
Niger	1976	15

[a]Energy generation of 5,217 gigawatt-hours (GWh) minus exports to Niger of 67 GWh, divided by population estimated at 79 million.

Within Nigeria itself, Table 10-3 reveals great differences in end-use oil consumption by the states. Compare this table with the political map of Nigeria in Figure 10-3, and the picture of the injustice among nations and within nations becomes clearer.

TECHNOLOGICAL OPTIONS IN THE CONTEXT OF JPSS

Now let us look at some important issues in the energy debate and examine how they relate to international justice and the concern of this Symposium for an efficient use of energy.

Conservation

As indicated earlier, the limits to growth debate brought out clearly the problems of pollution, ecological imbalance, and strain on nonrenewable resources. Parmar[8] puts the point of view of LDCs clearly thus: "The problems are the inevitable consequence of production and consumption patterns geared to an ever-rising standard of living. It is true that high rates of population growth,

Table 10-3. **End-Use Oil Consumption by Nigerian States, 1978**

State	In TOE[a] (x 10^3)	As % of total
Lagos	1,495	26.4
Sokoto	117	2.1
Niger	69	1.2
Kaduna	314	5.5
Kano	344	6.1
Borno	149	2.6
Bauchi	63	1.1
Gongola	63	1.1
Benue	97	1.7
Plateau	198	3.5
Kwara	174	3.1
Ogun	342	6.0
Ondo	148	2.6
Oyo	465	8.2
Bendel	513	9.1
Anambra	295	5.2
Imo	194	3.4
Cross River	189	3.3
Rivers	434	7.7

[a]Tons of Oil Equivalent.

Figure 10-3. **Map of the Federal Republic of Nigeria Showing the States**

characteristic of developing countries, contribute to the burden on resources. But here again, the inordinately high per capita use of resources in rich nations is more to be blamed for the resource disequilibrium that is assuming disquieting proportion." A look again at Table 10-2 demonstrates the great disparity in energy consumption patterns. Therefore, when we talk about conservation of energy, do we expect the United States to continue at a rate of 10,840 kilowatt-hours per capita (kWh/capita) while Nigeria stays at 65 kWh/capita? Or put another way, to what extent does the fact that the United States, Switzerland, the Federal Republic of Germany, the United Kingdom, and Japan would wish to maintain their average consumption of about 6,000 kWh/capita make it impossible for Niger, Chad, Benin, Gambia, and Togo to increase their present average electricity consumption of 32 kWh/capita?

It is clear where the initiative—or, shall we say, leadership—

must begin. Despite Parmar's observation that "over the last two decades much has been said about new structures of international political and economic relationships that will ensure a more equitable sharing of power and resources but very little has been done," it is the contention of this paper that energy conservation programs in the rich nations must be strongly stressed. It is in the interest of those of us in the LDCs that these conservation programs succeed, for if they do, there is less danger that the LDCs' energy resources will be taken away by force or by deceitful strategies such as obsolete technologies. In addition, it will provide a breathing space in which the Third World can plan realistically. Therefore, the principle of conservation is consistent with JPSS.

Self-Reliance and Indigenous Technology

The concept of self-reliance is well known. It does not mean self-sufficiency. It does not imply isolation. It means that a nation or group of nations must learn to make full use of their resources. However, it also means that a country such as Nigeria should not plan along the lines of the present consumption patterns of either the United States or the United Kingdom. Self-reliant strategy demands a development strategy in the LDCs that is based upon socioeconomic realities.

It is important for every nation to determine the minimum energy requirements that will guarantee an individual a meaningful quality of life. Next, the technology that would be applied in a self-reliant manner to achieve the above objective has to be determined.

It has been argued that indigenous technology is the key to self-reliance for a developing nation. Advocates of indigenous technology are to be found not only among the HDCs—such advocates are usually patronizing—but also among respectable leaders of the Third World. For example, Dr. Amadou-Mahtar M'Bow states[9] that "the Industrial World has profoundly influenced the advance of Africa primarily because the African[s] themselves have not succeeded in countering this pattern of development with an alternative model of progress rooted in their own tradition." In 1972, I advocated[10] the indigenous technology option. I drew from experience in the Biafra Science Group during the Nigerian Civil War. I had the opportunity to head a group of about two hundred scientists and engineers and about three thousand technicians/artisans for thirty months during that war. We were able to produce petroleum products such as petrol, kerosene, and diesel right from simple distillation on a batch process to the design and construction of a continuous flow system which processed 2,000 barrels per day of crude oil, in the bush. We also manufactured other things such as rockets, including antiaircraft rockets, bombs, and land mines. It is a pity that after ten years of war, Nigeria has yet to benefit from this experience of self-reliance.

One cannot go into details of the Biafran experience for obvious reasons. But one can say categorically from the Nigerian experience that indigenous technology and the principle of self-reliance cannot be achieved in the LDCs unless the *powers that be in the HDCs* think that this strategy is also in their *own interest*. Otherwise all efforts in this direction will be thwarted. However, my pessimism is giving way to guarded optimism. Due to the world energy crisis, it may be that powerful forces in the HDCs will support self-reliance and indigenous technology if it complements their conservation efforts. For an efficient energy future cannot be limited to twenty-five years, not even to fifty years. It must be considered from what we visualize the world society to be, in at least the next one hundred years and more. In this time perspective, conservation, self-reliance, and indigenous technology can be seen to be consistent with JPSS.

Technology Transfer

In defining self-reliance it was indicated earlier that this does not mean self-sufficiency or isolation. It was also pointed out that for indigenous technology to succeed, the cooperation of the powerful is required. Technology transfer implies the transfer of technological know-how from those who have it to those who do not. Many views have been expressed on the subject. But some of the views which are most attractive to me are those of an LDC scientist, Professor Animalu, who provides a model shown in Figure 10-4.

Important aspects of the model are:

- that the country receiving the technology and the industrialized country from which the technology transfer is to take place must be clear on the question of energy-related technology;
- the availability of funds;
- the bidirectional exchange of ideas;
- the choice of a suitable base (university, institute, etc.) for the transfer operation;
- the assessment of general political, social, and economic factors;

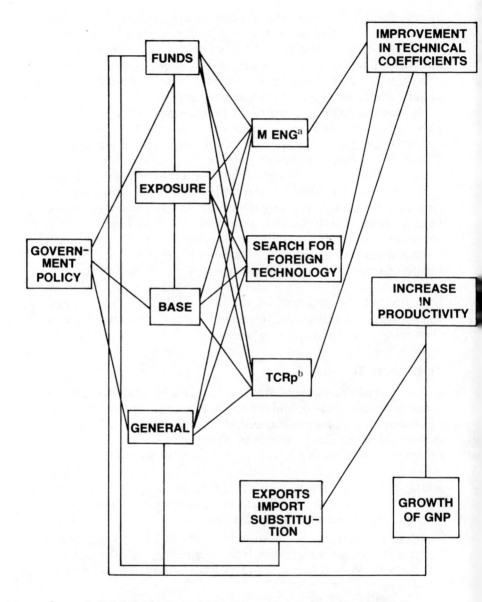

Source: A. O. E. Animalu, "A Model for Energy Research and Production in Nigeria," Proceedings of the Energy Policy Conference, Jos, August 1978, National Policy Development Centre, Federal Government of Nigeria.

[a] Mechanical Engineering.
[b] Technical Cooperation Representative.

Figure 10-4. **Model of Technology Transfer Feedback Mechanism**

• the search for relevant technology related to the solution of a particular problem (indigenous problems);
• the participation of representatives from both countries; and
• that the technology transfer must be mission oriented.

Animalu's model is based on the needs of a single country. It emphasizes bilateral relations between two countries. He presented these views again in the United States/Nigeria bilateral talks on technology transfer held in New York. Aspects of these talks will be discussed later under the theme dealing with practical realities of international cooperation between HDCs and LDCs. For now, it is sufficient to note that effective technology transfer as analyzed by Professor Animalu will promote *some* aspects of JPSS. Like most conventional papers on energy, it does not address itself to the principle of justice, nor does it set limits on the growth of GNP. It cannot be used on a global basis without modification.

OPEC IN THE CONTEXT OF JPSS

No discussion of present market determinants in the energy sector would be complete without some consideration given to the cartel known as the Organization of Petroleum Exporting Countries (OPEC). Closely associated with the role of OPEC in the energy market is the role of the multinational oil companies. There are many ways of looking at both roles, and I am sure these perspectives will be adequately presented during the course of this Symposium. However, the perspective of this paper is confined to the examination of the origin, objectives, and performance of OPEC in the context of JPSS.

Abdul Amir Kubbah presents an official OPEC view in the book *OPEC: Past and Present.* [11] He places on record that the immediate reasons leading to the formation of OPEC in September, 1960, was a reaction to the "challenge posed by the Multi-national Oil Companies in arbitrarily and unilaterally reducing the posted prices of crude oil in February, 1959 and again in August, 1960." He catalogues a number of direct and indirect factors leading to the formation of OPEC. These are:

• the inequity of the concession agreements,
• the posted price arrangement, "which was a tax reference price used for calculation of the companies' profits and did not always reflect market realities,"
• a growing awareness in the producing countries of "the importance and irreplaceable character of their countries' oil wealth," and

• political movements within the framework of the Arab League.

Most OPEC members depend heavily on revenues from oil for their development programs and were seriously affected by a unilateral cut in their revenues. It is estimated, for example, that the 1960 action resulted in an annual average loss, to the countries of Kuwait, Iraq, and Saudi Arabia, of $231,241,500.[12] On the other hand, the multinational oil companies made their decisions on market considerations. For example, in 1959 the United States decided, for security reasons, to impose compulsory import controls "in order to reduce dependence on foreign oil." The multinationals immediately reacted to this action. In turn, OPEC was formed.

When OPEC was formed, the multinational oil companies did not take it seriously, but events have proved this earlier attitude wrong. The companies' divide and rule tactics (by negotiating with individual countries), which were aimed at undermining OPEC's bargaining power, worked for a while until OPEC evolved a more effective bargaining stance. However, one must agree with Abdul Kubbah that "OPEC's first worry was (and still is) prices." (We shall discuss later the issues of prices in the context of JPSS and compare the price issue with other issues discussed earlier, namely, conservation, self-reliance, indigenous technology, and technology transfer.)

The rest of OPEC's history is well known. Its fortunes range from the period when oil was a buyer's market to when it became a seller's market. The Arab-Israeli war of October 6, 1973, was a turning point in the history of the oil market. To illustrate, oil prices that started early in 1973 at $2.591 per barrel for Arabian light marker crude rose to $11.651 for the Saudi marker crude effective January 1, 1974. This phenomenal jump shook the world economy. Things have never been the same again.

All members of OPEC are LDCs. The example of OPEC's successful pricing policy soon began to be quoted in LDC circles as an effective method of countering the international injustice perpetrated against producers of raw materials who lack the technical knowhow to convert their raw materials to finished products. Calls for more cartels to cover other raw materials were made. These calls do not concern us here. The issue before us is whether the objectives and performance of OPEC are consistent with JPSS.

Let us first note that this forum [the Symposium] is not a forum of the World Council of Churches. Even the churches admit that the Kingdom of Heaven is distinct from the kingdoms on earth!

Therefore, our definition of *justice* must imply only fairness. It is my view that OPEC satisfies the condition of fairness at least as its activities relate to the HDCs. OPEC members, however, have not shown enough concern for the energy-resource-poor LDCs. The funds allocated to them for financing their development programs are mere tokens. OPEC surplus capital is invested in HDCs based on traditional economic criteria of making safe and lucrative investments. But lasting international equity demands that OPEC change this attitude.

On the issue of *participation,* OPEC Resolution XXIV 135 emphasizes the principle of participation in the decisionmaking process of its member countries. Therefore, as a subset of the world, the principle of participation may be said to be satisfied. But to satisfy the principle of participation on a global level, new arrangements or forums have to be established to include consumers in both HDCs and energy-resource-poor LDCs.

Serious reservations arise when the OPEC performance is examined on the principle of *sustainability.* First, there has been too much emphasis on prices. Now and again OPEC members ask rich countries to make efforts at conservation and reduce their consumption of energy. But frankly, nothing in the lifestyle of the leadership in OPEC countries, including my own country, proves that expressions in this regard should be taken seriuosly.

Second, self-reliance as seen by OPEC countries means buying technology, appropriate or not, using the surplus cash available from increased oil revenue. No serious attempt is made to develop self-reliant technologies in association with member nations or in association with other relatively more technologically advanced LDCs. In fact, the technologies imported, particularly in military hardware, lead to further dependence on HDCs (both western and eastern). Take the question of crude oil refinement. Why can't OPEC nations, among themselves, commission the erection of refineries from scratch?

On the issue of *technology transfer* the problem is with the HDCs. Their attitude toward the poorer nations has hardened since the latest increase in oil prices. The recent United States/Nigeria bilateral talks revealed to me that, even for two countries with reasonably good political and economic relations, the technologically advanced countries are not really very serious in giving away their knowledge. Everything has to be paid for in cash.

OPEC is blamed for inflation. This blame is misplaced. Based on the price of crude oil in 1973, I estimated the cost of finished products from a barrel of crude oil (40 imperial gallons) to be about $100. The price of crude oil was then approximately $3. In 1974, the price of crude oil went up to approximately $12. The cost of finished products increased by 25 percent in 1974. The difference between the costs of finished products and raw material was $97 in 1973 and $113 in 1974. Who gained, the HDCs or OPEC?

Now, if we add to the above the crash efforts of the HDCs to develop alternative energy supplies—not necessarily to improve mankind's energy supply but to ensure that "Arabs and Africans do not dictate to us whether we shall drive our cars on Sundays"—it becomes clear that OPEC is not sustainable in the long run, unless its strategy is drastically modified to be consistent with JPSS.

ENERGY AND THE NEW LIFESTYLE

As a result of the rapid industrialization, population growth, and rise in per capita incomes and standards of living which have taken place during this century, there has been a phenomenal rise in the demand for energy. In turn, this has increased efforts to produce more energy. Associated with the production of conventional sources of energy, such as oil, gas, and firewood (forests), is environmental deterioration in the form of oil spillage, oil blowouts, gas flaring, defacement of large areas of land by practices such as strip mining, soil erosion, and the creation of deserts from formerly fertile lands. The production of nonconventional sources of energy such as nuclear power also has concomitant environmental problems such as the disposal of nuclear waste. In other words, all the significant methods of energy production, as we know them today, give rise to increased pollution of the environment.

Similar problems are associated with energy distribution. Bigger and bigger tankers are breaking asunder on the high seas, and transportation of coal over long distances in LDCs is associated with the disruption of social life. Long distance pipeline construction (e.g., the Alaskan pipeline) has led to massive intervention in the ecology of vast areas of land.

But perhaps the greatest threat to the environment lies in the consumption of energy. The environmental factor of critical importance on a global level is the heat release associated with burning fuel. It is a fundamental thermodynamic fact that all energy generated finally ends up in heat. This fact is true for all forms of energy. The rate of heat generation and its possible effect on the worldwide climate has been discussed by Randers.[13] Admittedly, there are conflicting views by experts. The same is true of the carbon dioxide question.

I think it is fair to say that if present trends in the production, distribution, and consumption of energy are allowed to continue, the global environment will be in great jeopardy. As a result, many have postulated alternative options for the human society that will reduce the need for excessive use of energy, particularly fossil-fueled and nuclear energy. This option is generally referred to as a "new lifestyle."

The interesting thing to note is that the advocates of this new lifestyle are mainly from the affluent nations. They link energy consumption with the demand for economic growth and the consequent increased use of all resources. Daly, for example, contends that there is a need to limit economic growth as a result of ecological and moral necessity.[14] He talks of "a long run aggregate biophysical constraint—something like ecologists' notion of permanent carrying capacity."

Most of these arguments are indeed impressive. However, to me and a number of the articulate citizens of the LDCs, there is a fundamental flaw in them. They do not take account of the realities of two-thirds of the world. I agree with Curien[15] that the discussions are "slogans that reflect a pampered minority." He gives the monthly per capita expenditure of people in rural India (80 percent of a total population of some 650 million) as 53 rupees, or approximately 7 US dollars. An educated guess is that a Third World average cannot exceed $21 per month.

Advocates of the new lifestyle cannot seriously expect people below the poverty line to limit their quest for a higher quality of life, which includes greater energy use. They are already as near to "nature" as possible. They already consume the lowest possible amounts of energy. Their quality of life demands improvement. Therefore, the call for a new lifestyle should be addressed to the rich nations and to the rich sections in the poor nations. The need for a lowering of the appetite of the rich is not necessarily expressed from a moralistic viewpoint. It is based on enlightened self-interest. The build-up of antagonisms between the haves and have-nots is more explosive than the energy crisis.

REALITIES OF NATIONAL ECONOMIC PLANS

The major difficulty in discussing the issue of the use of resources, particularly energy, is that economists are on the whole very conservative. As a result, national economic planning in all countries is based on tradition. Traditional planning implies planning for growth. We have argued that HDCs need to slow down so that the rest of the world can get their fair share. But this ethical dimension appears to be a nonexistent factor in national planning.

All nations plan for themselves and only take into account perturbations of their plans as a result of interactions with other nations. Consider, for example, Nigeria's national energy plans for the 1980s and beyond.

The study referred to earlier collated some interesting data on the correlation between GDP and energy consumption. It defines the ratio between the growth rate of energy use and growth rate of GDP:

$$\text{GDP elasticity} = \frac{\text{growth rate of energy use}}{\text{growth rate of GDP}}$$

This information is presented for Nigeria for the period 1960–78 in Table 10-4.

Table 10-4. **Nigeria's End-Use Commercial Energy/GDP Elasticity Coefficient**

	Average annual growth rate of end-use commercial energy (%)	Average annual growth rate of real GDP (%)	GDP elasticity of commercial energy
1960–70	5.8	5.4	1.07
1970–78	20.7	9.2	2.25

The GDP elasticity figures provide a simple way of demonstrating the relationship between economic growth and energy use. As can be seen from Table 10-4, the GDP elasticity coefficient during the 1970s was much higher than in the 1960s. This growth, of course, was due to the so-called oil boom. Using this encouraging development (which to my mind is quite fragile), the following forecasts were made on population, GDP, and energy demand, as shown in Tables 10-5, 10-6, and 10-7.

The authors of the study state that "projections of domestic energy demand in Nigeria entail large difficulties and uncertainties especially given the still very low per capita energy use on one side and the very large potential for future increase in energy requests due to economic development and population growth on the other." They also point out that political decisions can change their projections.

The accuracy of the projections is not the main issue in this paper. The main concern for the *future* is what will happen if the

Table 10-5. **Forecast of Nigeria's Population, 1979-2010**

	population (x 10⁶)	annual population growth rate (%)
1979	80	2.5
1980	82	2.5
1981	84	2.5
1982	86	2.5
1983	88	2.5
1984	91	2.5
1985	93	2.5
1990	104	2.2[a]
1995	115	2.0[a]
2000	126	1.8[a]
2005	136	1.6[a]
2010	146	1.4[a]

[a]Annual rate also applies to each year in the preceding five-year interval.

rest of the world (particularly the LDCs) have similar targets. For example, from 1978 (actual) to 1985 (projected) we have a quadrupling of energy demand. On the face of it, this is not an unreasonable target for Nigerians. But how do these targets affect Chad, Benin, and Niger? Put another way, can the earth's energy resources, including possible technological breakthroughs, sustain a worldwide average energy consumption using Nigeria's projections? We do not need to look into a crystal ball to make some deductions. Already, one project—namely, the 2000-

Table 10-6. **Forecast of Nigeria's GDP, 1979-2010**

	GDP at 1973–74 factor cost (naira x 10⁶)	Annual GDP growth rate (%)
1978 (actual)	17,182	5.5
1979	18,740	9.1
1980	20,245	8.0
1981	21,983	8.6
1982	23,652	7.6
1983	25,579	8.1
1984	27,941	9.2
1985	29,897	7.0
1990	41,932	7.0[a]
1995	56,114	6.0[a]
2000	73,339	5.5[a]
2005	93,601	5.0[a]
2010	119,462	5.0[a]

[a]Annual rate also applies to each year in the preceding five-year interval.

Table 10-7. **Forecast of Nigeria's Energy Demand, 1979-2010,
Based on International Sectional Analysis**

	Energy demand (x 10^6 TOE)
1978 (actual)	6.2
1979	14.1
1980	15.5
1981	17.3
1982	18.9
1983	20.9
1984	23.4
1985	25.8
1990	39.5
1995	57.2
2000	77.4
2005	105.4
2010	144.3

megawatt Lokoja Dam—has run into serious problems. The large hue and cry raised by citizens of the villages that will have to be uprooted and resettled has forced the government to put the project in the cooler.

THE NUCLEAR POWER OPTION

Estimates of available fossil fuel resources vary widely. The figures given are not usually very objective. The information provided goes through several kinds of "color filters." The multinational oil companies, the oil-producing countries, the nuclear power plant suppliers, the environmentalists, and so forth all give estimates that strengthen their vested interests. However, it is self-evident that since the fossilization process took place millions of years ago, we cannot expect to replace fossil fuels within the time frame which we refer to as the *future*. Therefore, the use of fossil fuels for energy is limited.

Since we must use energy as long as mankind exists on earth, the main concerns for the future relate to the issue of alternative energy sources to replace fossil fuels. At first glance, the answer is simple; namely, a renewable nonpolluting energy source. Solar energy is immediately considered the best candidate. Other candidates advocated include biomass, wind, geothermal, and tidal energy.

Solar energy has become so popular in Nigeria (in academic

circles) that we have a Solar Energy Association of Nigeria. Views of budding Nigerian solar energy experts have been adequately presented in the document "Towards a Comprehensive Energy Policy for Nigeria."[16] I agree with most of the views expressed by the advocates, particularly Sulaiman and Adebeyi. They emphasize the need for Nigeria to begin immediately to involve itself in essential research and development work on solar energy. But I do not subscribe to some other views which border on oversimplification (an attitude not peculiar to LDC scientists) on the prospects of solar energy availability within the near future. In fact, Ofodile states[17] that "there is indeed very little technology required for diverse utilization of Solar Energy; and by applying such technology to capturing of Solar Energy, Nigeria would be on the way to reducing consumption of fossil fuel resources, vital for other uses." It is this type of oversimplification that forecloses meaningful discussion of the nuclear power option for LDCs.

I think it is important to repeat my endorsement of *solar energy as an important energy of the present as well as the future.* It is not easy to quantify the amount of energy provided by the sun in preserving food (e.g., drying fish, meat, etc. in the sun), drying clothes (instead of using electrically operated dryers), and so forth. But this amount of energy is significant in any evaluation of the total end-use energy of the rural populations in LDCs. But the real situation is that solar energy contributes significant energy only to those living at subsistence levels. There remains, however, the need for electricity to reach the rural areas. Solar energy can provide this clean electricity, and authoritative studies reveal that photovoltaic electricity will begin to make a serious impact on a global scale by the year 2000.[18]

It is important, therefore, that scientists and engineers in LDCs should work hard to provide accurate environmental data on solar radiation, and, in collaboration with technologically advanced countries, develop and test new devices under local conditions. This is an area where self-reliance and technology transfer models are not ambiguous. I think there is general agreement on this issue.

But when the issue of the nuclear option is raised, we have fireworks! To the average literate man, either in the HDCs or the LDCs, nuclear power means the nuclear bomb, and for some it means still worse: Hiroshima!!! This situation is unfortunate. An operational atomic bomb preceded an operational atomic power plant. We cannot get away from this fact of history.

The contention of this paper, however, is that LDCs should not foreclose the nuclear power option. They should proceed in the same manner as is advocated for solar energy. There should be research and development, which should include intensive manpower development. It will be a tragedy if, by the year 2010, there is a significant breakthrough on nuclear fusion but there are no scientists and engineers in the LDCs to apply the knowledge so gained in their respective countries.

It has been argued that providing nuclear technology to LDCs is like giving matches to a small child. It does not occur to such advocates that Hiroshima was not the act of an LDC. What upsets one most is that a number of LDC scientists are even more brainwashed on this matter. In the policy paper referred to earlier, a Nigerian scientist had the temerity to say that "from the analysis, a lot of recommendation [s] are drawn and a proposal for shifting from conventional to nuclear is strongly advocated provided the Nigerian can show a great sense of maturity and responsibility." A sense of maturity and responsibility indeed. I ask, how can Nigeria install and operate a 2000-megawatt dam if Nigerians cannot show a sense of maturity and responsibility?

Debate on the specific issues of nuclear profileration reveals the state of international injustice and self-centeredness in the world today. The nuclear nonproliferation treaty is about the most unjust treaty ever sponsored by a United Nations agency. However, for the record it must be said that the International Atomic Energy Agency (IAEA) has tried to serve the Third World, given all the constraints of the treaty, although the pressures of the big powers have led to the imbalanced funding of programs by IAEA regulatory and promotional functions. But the point that must be made is that due to the intensity of the nuclear armaments race, both in quantity and in quality (vertical proliferation), there is no justifiable reason to deny the LDCs *access to nuclear technology*.

Another argument raised against the nuclear power option is the one arising from problems of safety and waste disposal. Experts such as David Rose have made comparisons over the years about the relative safety of civilian nuclear power as compared with power-generating systems based on fossil fuels.[19, 20, 21] For someone favorably disposed to nuclear power, the arguments are straightforward and convincing. But let us admit that risk analysis as a science has yet to attain maturity. Therefore, questions on nuclear power plants' safety and so on are not to be dismissed. But the argument that is not strongly emphasized is the one which compares the safety and hazards of the silos beneath the ground with the safety and hazards of submarines beneath the oceans loaded with thousands of megaton bombs! Furthermore, what is the level of radioactive waste accumulating from the production of these megaton bombs? In short, it is extremely hypocritical to talk in terms of the safety and hazards of civilian nuclear power if nothing is done about military nuclear power programs.

The world needs the nuclear power option to generate energy for peaceful purposes. We need the present fission reactors to bridge the gap between the medium term and the long term, when either solar energy or fusion can make a significant contribution to world energy needs. In the short term and the medium term, LDCs not only must exist, they must lift the majority of their citizens above subhuman levels. Energy is required for them now. The nuclear power option should not be excluded. Not all the LDCs will be in a position to buy and operate nuclear power plants. Those who desire them should have access to them, provided these countries make efforts on their own to produce the necessary trained manpower. The HDCs should not stop their civilian nuclear power programs, because they then will be forced by political pressure to seize the resources of the weaker nations, leading to grave global instability.

CONCLUSION

In the preceding chapters we have tried to discuss conceptual issues on a whole range of energy-related problems. We have also treated some specific problems and done some analysis of these specific problems. It is not the intention in this concluding section to summarize the foregoing individual sections. Rather an attempt will be made to present, for discussion, a model which will link up these ideas. The objective is to promote debate.

The first assumption is that only when ideas and concepts are reconciled with political, social, and economic realities do we have the possibility of action programs.

The second assumption is that the concept of a Just, Participatory, Sustainable Society is a concept which tries to take into account the LDC concern for international *justice* in the discussion of *energy-related* problems of the *present* as well as the *future*.

Let us start with a simple model:

- Efficient Energy for the Future (EEF)
- Just, Participatory, Sustainable Society (JPSS)
- Political, Social, Economic Realities (PSER)

This model is shown in Figure 10-5.

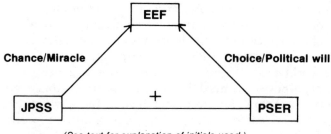

(See text for explanation of initials used.)

Figure 10-5. **Model I**

On the basis that we would prefer *choice* to *chance,* we attempt
to construct a more detailed model. It is predicated on the fact that
the major component concepts of JPSS as they relate to energy are:

Justice
Minimum-Maximum Energy Requirement	MMER
Redistribution of Energy	ROE
Fair Price for Energy Resource	FP

Participation
Participation in Market Decisions	PIMD
Choice of Quality of Life	QOL
Energy Technology Sharing	ETS

Sustainability
Disarmament	DIS
Conservation	CON
Global Carrying Capacity	GCC
Dynamic Equilibrium in Economic Growth	DEIEG
Self-Reliant Technology	SRT

The component concepts of JPSS are essentially ideals. The world
in which we live is based on the realities of greed, selfishness, the
desire to dominate others, and the disposition to waste resources.
These are *natural tendencies* which are universal and cannot just
be wished away. *Therefore, in our search for an Efficient Energy
Future we have to establish new and strengthen existing structures
to counter these natural tendencies.* To do this, we propose some
structures:

Global Security Commission	GSC
Global Ecological Monitoring Commission	GEMC
National Energy Council	NEC
Regional Energy Council	REC
Global Energy Council	GEC

—and a final model, as shown in Figure 10-6.

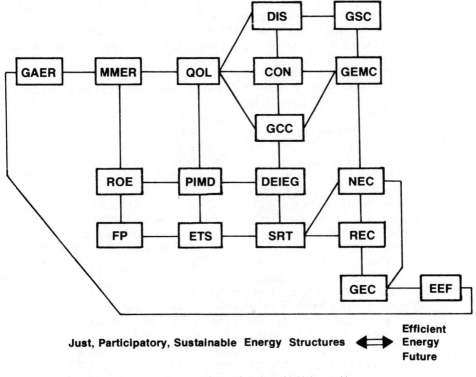

Just, Participatory, Sustainable Energy Structures ⬌ **Efficient Energy Future**

(See text for explanation of initials used.)

Figure 10-6. **Model II**

NOTES

1. B. C. E. Nwosu, "Is there a Future for Science in Africa," *Anticipation* 1 (Geneva: WCC Press, 1972).

2. See study commissioned by the Federal Republic of Nigeria, Ministry of National Planning, "Energy Study for Fourth National Development Plan 1980–1985," Vol. 1, Main Report (Draft—March 1980).

3. B. C. E. Nwosu, "Scientific Technology and the Future of Africa," *Ecumenical Review* XXIV (Geneva: WCC Press, 1972), pp. 289–300.

4. *Faith and Science In an Unjust World*, Report of July 1979 WCC Conference held at MIT (Geneva: WCC Press, 1980).

5. J. Randers, "The Carrying Capacity of Our Global Environment—A Look At the Ethical Alternative," *Anticipation* No. 8 (WCC Press, 1971).

6. Ibid.

7. "Science and Technology for Human Development," 1974 WCC Conference Report, *Anticipation* No. 19 (WCC Press, 1974).

8. S. L. Parmar, *Anticipation* No. 18 (WCC Press, 1974).

9. A statement in June 1979 credited to Dr. Amadou-Mahtar M'bow, Director General of UNESCO. See p. 159 of reference cited in note 4 above.

10. See reference cited in note 3 above.

11. Abdul Amir Kubbah, OPEC: *Past and Present* (Vienna: Petro-Economic Research Centre, 1974).

12. Ibid.

13. J. Randers, op. cit.

14. See H. E. Daly, "The Ecological and Moral Necessity for Limiting Economic Growth," pp. 212–20 of reference cited in note 4 above.

15. See C. T. Curien, "A Third World Perspective," pp. 220–25 of reference cited in note 4 above.

16. See the papers by Drs. A. T. Sulaiman, O. A. Bamiro, G. A. Adebeyi, E. I. E. Ofodile, Olu Odeyemi, A. M. Salau, and B. A. Ajakaiye in the chapter on solar energy (pp. 365–448) in the Proceedings of the Energy Policy Conference, Jos, August 1978, National Policy Development Centre, Federal Government of Nigeria.

17. E. I. E. Ofodile, ibid.

18. See the study entitled "Principal Conclusions of the American Physical Society Study Group on Solar Photovoltaic Energy Conversion" published by the American Physical Society, January 1979.

19. D. J. Rose, *Science* 184 (1974), p. 351.

20. D. J. Rose, P. W. Walsh, and L. L. Lesrovjan, "Nuclear Power Compared to What?" *American Scientist* 63 (1976), pp. 291–99.

21. D. J. Rose and R. K. Lester, "Nuclear Power, Nuclear Weapons and International Stability," *Scientific American*, April 1978, pp. 45–57.

CHAPTER ELEVEN

Selected Comments

LUIS SEDGWICK BAEZ

I had the opportunity to attend the 11th World Energy Conference held last month in Munich in the Federal Republic of Germany, and one of the main conclusions reached there was, and I quote, "International energy cooperation is not only necessary but must be closer than in the past."

We in Venezuela believe that one of the solutions to the energy problem is regional cooperation. The Latin American developing countries possess enormous potential energy resources but lack the financial capacity and technical know-how to utilize them fully. The utilization of this energy potential would help to accelerate the developing process of these countries, and at the same time it would permit reduction of the dependence that many of these countries currently experience because of oil imports and their effect on the countries' balances of payments.

Venezuela, blessed with natural energy resources and oil revenues, has an obligation to help the less developed countries of its region. Venezuela has recently signed, with Mexico, a joint cooperation program put forward eloquently at this Symposium by The Honorable Minister of Energy of Costa Rica, Fernando Altmann Ortiz [see chap. 7]. Moreover, Venezuela is in the process of creating a continental program of energy cooperation in coordination with the Latin American Energy Organization (OLADE), with participation by all its member countries as well as by the United States and Canada. The main objective of this continental program of energy cooperation is to develop energy resources and improve energy efficiency in the region. This would help to reduce energy restrictions on the growing economic development of the region's countries; free more conventional resources in the region such as oil for those uses where substitution may be difficult (as in the petrochemical industry and transport); contribute to the energy self-sufficiency of those countries in the region whose oil consumption has been increasing rapidly; and stress energy conservation, the most important source of energy. These considerations would, at the same time, help promote commerce and economic development in the region.

EDWARD LUMSDAINE

My comments relate more to this morning's session, so I won't ask any questions but will make comments. Not being a policymaker but an engineer who directs an institute that attempts to bring solar technology to commercialization, I must say I feel rather out of place making comments on specific items in a field filled with generalizations. But I do want to point out some specifics that affect my job (or our institute) in our frustrating attempt to pull solar energy technology toward commercialization.

The key word is "education" in the broadest sense. I am particularly surprised that the previous speaker, Dr. Nwosu, who is Chief Education Officer of Science in Nigeria, did not mention the importance of education in the broad context. For example, we find that despite years of demonstration of solar facilities, there's very little consumer understanding of and confidence in solar products. Perhaps the choices in energy policy are not technical, political, ethical, or economical but instead are philosophical. In this case, they can only be changed through education, and that is a universal need, if we want a soft-path future or any future at all.

Although as an engineer I am quite sensitive to cost effectiveness, I am also quite suspicious of the notion that, personally and as a society, we should always take the most cost-effective path. Many decisions we make are not based on cost effectiveness, or who among us would have children? The fact is that, in recent decades, we have been receiving energy in a dependent way. In this dependency we have taken first-cost options rather than those options that give delayed gratification. In the United States, for example, we are a credit card society; we want to enjoy the benefits of goods before paying for them. In fact, on the rebate system, they pay you to drive the Petropig before you even start making payments. We are dealing with a society that is conscious of first costs but not of life-cycle costs. Having been involved in the design of many passive solar homes and passive solar retrofits in the last few years, and in solar in general for more than fifteen years, I have yet to find a solar house that would cost the same as a conventional house. So, why should builders or architects design or build houses that have higher first costs, since they do not have to be saddled with paying the fuel bills?

What the soft path requires is that people in our society become active participants in long-term investments and power generation and become very conscious of life-cycle costs. If you consider that in a recent survey conducted in the United States, nearly 60 percent of the people surveyed did not know what a solar hot water system is, then requiring people to raise themselves to such a level of sophistication is really a formidable task. A passive solar home, for example, requires some participation by the occupants, which means some understanding of heating and cooling. Turning on a switch does not require this understanding. When you ask people to invest in passive solar homes, they know immediately, by casual comparison, the extra first cost. This is the equivalent of asking them to pay for all their incremental future fuel bills at the present, whereas traditionally we depend on the power industry to make this investment and we pay for the energy at unknown although fearfully high and escalating fuel rates. To supply power by nuclear, oil, and coal requires only a few people talking to each other and making decisions in a small gathering, whereas passive solar requires the energy education of the masses.

I am quite surprised that in this Symposium the issue of energy education has not been mentioned at all. The most ignored yet potentially the most important subject in terms of energy productivity is education.

MARCELO ALONSO

I am going to make some comments about what I gathered were perhaps the highlights of what has happened today in Sessions II and III. I think it is very clear that the issue, not precisely of conservation but of increasing energy efficiency—making the best possible use of energy—has come up several times. But I think at the same time that—and this was put very clearly by Dr. Nwosu—we cannot rule out a priori any particular energy, so regardless of one's opinion about the different energy alternatives, one has to consider all of them as essential in planning energy development.

I think that in addition, as Dr. Nwosu said, there is a problem with nuclear energy in particular. We should try to separate entirely the use of nuclear energy for power generation from the problem of nuclear weapons proliferation. I think it is terrible that mankind has tied up an energy source whose use is essential under many circumstances, but particularly in industrialized countries, with the issue of proliferation, for this issue, although a terrible threat, is not directly related to the use of nuclear energy for power generation. So we should try to see how we can separate, or uncouple, those two issues. But at the same time the problems of nuclear energy point to the tremendous importance of having people, especially those in countries that need nuclear energy, understand the issues of this energy source.

Furthermore, policymakers—the people who have to make decisions about energy—are essential for every country's planning. It thus is very difficult to expect that a country will tackle its energy problem in the correct way if it cannot prepare people who are competent in the field of energy policies. For that reason, international cooperation is very important—not only in research, development, financing, and so on, but in training cadres to provide leadership to deal with energy policies. .

The importance of international cooperation is related also to the problem of technology transfer, because that problem involves the production of energy itself. One of the predicaments in which the developing countries find themselves is that their industrial production methods essentially have been imported from the more developed countries and are therefore in a sense a duplication of the latter's systems. The less developed countries have a very limited flexibility to change their production methods and become more energy efficient, so unless the more developed countries modify their exported production methods and also begin to orient people toward more effective energy use, the developing countries are going to find themselves in a very difficult position— particularly those that are not oil producers, that are not oil exporters. So the more developed countries have a tremendous responsibility.

In addition, the developing countries that must substantially increase their energy consumption to achieve their development goals have very limited possibilities for conservation, while the more industrialized countries have greater possibilities. And I think that the energy problem has also been complicated by the combination of, on the one hand, the production of oil by a rather small group of countries and, on the other, the predominant consumption of oil by another relatively small number of highly industrialized countries, basically those in the Organization for Economic Cooperation and Development (OECD). It is those latter countries that have to make a special effort to increase their energy efficiency, their energy conservation efforts. So from this point of view, international cooperation (as somebody said earlier today) is a two-way street: going from the more developed countries to the less developed countries, to help the latter to change their industrial methods and adopt new technologies for energy use and generation; and at the same time, going on within the developed countries themselves, by becoming more efficient in their own energy use and thereby setting examples.

B. C. E. NWOSU

Basically, I agree with his statements, and to save time I won't add any more.

MILTON KLEIN

I am grateful to the two speakers in Session III, who I think did an excellent job of putting certain aspects of the energy problem into perspective—different perspectives, certainly, but nevertheless both useful. Professor Goldemberg was cautiously constructive in describing the potential advantages and the problems of a particular promising technology. Dr. Nwosu, of course, gave the perspective from the poorer of the developing countries. It leads me to a comment that applies here but also relates to this morning's session particularly, and that is the importance of these perspectives—the importance of helping the public understanding of the energy problem.

In the United States, and I think in other industrial countries to a degree, one of the principal impediments to making progress in energy matters is, in my opinion, a lack of understanding—a lack of any kind of consensus whatsoever. I believe Mr. Landsberg alluded to this in a different way in his introductory remarks to Session III, and I think that it is incumbent on all of us, particularly those of us who dabble more directly in energy matters, to go out of our way to be responsible in describing the pros and cons and to put into understandable perspective everything we do.

And frankly, that's one of the problems I have with the second paper of Session II [chap. 6]. It is difficult to comment on it, as

Jane Carter said in Session II's discussion. Amory Lovins can speak and write faster than anyone can think, but I know that he knows that many have challenged his numbers. Other people come up with very different numbers than he does, and there isn't time to refute them. But the real difficulty I see is that this kind of a discussion tends to put things into an "either/or" context. For example, either we go only with conservation, or we go only with supply. I don't think that's real. It is clear that we must do everything we can in the way of using energy more effectively, particularly in the United States, because, as several speakers have indicated, the US problem is an international problem—the United States is such a major factor that what its people do has tremendous consequences internationally. And there again, the need for the public to be led to reasonable understanding rather than to be confused is a great responsibility. An essential additional factor is that energy is not one dimensional. As others have said, it is one of many tools for reaching social objectives. Dr. Nwosu pointed this out, and others have as well. I didn't hear much in Amory and Hunter Lovins's paper this morning about that aspect—a few words and some arithmetic, but not really much more than that about the effectiveness and the purpose served by energy.

Finally, since I am involved with electricity I ought to say a good word about electricity, which didn't get much attention in Session II. It misrepresents the industry to imply that we feel that electricity is the answer to everything. Certainly nobody I deal with has that view. Electricity, we believe, should be used where it is economically and socially effective. Conservation measures are and should be promoted by electric utilities. However, electricity does have important economic and social benefits. It's particularly easy to see that here in the Tennessee Valley—I think those who have watched the development of the Tennessee Valley will understand that better than many. The situation here is not parallel to developing countries in every respect, but it is not without some lessons.

AMULYA KUMAR N. REDDY

Dr. Nwosu's paper has stressed the importance of justice and equity between countries. This is absolutely necessary but not sufficient. Energy justice and energy equity *within* countries is also crucial. The point is that most developing countries are dual societies—small elites among large masses of poor, islands of affluence amidst vast oceans of poverty. Since these dual societies are the antithesis of development, we have to ask the question: Do we want energy policies that perpetuate dual societies? Or do we want energy policies that promote development? We must determine whether the perspective is that of the rich in poor countries or the poor in poor countries. So what is needed is not only a new international energy order, which he rightly stressed, but also new national energy orders, and herein lies the path to solving the energy problem, as I propose to demonstrate in my paper tomorrow [chap. 14].

My second comment is with regard to Professor Goldemberg's paper. Ethanol technology is *a* solution, but it is not the only solution. In high-population-density developing countries other solutions may be necessary, for these countries have several objectives. Specifically, these are food, fodder (inasmuch as animal energy is the basis of their agriculture), fuel, and fertilizer. Hence, we should not be so obsessed with energy as to forget food and fertilizer, and we need decision rules to deploy land, which is the basic resource and constraint.

One possible set of decision rules is presented in Figure 11–1's diagram. In high-population-density developing countries, we should not follow the route of going down the left where arable land is used to produce alcohol; instead, nonarable land should be used to grow trees, and from the resulting tree biomass, only the parts that are not rich in nitrogen and phosphorus (that is, the ligneous or woody parts of the trees) should be used to produce fuel. With regard to that fuel, there are two options: there is the methanol option and the gas production option, but the choice between these two options has to be based on other considerations.

B. C. E. NWOSU

In my attempt to rush through the presentation of my very long paper, I may not have stressed sufficiently the point about justice being both between countries and within countries. That point was in my paper, if you read it. But I think that in my presentation I did show one of the few tables in the paper, to demonstrate that in Nigeria there's a wide distribution in the use of energy as of now. So I subscribe to your view. Somehow, in my African or Nigerian English, or my speech, or both, this point must have missed you.

JOSÉ GOLDEMBERG

I couldn't agree more with Professor Reddy on the idea of dual

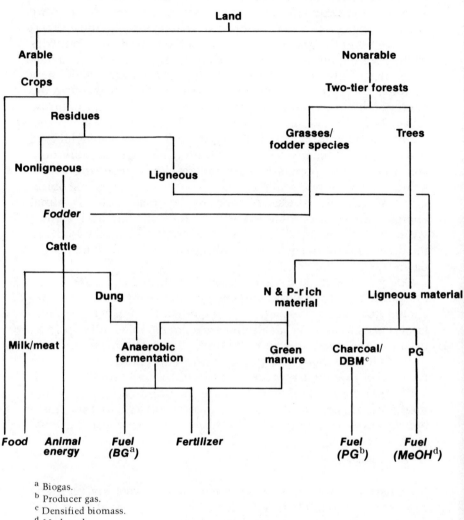

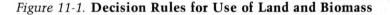

[a] Biogas.
[b] Producer gas.
[c] Densified biomass.
[d] Methanol.

Figure 11-1. **Decision Rules for Use of Land and Biomass**

societies. That duality is very, very obvious to anyone living in a less developed country, and what I tried to show is that even in highly urbanized, less developed countries such as the Latin American countries (and some other countries around the world), solutions are possible, hinging on renewable sources. The reason I prepared my paper was to show that I am becoming slightly impatient with a large number of the solutions that have been proposed to solve the energy crisis in the developed countries—solutions which are really inapplicable to our countries. All the questions that Mr. Landsberg raised about economics, suitability, and so on have been discussed for years and years, but what was done in Brazil—and I think this is what makes it interesting—is that the country has moved toward adopting a new solution based on ethanol, which is a renewable source. The solution has limitations, but we are coming to understand these limitations now, and I think it is a live example, a vivid example, of what can be done; it shows that conservation is not the only thing you can do in the world. Of course the modern, urban sector in the societies of less developed countries will benefit from whatever conservation measures are adopted in the affluent countries, but the rural sector—and some of the urban sector, too—will need renewable resources, and they can be developed.

MIGUEL S. USSHER

I will make a few remarks on the subject of international cooperation. This subject is very poorly understood and very badly put into action.

First of all, technology is considered a magic word. Technology is not a miracle. Technology means people who have to get a certain knowledge. And if we transfer something, we transfer the knowledge to people who in a few years will have to put that knowledge into action and promote other actions to solve the problems that have to be faced.

Second, we must realize that the issue of international cooperation must be divided into two categories. The first category is the interchange of technology between countries that are on the same level, like Russia and the United States—that is not a problem. But the second category is the transference of technology from a very developed country to a country which is not so developed—in that case, we must first know where and why we are going to transfer technology. To do so, an energy assessment must be run to spot the places, or areas, that require technology transference. If this is not done, the transference may be useless work because sometimes what is being transferred—for example, solar technology—will accomplish nothing in the recipient country. Both the transferrer and the recipient should understand that this assessment is a prerequisite.

Furthermore, the time factor must be considered. There is not enough time in the world to discover new technologies country by country before the oil runs out, especially in less advanced countries that are not in a condition to establish new labs, put in more money, and do all that has to be done to discover new technologies or apply existing ones. Thus, in my opinion, the very advanced countries that share certain technologies should try to pool those technologies with countries that need them.

For example, if through an energy assessment we know that a certain less developed country has a good deal of undeveloped oil shale, its technicians should be invited to countries where oil from shale is currently being produced to allow those technicians to see, or personally participate in, what is being done. They should be allowed to learn, investigate, and copy all the trade secrets, and the host countries should commit themselves to disclosing and transferring their experiences and findings to the visitors. Since an image is worth a thousand words, this approach will pay better dividends than one which, by transferring theoretical details through government agreements, causes everything to be slowed down or left on the shelf.

If the former apporach is used with many countries that have the same problem, the effort will be worth it. For we have to consider that since the need for new technology options is imperative, the amount of money involved enormous, the current number of technicians too low, and the technology's results useful and vitally needed by everybody, then it's worth it to pool these efforts. Of course, there are matters of national security in certain technology transference areas. Let's leave those areas apart, but let's at least work on those areas that can be worked on, for then something practical will be achieved. If we don't do this the whole subject will be all talk, and this is what has been happening up to this moment.

B. C. E. NWOSU

I would like to reinforce this point of view, even though I haven't done so in the other cases. Now, when I was talking about international technology-sharing, I had this in mind: that we were talking about any energy-efficient feature, and that technology is a resource—something to which you add on. And I want to support the point made by Dr. Ussher from Argentina, for it is important to utilize the human resources in the less developed countries because doing so will increase the pool of energy by increasing productivity and efficiency and so forth. Technology-sharing should be done not only on humanitarian grounds but also from the standpoint of justice and enlightened global self-interest (if we reach that stage). Thus, if Nigeria or Ghana or Niger is able to utilize solar energy effectively because of American know-how or sharing of that know-how, a contribution to the global energy resources as a whole can be made. If it is looked at that way, there is room for guarded optimism, as I say in my paper. But if it is looked at from other angles, it will be mere talk, as Professor Sadli has said—talk which we have had for more than twenty years, at least.

ISHRAT H. USMANI

Just last week I returned from Brazil, which in size is not a country but a continent. And a short calculation shows that if the population density of Brazil equaled the population density of Bangladesh, the entire world could be accommodated in Brazil, and the rest of the earth would be totally devoid of people. Because of its vast territory, I want to ask Professor Goldemberg why oil has not yet been discovered in that continent called the country of Brazil? When did Brazil start exploring for oil, or was no importance attached to it because it was selling at $2 per barrel in the 1960s, and so forth?

That was my first question. My second question is: Why should a small consumption of imported oil—about 1.1 million barrels per day—agitate the national energy policymakers of Brazil so much? Take lessons from the United States, which in area is as big as Brazil, but, being highly industrialized, is living with about 6 to 7 million barrels of imported oil per day. How the United States does it, of course, is by exporting armaments and food, both of which are energy intensive. Just as the prairies of the United States are the "breadbasket of the world," where wheat grows naturally without any irrigation, so the climate of Brazil is such that in many of its arid parts and in the great Amazon basin, hydrocarbon-yielding palms and other biomass could be harvested to produce liquid fuels which could replace imported oil. I hope the Brazilian authorities are aware of the exciting research and development work that is being done by Professor Calvin in California on hydrocarbon-yielding plants. So I am not pessimistic and feel that, with proper planning, the oil problem in Brazil can be solved.

Now, having said this, I want to know the following: Why is the Brazilian government concentrating on the production of ethanol from sugar cane in centralized distilleries? Because of the vast distances involved, the cost of transporting ethanol from centralized distilleries to rural areas scattered all over Brazil must be very high, and therefore, perhaps, a decentralized approach would be more appropriate if the whole country has to be covered. The cost per kilometer of transporting a liter, gallon, or cubic meter of ethanol is not given in Professor Goldemberg's paper. Finally, exactly what is being done to treat the molasses, and how is the problem of pollution caused by distillage being solved? I think the distillage could be subjected to anaerobic fermentation, which could produce biogas as methane fuel and nitrogen-rich slurry as liquid fertilizer. I would like to pose these specific questions to Professor Goldemberg.

JOSÉ GOLDEMBERG

Well, it is very good to see someone optimistic in this world, and you should come more frequently to Brazil; oil has been looked for there since 1950, and the amount of drilling has been reasonable by any standards. Brazil produces about 200,000 barrels of oil per day, but no important fields have been found in the last few years, although exploration has been going on. Actually, the laws of the country have been changed to allow foreign companies to come in. They were kept out for about twenty years but were allowed to come in a few years ago, and they have now been there for a couple of years. They are not numerous, and they did not have much success, either. So it seems that geology is just playing some tricks on Brazil.

Now, on the extent of the land, I would like to answer Dr. Usmani by mentioning that the amount of arable land in Brazil is not very large—there are 40 million hectares being used today, and apparently the total amount of arable land is close to 100 million hectares. The country is very large, but a good part of it is covered either by the Amazon forest or by lands of low quality, so it's no use throwing grains of wheat around, for they will not grow.

As far as distilleries are concerned, I did mention in my paper that these distilleries are relatively small in scale. They produce about 600 barrels per day, and they are spread out all over the areas of the country in which sugar cane grows.

This leads me to the next question. Why don't we use molasses and other crops? They have been tried. Cassava has been tried as the next candidate, but what happened to cassava is that the agricultural problems involved in growing large quantities of cassava on an industrial basis—on an agri-industry scale—have not been resolved. I think this is a great shame, but they have not.

Brazil is the greatest producer of cassava in the world, but in very, very small lots. There are no problems in the technology of producing alcohol from cassava, but there are problems in growing cassava in large amounts.

As for Calvin's ideas, we have had many visits from Professor Calvin, who has been trying to impinge on us with his brilliant ideas. My answer to him—I will say it in public because I told him personally—is that we'll let him develop petroleum trees in California while we produce alcohol. It might not be the most effective way of producing it, but it harvests the sun's energies with an appreciable efficiency. He has not come out with a feasible scheme, but we are following his work rather closely.

AMORY B. LOVINS

I will try to be very brief and start with Milton Klein's comments. I am sure Mr. Klein knows the difference between historic and marginal utility. There's no doubt at all that the first kilowatt-hour, or the tenth, or the hundredth, or even the thousandth, perhaps, produced great social benefits. We are arguing about the next kilowatt-hour, not the first one.

There are about thirty major exchanges in the literature—critiques and responses—on my numbers, and I think those who read the responses will find that people who disagreed with my numbers did not show the numbers were wrong; they showed rather that they had failed to verify the references. I do try to document where the numbers come from, and if you find a wrong number, I'd be delighted to know where it is.

It is quite true, as several speakers remarked, that we based much of our argument of what is possible (in technical performance or in cost) on existence proofs. Boulding's First Law: Whatever exists is possible. This is a reasonable line of argument for a physicist. It is not logically airtight. I don't know any argument that is. Måns Lönnroth has said, "Okay, you can build such and such a solar system in that place at that cost. That doesn't prove that you can build it in *my* country at that sort of cost, because there may be institutional barriers in my country that you don't know about." That's perfectly true, but I still think that it is the best form of argument around, and one which is often used, for example, by advocates of fast breeders.

One brief comment and a very brief question to each speaker, if I may. Professor Goldemberg's paper, I think, casts more doubt than is warranted on urban solar energy. Urban density actually improves solar economics, and we have had some careful aerial photogrammetric studies of the area, orientation, and shading of roof and wall space in places like several European cities, Denver, and Los Angeles. It has turned out that there is plenty of collector area to go around, even without much efficiency improvement, and if we go in for photovoltaics and urban forestry, one can even envisage the city being very largely self-reliant in energy. The marginal land use, if soft technologies are properly done, is approximately zero—it's not very large.

A question for Professor Goldemberg. Brazil has a lot of lateritic soils. There is also something called the Munchkin Principle: If you leave a lot of free energy lying around, something will come munch on it. That is why, for example, the net primary productivity of the rain forest is practically zero, even though its gross productivity is very high. In the Brazilian climate, the munchkins can really go to town. If high-productivity crops like sugar cane or cassava in large monocultures were ecologically sustainable, why aren't they there already? In other words, how sustainable is the sugar cane crop, and how do you know?

And finally, a question for Dr. Nwosu, who talked very appropriately about salesmen who come to sell technology to developing countries. What would be the reaction in your country if people came in from, say, the United States, trying to sell you soft technologies as a way to increase their business rather than as a way to promote self-reliance by making these technologies yourself? This is a major concern, because many developed countries are trying to promote renewable energy as an extra export for their own industries rather than as a way to promote world equity.

JOSÉ GOLDEMBERG

In answer to the question on sugar cane—I am not an expert in these things. The only thing I can tell you is that sugar cane has been a very important crop in Brazil for about three hundred years; apparently the Dutch came to Brazil because that was the way of beating the Central American countries, which had a monopoly on the production of sugar. We have been able to keep this production going; probably the future will tell if the present way of producing sugar cane—with mechanized equipment and all that—is really destroying the land, but there are no clear indications of it yet. People have been watching it rather carefully.

B. C. E. NWOSU

First, the criterion for appropriateness of technology transfer, whether it be high technology or soft path, is the criterion of replicability. Therefore, if, when the salespeople bring these soft technologies or solar energy technologies, the Nigerians can replicate them or improve on them or participate in their use and maintenance, that is of course preferable. But the mere fact that the technology is very tiny or small, a microprocess, does not make it any better for our purposes. Does that answer the question?

AMORY B. LOVINS

I was wondering more about the psychology of salespeople coming in. Obviously, if our country—the United States—sets a bad example, and we say, "Soft technologies are great, but we are not going to use them ourselves," we can't expect countries lacking our advantages to come to a different conclusion. We send high-tech salespeople all over the place pushing hard technologies and giving concessionary loans and so on. But what if we view soft technology as a way to boost our own industries—if we are not really interested in energy for development but only as another export—does that turn off developing countries?

B. C. E. NWOSU

Well, the time of the so-called iron curtain or bamboo curtain is over, I think, because everybody's been penetrated now. So I think it is a fait accompli that we are going to have salespeople, and we are not turned off by the salespeople or the people trying to promote their industries. We assess what they're giving us based on what real advantages we get. So there is no psychological barrier to somebody trying to sell something to you.

MÅNS LÖNNROTH

I would like to pick up a point that Amory Lovins had and elaborate on it. Our studies show that in the Swedish climate, solar heating has its highest benefits in the rather densely populated suburban areas—not, perhaps, in the city cores. And the disadvantage lies in the rural areas, where distances between houses are so great that a common distribution system is too expensive. You cannot really make any generalizations where the economics of solar heating lies.

My major point was actually another one. I'm struck by the Brazilian ethanol case for a slightly different reason from most people here: the fact that the Brazilian government apparently has forced its car industries to use a fuel which is not used in any other country. The reason why I am struck by this is that we have had some similar problems in Sweden. We have had long discussions about using methanol as a fuel in cars and have been opposed by a coalition of oil companies and car industries that we first had to fight down. Now, I would very much like José Goldemberg to tell us: How did the Brazilian government fight down its own car industry? This question is linked to a discussion that we had this morning about "big government." I found it odd, when I read both the Lovinses' paper and John Deutch's paper, that there is an obsession in the United States with what they call "big government." It seems to me that the major problem in most countries is not that there is too much government but that there is too little government in the face of an already too great level of corporate power. And, at least in my own country, I don't think we can solve our problems unless we have a stronger government in relation to corporate power. I would thus like to hear José's view on the Brazilian government and the Brazilian car industry.

JOSÉ GOLDEMBERG

I think this is a very interesting subject that deserves analysis by some political scientist, but I will answer as well as I can. The point is the following: when the alcohol program started four years ago, everybody was against it. The state-owned enterprise, Petrobrás, which has a monopoly over oil distribution, was against it; the automobile industry was against it; the sugar growers were against it. The only people in favor of it were the intellectuals—who are considered to be ineffectual in general but were not in this case—and some but not many government officials, who were championing the alcohol program in one of the ministries. Now what happened, slowly at first, was a large amount of discussion which I think captured a very large audience.

As a brief digression, I will use this moment to answer Dr. Lumsdaine, who made a very interesting comment about energy education at the beginning of this session's discussion. My answer: education on energy is being done all the time. This Symposium is an occasion in which the people in Tennessee—and probably a wider audience—are being educated on energy matters, and this is going on all the time. These conferences, of course, are too frequent for most of us to attend all of them, but they implicate other people, and it is a way of extending what we do into the classroom.

Turning back to my answer, education was an important factor, but what really turned the automobile industry was the danger of rationing gasoline, which was a very, very real danger. The sugar growers joined the program as soon as the price of sugar in the international market went down. The fear of rationing was the major force that drove the automobile industry to face their home offices. (All of the automobile companies in Brazil—Volkswagen, General Motors, and so on—are foreign owned.) Their cars are produced only for the internal market, but this internal market is big enough to absorb the production of the companies. Now people are beginning to talk about exporting these cars, and I think that is going to be a new crucial point. Now everybody is very enthusiastic, and of course the government gives easy credits on cars that run on alcohol. So a combination of threat, rationing, and some carrots which included easy terms on automobile financing did it.

SHEM ARUNGU-OLENDE

My comments actually relate to two sessions—Sessions II and III. I find, after listening to the speakers of both sessions, that there is a gap regarding technology perspectives, a gap that I hope will be filled at the next round of meetings. The perspectives should, in my view, give us a whole spectrum of available technologies and a methodology for comparing them so that decisionmakers who want to use them can make decisions knowing the range of the technologies' applicability and effectiveness. The discussions so far have concentrated on one or two major technologies without giving due consideration to the other technologies. The next stage of the discussions is the most difficult one, for it involves synthesis: trying to harmonize all the structures of different technologies and different methods of producing energy. But it is, in my view, one of the most important and crucial tasks, and I hope that when it is undertaken it will take into account the differences in various regions and parts of the world.

The other point I should like to make regarding Session III is that we should give more thought to constraints, conflicts, and critical paths. Here again, some constraints have not received enough attention and yet are very important. They entail financial constraints; research and development, including transfer of technology; education and training; information flows; political will—already mentioned; understanding the nature of the problem—a main topic of this Symposium; justice—already mentioned; social acceptability, which I think is very important; some social and environmental problems that have been touched on in passing; institutional problems and barriers; problems of weak infrastructure in developing countries; and of course the problems of other resource constraints. I believe that these are some of the issues we need to address if we are going to resolve the energy problem.

JAN K. BLACK

I have a wrap-up comment. I believe that the energy crisis has been seriously overrated. So many of us start out talking about the problems of energy and end up talking about injustice and inequality. Energy problems are only one aspect of global inequity, and a fairly minor one when you consider all the other problems in that regard. In the United States, the energy crisis, as it has been presented by government and by the companies that have the most to gain, has been something of a sham, whereby the exploitation of the consumer and the taxpayer has been raised to a new order of magnitude. But I know that certainly is not the case in a great many other countries. Brazil in particular needs a miracle.

GUY J. PAUKER

I would like to introduce a slightly different perspective to some of our discussions. Professor Sadli told us that only about 10 percent of Indonesia's villages are tied in to that country's electricity distribution network, in contrast with about 50 percent in India. In the United States, electrification only started about fifty years ago and was completed very rapidly. I think that it makes a lot of sense to look at rural electrification as a major global cultural process that will continue, regardless of our views about alternative paths.

It took the agricultural revolution about three thousand years to spread from the Middle East to northern China. We are only at the beginning of the third century of the industrial revolution, and of course the use of electricity is much more recent than that. I find it very hard to believe, using Wolf Häfele's time frame, that fifty years from now there'll be any village in the world in which the population will accept being deprived of electric light, access to tele-communications (television in particular), and a certain amount of food refrigeration. I don't think villagers will revolt if they cannot drive Cadillacs, but I doubt very strongly whether the political pressures that would be generated in those parts of the world still lacking electricity in 2030 would be tolerable to any government, no matter how tough-minded.

A brief back-of-the-envelope estimate is quite illuminating. Statistics are misleading. It's not true that Nigeria uses 65 kilowatt-hours per person annually and the United States over 10,000; most Nigerians have absolutely no access to electricity but will not tolerate this situation forever. Let's take some arbitrary figure: assuming that a 0.5-kilowatt generating capacity per person was a minimum goal for fifty years from now, if the population by then was on the order of 8 billion—which is a plausible figure—total requirements, just to provide a minimum of electricity to humankind as a whole, would amount to at least 4,000 gigawatts. This is the equivalent of not fewer than 4,000 power plants of 1,000 megawatts each which would have to be available at that time, using coal, hydropower, nuclear power, biomass, and any other source of energy available.

Although John Deutch told us that it is not wise for a policymaker to start from the goals, I think that from a global management perspective one cannot avoid asking the question: How will this goal of electrification be achieved? How will this process be accomplished? What paths will be followed? I am sure it will be a very eclectic process. It will be a combination of every method that humanity will be able to devise in order to give everybody access to electricity, regardless of our current ideological debates.

B. C. E. NWOSU

I would like to thank the last speaker because he has given a bit more impetus to my minimum-maximum energy requirement (MMER) concept. This idea of rural electrification, for example: if we think (as I believe and as I stated in my paper) that the average person should have electricity as part of his quality of life, as part of his standard of living, then what implications does this have for our choice of our alternatives and so forth? And Dr. Pauker gave a small figure for kilowatt-hours per person—I don't care about the figure, but it's a good starting point.

Finally, I would like to speak to the point that Dr. Lumsdaine made earlier: that I did not take up the issue of education. I have done teaching all my life except the last five years. I've been a bureaucrat for five years; the rest of the time I have been a teacher. I was also glad somebody—I think my colleague—helped me out by saying that this is an educational process. I don't want to spend more time on this, but I would like to say that I did not reach the conclusions of my paper without education by interaction. It took ten years of education by interaction with other scientists in the World Council of Churches before I reached some of the things I said in my paper; otherwise, I would just have been a plain physicist teaching physics. So I think education is important and thus the whole subject of this Symposium, and so forth.

P. C. ROBERTS

I have just a brief comment to make, and it refers to the discussion of biofuels. The point that I want to make with respect to biofuels can be made with a handful of numbers.

If we spread out all of the world's current population over all of the available land area, the actual number of people per square mile is about 90. So, if we allocate the land equally to each of the individuals who live on the earth, each of them has an area equal to only slightly more than three English soccer football pitches. (As soccer is now played in the United States, I presume that most of you are familiar with the size of a pitch.) Now, the point I have to make is that the first two of those pitches are covered with mountains and deserts and ice and forests. It's only about one-third of the third football pitch which is actually usable in its current state as agricultural land. Having made that point, I will tell you how much energy is used, first, for feeding one individual, and second, for driving one motor car. To sustain one individual on an ordinary Western European or North American diet requires about 150 watts of food energy per year, but to sustain one European motor car driving an average distance per year requires about 1,500 watts, or ten times as much. I want to ask this question: Does anyone seriously think that we can sustain the world's motor cars by producing biofuel? We can do it, obviously, in one or two places where there is a very low population density, as in Brazil. But can anyone seriously maintain that in the long term we could sustain the world's motor cars on biofuel?

JAY H. BLOWERS

I have comments to make about the topics of both of Session III's papers.

In his paper, Dr. Nwosu called for a just and equitable energy-efficient future for all people. One Symposium participant suggested that a figure—0.5 kilowatt-hours per capita—could be chosen as a bottom line for all. It is suggested that an initiative could be taken in the United Nations to produce a "Declaration of Energy Rights." We know the quantity and quality of food necessary to sustain life. A similar figure could be established for energy. The upcoming [August, 1981] UN Conference on New and Renewable Sources of Energy in Nairobi, Kenya, might be an appropriate forum for this initiative. "Declaration of Energy Rights" could also be a topic for the Symposia to follow.

Professor José Goldemberg described in his paper Brazil's program to substitute, for vehicles in that country, ethanol produced from sugar cane for fuel produced from petroleum. Professor Goldemberg sketched the negative aspects of relying on a monoculture; for example, the possibility of attack by disease or insect vectors, the problems of using arable land for nonfood production, and so forth, and he suggested that countries with wood resources should utilize these resources for energy.

An international workshop on biomass for energy sponsored by the US Man and the Biosphere Program concluded that the use of forest resources for energy is feasible and has the following benefits: a dependable supply based on the renewability of the resource, an even spread of developmental activities by afforestation of ecologically or economically marginal land, the generation of employment opportunities, and so forth. However, care must be exercised in deploying this resource because knowledge of the functioning of forest ecosystems is incomplete. Research must go hand in hand with utilization.

KERRY McHUGH

In Australia we have been intensively examining the potential use of ethanol for fuel. Our results indicate the following: that the production of fuel ethanol from either sugar cane or wheat is still uneconomic compared with fuel from petroleum; that ethanol is highly energy intensive in its factor inputs (mechanical planting, harvesting, transporting, crushing, and so forth); and that the land input needed for ethanol fuel production is especially high—for example, to meet *20 percent* of Australia's transport fuel needs would require a *doubling* of the land presently used for the production of wheat and sugar cane. It is questionable, given the shortage throughout the world of both arable land and food, whether the large-scale production of ethanol via biomass represents an optimum approach to solving the world's fuel transport needs.

DEE ASHLEY AKERS[1]

The concern expressed at more than one session that new forms of energy might not be affordable by developing countries was treated partially by a discussion of soft path alternatives. Such alternatives have great significance to the overall energy picture but leave much to be done relative to meeting the industrial development ambitions of developing countries in the near term. This need will obviously not be met by complex technologies that require several years to develop and typically involve low conversion efficiencies, but it can be met quickly by a series of current, less complex processes that tend to be simpler or less costly, and usually both. Included in these processes are coal-oil mixtures, fluidized-bed combustion, low-temperature (low-pressure) pyrolysis, low-Btu gasification, decentralized power generation with slow-speed engines fueled with coal-oil mixtures, coal-bed methane recovery, waste heat recovery, cogeneration, waste coal recovery, and municipal and industrial waste combustion. Methanol-from-coal and magnetohydrodynamics are also processes that produce low-cost clean energy. Generally speaking, these processes do not require large investments of time and money and are more efficient than many of their more complex counterparts. They constitute alternative fuel systems that are affordable in comparison with today's energy prices.

PAUL DANELS

At the close of Session III, I would like to pose two questions: What has Nigeria done to address the inequitable and unjust distribution of resources within Nigeria? Can these efforts be seen as serving as a model for the redistribution of energy resources among less developed countries and highly developed countries to achieve the just, participatory, and sustainable society to which Dr. Nwosu alluded?

NOTES

1. Dee Ashley Akers; Director, Government Law Center (Technological Law); University of Louisville; Louisville, KY.

For identification of the other discussants, please refer to the list of participants in Appendix II.

CHAPTER TWELVE

Summary

David J. Rose
Professor of Nuclear Engineering
Massachusetts Institute of Technology

In my city of Cambridge, Massachusetts, a cashier at the express checkout at the local supermarket, where the sign stated "Eight Items or Less," said to the man who wheeled up a whole shopping cart of groceries: "You're either from MIT and can't read or Harvard and can't count." And so we have different backgrounds, abilities, points of view, attitudes, which tend to persist. The experience of listening to the energy debate shows this; less than ten years ago at an energy meeting, each participant would talk technology and throw rocks at other speakers for introducing apparently simplistic and largely irrelevant social issues. I have been to plenty of those. Today we partly see the inverse, where they talk social purpose and point the fingerbone of scorn at simplistic technologies. We need a combination of many of these ingredients, as one of the popular television commercials would say.

I came across the following statement the other day, the opening sentences of chapter one of a book. Those very first words were: "Man's search for energy to meet his needs has proceeded since the dawn of civilization. At least some of our international friction is directly traceable to disputes over possession of the areas favored with abundant energy reserves. Classical conquest and exploitation rarely are undertaken for barren areas alone. The acquisition of land endowed with energy and material resources motivates much of our present strife." That was written in 1960 and applies today.

Is anything fixed in this fast-moving discussion besides the basic problem itself? I find two great statements of faith shining through it all, which rise up to heaven from the debate as smoke from the incense in cathedrals of learning, or the burning of prayer papers—

sanctified by convention and tradition. The first one is: *Non mea culpa;* don't blame me. The second is: Maybe wrong then; never wrong now.

You've probably noticed that all three sessions of this Symposium so far, and the fourth to follow tomorrow, all deal with the same thing, and we could have shuffled the titles around. The program committee realized that and planned it that way, because the sessions were intended to stimulate discussion rather than to limit it. The four titles were chosen in order to stimulate different views and the developments of one holistic theme. Thus we see in each of the sessions a repetition of the same general motif, but different development of it—like Mozart or Beethoven sonatas.

We take, as some *modelers* would say, a cross section of the problem—this way, that way, and the other way. But beware of just taking cross sections and imagining that all the dimensions of the problem have been included. Some essential directions might be missed. One could take a cross section of an elephant—say, 1 millimeter thick—paste it on the wall, and look at it. It would look something like an elephant, and you could maybe see more about some of the internal workings of that elephant than you really wanted to know. But all you would have is that 1-millimeter-thick slice on the wall, and the two halves of a dead elephant.

After that preface, I now approach the task of integrating today's excellent talks, thanks to Professor Goldemberg and Dr. Nwosu, into a more general discussion, and will paint on a very large canvas. I've been listening to all the comments during this session and have developed some notes on them. These notes are all my own, and I take responsibility for them.

TEN GENERAL STATEMENTS ABOUT ENERGY

Here are ten general statements about energy.

1. Oil is increasingly scarce, expensive, and out of reach. Everyone has said that.
2. There are four and four-tenths billion people now (2 persons per hectare on good agricultural land, as Professor Goldemberg said), and more people coming later.
3. The energy route of presently industrialized countries (ICs) cannot be followed by less industrialized countries (LICs) under present distribution circumstances. (I prefer these terms—less industrialized countries and industrialized, or more industrialized, countries—as more accurate descriptions of what we are talking about. The old words, "less

US-European route is necessarily better. In some cases it may be, in others not. So let's not get into unnecessary semantic difficulties, and let's describe what we mean as accurately as we can.)

4. There's lots of coal (and oil shale, and tar sands).
5. But what hasn't been pointed out at this meeting, except almost in passing, is that coal, oil shale, and tar sands bring some very big problems, as do all other fossil resources, especially in the long term. Here I have global environmental problems particularly in mind, such as a carbon dioxide problem, but regional and local ones exist also.
6. Consumption patterns can change, both in the LICs and the ICs, and this will be especially important as we consider what people call energy conservation. But these changes come slowly, as I'll describe later.
7. The energy problem isn't unique, as was described in this morning's session [Session II] and also the present session. Rather, it is typical of many other large societal problems—food, urbanization, environment, water quality, and so on.
8. Different responses to the energy problem are attuned to different interests, and perspectives clash. (The energy problem is different from solving problems in energy. Here is another important semantic difficulty which has led to a great deal of trouble. For example: a problem in energy would be how much energy it takes to put some Assistant Secretary of Energy into orbit. You can calculate that. But the energy problem is something different. It is a socio-technological problem that we didn't want, or having it, we wish to ameliorate. And so you have to ask in regard to this Assistant Secretary of Energy, "Was that a good thing?" or "Compared with what?" The same comment could apply to and help illuminate the environmental problem, as distinct from solving problems in the environment. The former transcends the latter.)
9. A century from now, it will all be solar (in many different forms—photovoltaics, hydro, or biomass) or nuclear (fission or fusion)—probably some combination. It's all that God gave us.
10. Who's minding the shop? The problem is global, but the international organizations are weak. Here's a big mismatch. National and local forces dominate, as we have seen through all the discussions, and the real problem remains.

SELECTIVE INATTENTION AND TIME PERSPECTIVES

Two important things can be strained out of all this. The first is *selective inattention* (or selective attention—call it what you want). The problem is vast; the issues are too broad, too many. So we focus on just one thing or a few things to the exclusion of all the others. For example, we have villain theories (most of which are wrong): it's the oil companies or OPEC or whatever. We have single-issue solutions: it's all coal or solar or conservation. Selective inattention, carried to such extremes, loses security rather than gaining it.

That is the same point, incidentally, as the Buddha's third noble truth, where he says people tend to concentrate on just one thing and get into trouble by not looking around them. The idea is 2,500 years old. It leads to clashes of special interests. I have described the nuclear power debate as being like a duel in the dark with chain saws, no matter which side you're on.

The other important aspect of these problems is *time perspectives*. If everybody said what kind of time horizon they had in mind in their debates, then the debates would be much clarified. People talk about the short term, medium term, and long term but mean different things. To some the short term is oil next winter, the medium term is the next election, and the long term is the election after that. Let us search for better ways to describe the idea of time perspectives, in terms of events relevant to energy.

The economic rate of return that businesses expect suggests a time perspective of, say, five years—rarely much longer. But developing new options, from early concept to early deployment, takes more like twenty-five years, sometimes longer. Controlled fusion research started twenty-five years ago, and its viability is still undetermined. Modern coal technology has been developing for a decade, using a German World War II technological base, and it will need a decade more to become fully modern. As a result of this time mismatch, the business or political sector may want to keep on boring holes in the ground as long as disaster is much further away than five years, to the neglect of preparing longer term options.

Social inertia is often high, and social rates of change must be slow if disruptions are to be kept to a minimum. Cities last a long time; once they're built, it's hard to change their technological or structural base. Witness the Washington, DC, subway: 5 billion dollars for its construction, largely because it had to be excavated almost by hand.

Actions designed for different time perspectives often clash. It is often said that in the long term we have many options, but in the same breath it is said that of course the pressure of present events forces us just now to look only at the short term. And the flexibility of the long term remains a mystical goal. It takes foresight and fortitude to decide on a combination of activities designed (with some essential redundancy) to cover all these time perspectives.

Another deceptively simple question: How long do you want the civilization to last, in the sense of being willing to take positive action to realize its sustainability? Neither the question nor the answer is easy, but they are important nevertheless.

A FEW SELECTED ISSUES

I could list many, but here are four that strike me as relevant to our discussion today.

1. *Energy is more than technology,* or technology plus economics, or plus other deterministic additions. It's also the North-South debate (as some call it), the social purpose debate, the high-low technology debate, and so forth.
2. *There are alternative energy sources, alternative uses, and alternative other things.* There are alternative lifestyles, including relatively pleasant ones that tend to be energy-frugal (but I do not intend to debate the question of lower plausible energy-use limits here).
3. *Conservation.* It's been a good workhorse word up to now, but I suggest that we retire it with thanks, in favor of "rational and effective use." Semantics got in the way again. Preaching of conservation by the industrialized countries sounds to many LICs like curtailment of LIC energy. *Sie pregen Wasser und trinken Wein* ["They talk out of both sides of their mouths"], so to speak. Practice conservation yourself, say the LICs to the ICs, and we'll all be better off. So we will, but the phrase "rational and effective use" describes much more accurately what we are trying to do, and properly suggests that it is something which concerns everyone.
4. *Selective inattention*—here applied to various long-term problems that, if ignored, are likely to lead to global catastrophes. A good example is the problem of carbon dioxide's effect on the climate—a problem which, I hope, will be an exception to the rule of neglect. We at the Massachusetts Institute of Technology and others have been working on that problem during the past several years. If the various national and global scenarios presently fashionable actually come to pass, the use of coal and other fossil fuels will, in about fifty years, bring about the conditions for unavoidable temperature and climate changes, the effects of which will be mainly

negative. The bullet will be on its way, so to speak, and it will be too late to move out of its path.

Some say, regarding this carbon dioxide problem, that we shouldn't worry—new species of corn can be developed and so forth. But the forest biome takes a long time to adapt, and soil conditions in presently nonproductive lands take time to change, even if the temperature and rainfall patterns become more favorable for them. Some countries will see benefit, and so international conflict will grow. Even if the probability of very serious climatological change is as low as 1 percent—I think it's much higher—it is still a real cause for worry.

SESSION III PAPERS

Now to the two papers presented at this session, which were both very good. The first by Professor Goldemberg poses an exquisite dilemma—a dilemma of conflicting interests, goals, time perspectives, and so on, because there are obvious benefits, all well described, from reducing oil imports. The dilemma relates to the various costs, both economic and social, of ethanol production: the disruption of farming patterns and food supply, the problems attendant with monocultures, the inequality of having benefits go mainly to minority elite groups, the exacerbation of income disparities, and so on. How big are these costs? How does one compare them with the benefits of reduced oil imports? How does one decide? That's the dilemma, which often tempts people to apply selective inattention in order to force a simplistic solution by ignoring large parts of the problem.

While I don't know enough about Brazil to make a final judgment about ethanol production there, I do know something about the US situation. To replace 1 percent of our US gasoline with ethanol made from corn—for which every tax incentive known to man or woman has been offered—would take enough corn to give some 2,000 food calories per day to somewhere between 10 and 30 million people per day, depending on technology. Were I a member of the something or other party in some Country X wanting to stir up people against the United States, I would have a perfect argument by saying: "Are you hungry?" "Yes." "Do you want corn?" "Yes." "What does the United States use its corn for?" "To run its cars." "Do you like the United States?" Yes?

Scrap or even dedicated forest products may help here, for producing methanol, but also remember the present pattern of deforesting tropical lands. The dilemma persists. Professor Goldemberg has been careful to enumerate the various pluses and minuses of the sugar cane route, and he has laid a basis for similar consideration of forest biomass.

Dr. Nwosu and Professor Goldemberg both were concerned with government policy in which the main initiative is governmental. This morning Professor Deutch emphasized private sector roles and the critical importance of the action of the marketplace. We also might ask what the proper role of multinational corporations is. They can help, but some would ask: Can they be tamed?

Among the many topics in Dr. Nwosu's fine paper, I was very interested in his references to the just, participatory, and sustainable society. I worked with Dr. Nwosu for some years in the World Council of Churches (WCC) on that global topic; maybe one-third of my writing in the last five years has been for the WCC. The debate did not have to do with bad hymns. We've had some better and more realistic energy debates in the WCC than many seen in other places. That was because the WCC didn't make the foolish mistake of involving only church people, but instead got all kinds of people—sociologists, administrators, environmentalists, engineers, and so forth—and all with a real interest and willingness to work hard.

Dr. Nwosu's paper raises questions about the joint responsibilities of the industrialized and the less industrialized countries. He asked how confrontation can be avoided and what mix of energy strategies can contribute to a just, participatory, and sustainable society. These three qualities reinforce each other. A "peace" in which the world is divided ever more rigorously into haves and have-nots is neither just nor very sustainable, whether the basis for division is social, economic, or seemingly technological. Such a division not only defeats itself in the long run; even worse, it is morally wrong. This applies both between nations and within each country as well.

In the opening address to this Symposium, John Sawhill said he wanted to look at the future rather than the past. However, we need to look both ways. The historian George Santayana remarked that those who do not study history are condemned to relive it.

We can learn from history. Edward Gibbon in his *Decline and Fall of the Roman Empire* wrote that Odoacer, viewing from outside the empire's boundaries how the center had grown complacent and uncaring and how the empire tried to use mercenaries from beyond one end of it to protect the interior from invasions at the other end, finally despaired of the Roman government ever doing better, sacked Rome in the year A.D. 476, and threw over the entire European civilization. If this present world becomes too unbalanced and redress seems impossible, the same thing could happen again, this time with no real winners but only losers in different degrees. It's much better that everybody get on the train of civilization, rather than leaving people standing by the tracks, watching the world's goods and qualities carried away.

Such general ideas have application here and now.

The just, participatory, and sustainable society raises questions about the redistribution of resources. Improvement of energy use efficiency, everyone agrees, will contribute to this redistribution, as will efforts to stimulate self-reliance. We need also interdependence based on cooperation. There is the question of the extent of the coupling of the economies of the LICs and the ICs. For example, what about the multinational corporations, some of which evidently exploit the less industrialized countries shamelessly? Some say they should be scrapped. Others point out that true global collaboration requires some global effort to civilize them, so that their benefits can be made available to all, not just a few.

All of these provocative questions and issues of Dr. Nwosu have led me to think about what might be a strategy for the industrialized countries. Here it is. Industrialized countries should curb the use of fossil fuels, especially oil, much more strongly by rational and effective use and by adopting alternative technologies—not to spite OPEC, for example, but as part of a global cooperation. This would result in the following positive results:

1. Better availability of fossil fuel to LICs in their difficult coming transition.
2. Better match of time perspectives in fuel use, to avoid the carbon dioxide climatological catastrophe, and so forth.
3. Reduction of the probability of global conflict.
4. Strengthening of international collaboration on a basically global pattern.
5. Development of a foundation for global collaboration on rational and effective use of energy and other things.
6. Stimulation of local initiative, as each sector and region considers more carefully its own opportunities and responsibilities, knowing that it is no longer alone, and that its efforts will not be frustrated by others.

Regarding this last point, one of the Symposium papers implied that more rapid use of oil would help the OPEC countries; another one said a lower use would help the OPEC countries more. The question is indeed somewhat region-specific, but I tend to subscribe to the latter view because what they're trying to do is turn a one-time resource into a long-time permanent benefit. The longer time perspective and slower rate of withdrawal fits better with peaceful rates of change in society.

SESSION III DISCUSSION

The discussion following the two principal papers of this session has been included in these proceedings. However, in conclusion I wish to identify some important points not sufficiently stressed so far in my summary.

Several persons (Pauker, Reddy, Nwosu in response, etc.) remarked that the better energy programs we envisage have not only international aspects but also national ones. What lasting good, one can ask, will come from an energy program when meanwhile in some countries the urban rich get richer and the rural poor languish in want and neglect? Thus, several persons spoke to the need for rural electrification, the replacement of rural energy-inefficient methods and devices with better ones, and so forth. These require capital investment and offer opportunities not only for innovative devices and networks but also for true collaborative programs between the more industrialized and less industrialized countries, to everyone's benefit.

A second major discussion topic (Alonso, Arungu-Olende, and others) related to the need for doing much better assessments of energy options, bearing particularly in mind the different conditions and opportunities in different parts of the world, and using different time perspectives. These should not just be narrow technological-economic assessments but should include the full panoply of major tradeoffs, so that we get an understanding of and even a sympathy for the need to live in a world of tradeoffs, in a world where paradox itself will remain forever an essential part. This cannot be done with only the involvement of selected small groups (though leaders are sorely needed) but will necessarily require the education of ordinary people everywhere—a task I think quite possible to accomplish.

A more specific suggestion was made (Usmani and others, especially in corridor discussion) that the world is very unevenly explored for gas, oil, and coal. Thus the feeling exists that there's a lot out there, especially in the less industrialized countries, and that the discovery of even a modest part of it would be a great boon, if used for and by LICs and not just to fuel the engines of the present large users. Here is something valuable that the industrialized countries can do: assisting the LICs to explore likely energy resource places, for the LICs' benefit. Such an activity would not conflict with my worries stated earlier about global carbon dioxide build-up—the LICs need to increase their use of fossil fuel during the next few decades by all practical means; their present total energy use is small compared with what the industrialized countries use now.

POSSIBLE TOPICS FOR FUTURE SYMPOSIA

A number of institutional and other nontechnical issues were raised here which it would be good to follow. One was related to the establishment of a more just, participatory, and sustainable society. Everyone would agree to this, but how do we go about it? There are several suggestions. This year's Symposium and discussion often dealt with individual nations and their problems, options, and so on. This was essential, of course, because the national organizations and incentives are now usually much stronger than the global ones. However, that circumstance tends to bias discussion toward options where each country goes its own way, imagining that the global costs can be ignored or will be borne elsewhere. This is the classic tragedy of the commons, wherein each partner contributes to a set of activities that causes the system as a whole to collapse. The cure involves development of understanding and of arrangements at the larger, more holistic level. So, in the future Symposia, let us have sessions on possible new international arrangements: A tax on internationally traded fuel, to help develop new options? New international arrangements so that the benefits of transnational corporations are more evenly shared, and (importantly) perceived to be shared? An organization devoted to the development of new options in using energy more effectively all over the world, at all levels? (As the LICs use more energy, let us hope that we can be of help to them and to each other, enabling them *not* to repeat energy-wasteful habits of yesteryear.) Do such organizations belong in the United Nations? Should they be started bilaterally, or regionally?

A second major theme could be put as a confrontation, so to speak, to highlight the differences in global views between the benefits of large near- and medium-term energy supply increases and the long-term difficulties that might emerge from these increases. One example is the global trend to coal versus the predicted "greenhouse" effect of burning fossil fuels; much has been done on both parts, and the parts are very much out of joint. Another example is the global acceptance (or nonacceptance) of nuclear power and the prospects of solar power—in other words, the principal eventual "alternatives." This brings in nuclear proliferation—whether it is germane, whether it can be brought under control, and so forth—and solar power's prospects—when it can be expected, in what form. (The vibes for solar power are good, for some countries, at least.)

A third major theme could be energy for cities. The world is becoming more urbanized, both in ICs and LICs. Although energy supply and use patterns differ between city and country, between one country and another, it is a common concern. Right now, the Europeans are doing a lot on it (e.g., the German Marshall Fund papers), but more needs to be done.

Finally, many of these issues involve the "North-South" debate. This debate is something which must be recognized, acknowledged, understood by all, and worked on—as must be all the issues mentioned here and elsewhere during this session. For energy problems will not go away by themselves, or by being dealt with unilaterally. A *global* perspective is needed, one which takes into account the complex timing and multiplicity of the issues and the interrelatedness of the energy problem as a whole.

Alternative Policies for Improved Energy Productivity and Production

Introduction

Marcelo Alonso
Executive Director
Florida Institute of Technology
Research and Engineering, Inc.

Since this first Symposium is preparatory to the 1982 World's Fair and has as its main theme energy production and productivity, it is very appropriate that it concludes with a session dedicated to alternative policies to improve energy production and productivity. In fact, no matter what alternative energy sources are available and which technologies exist for exploiting such energy sources and improving the efficiency of their use, appropriate policies are a key factor for moving toward a more acceptable and even feasible energy world. These policies require an understanding of the energy situation in the world as a whole and in each country in particular—an aspect of what might be called the geopolitics of energy—as well as a matching, at national levels and over a certain period of time, of the energy requirements with the energy possibilities. Both aspects are still not well understood, and a good deal of research and analysis is required. But one thing is clear: no single solution exists which can be applicable to all countries without regard to their sizes, their resource endowments, and their degrees of development.

Broadly speaking, there is a policy conflict between the more industrialized countries and most of the developing countries. For the first group of countries, the main goal is how to cope with a situation of increasing prices and uncertainties in the supply of energy without affecting the countries' economic outputs and their populations' relatively high standards of living. In addition, security may be an overriding factor in defining their energy policies. For developing countries, on the other hand, the problem is how to increase energy consumption to meet development requirements and to improve the quality of life of a vast majority of their populations. This dilemma is what Professor Reddy very

267

clearly states in his paper as "growth"-oriented versus "development"-oriented energy policies. Obviously, it is not easy to reconcile these situations. There is, however, one aspect common to both groups of countries: they both have an imperative need to decrease their dependence on oil as a primary energy source. And this is the major issue faced by the whole world at present and for the immediate future—let us say, until the year 2000.

Even moving away from oil does not mean the same thing for the advanced industrialized countries that it does for the developing countries, and this is another important issue. For most of the industrialized countries with perhaps the sole exception of Japan, oil constitutes only a fraction of the total energy used and is shared with gas, coal, nuclear, and hydro, but these countries are the major users of oil as a primary energy source. For the developing countries, however, oil is the major source of commercial energy for historical reasons, even if their total oil consumption is relatively small compared with the more industrialized countries. Thus, the industrialized countries should have more aggressive policies to restrain—particularly through conservation—their growth in oil use and to accelerate oil substitutions. On the other hand, the policies of the developing countries probably have to be oriented in the short term toward how to finance their needed oil imports, and these countries will move away from oil at a pace that is slower because of their circumstances and limited feasible alternatives.

To reduce oil consumption requires a delicate interplay of government and market forces—an issue that, again, has no unique solution and is different for industrial and less developed countries. It appears that in the industrial countries market forces can play a major role with perhaps a minimal government involvement related to pricing and deregulation. In the less developed countries, governments may have to intervene more directly in the management of the energy problem, allocating resources and adopting strong regulatory measures, such as limiting the size of cars.

An issue peculiar to many less developed countries is how to diminish their heavy reliance on noncommercial energy, especially for household use, without replacing it with oil-based fuels. Besides the undesirable use of animal dung, firewood, and charcoal, there is a marked trend in rural areas toward the use of kerosene, which again places a heavy pressure on oil. One possible alternative would be massive rural electrification, as long as the electric power is not generated by burning oil. In this respect, small hydroelectric units and wind-powered systems might offer a sound alternative. Biomass and biogas should also play an important role.

A method that could save oil might be to gradually change the industrial structure, phasing out those industries that are energy intensive by transferring them to energy-producing countries, or replacing the industries' primary heat source by shifting to nuclear or gas. Either solution is capital intensive and involves risks and difficulties as well as lead time for implementation. For example, based on Professor Reddy's data, it might be desirable in India to shift the long haul of goods and passengers from roads to railways, but in turn the railways should change from diesel to coal to the extent feasible. But these changes face severe social and political resistance and are unlikely to occur, even if the capital to make changes was available. An interesting case worth mentioning here is Korea, which before 1973 was pushing investment in heavy industry, resulting in a sharp increase in oil use. As a consequence, Korea now has under consideration the restructuring of its manufacturing industry, particularly those industries having an energy input ratio larger than 10 percent.

Oil substitution means considering all possible alternatives. One such alternative available to industrialized countries is nuclear energy. It is expected that by the year 2000 the Organisation for Economic Co-operation and Development (OECD) countries will have increased fivefold their nuclear capacity. However, for most developing countries the alternatives are limited and nuclear energy may not be one of them, mainly because of technical reasons. In fact, most developing countries capable of using nuclear energy have already decided to use it, while for the rest its use is in the distant future. Similarly, the OECD countries will increase their use of coal three times by the year 2000, but for those developing countries not having coal—and they are in the majority—coal will remain a remote possibility. In conclusion, the big issue of national and international energy policies is how to achieve a more equitable energy use which will allow a real, equitable access to goods and services by all population sectors, while in this transition period gradually replacing oil with other sources.

CHAPTER THIRTEEN

Policies for Energy Production and Use

David Sternlight
Chief Economist
Atlantic Richfield Company

The purpose of this paper is to discuss policies for energy production and use. Following a brief overview of the current world oil price situation, a set of general principles for energy policy is presented. Overall energy policy objectives are discussed in the context of supply, demand, and price, followed by discussion of related policy principles and major levers that influence energy decisions and outcomes. The strategic translation of energy policy principles evidenced by the International Energy Agency (IEA) is then described to complete the analytical framework. This framework is used to examine choices made by individual IEA member countries. The paper concludes with a discussion of major strategies for the future.

INTRODUCTION

The Iran revolution caused a major decline in Iranian crude oil exports, partially compensated by expanded production in other countries. At the same time, panic stockpiling took place to fill low stockpiles at a time when the magnitude and duration of the oil cutoff was uncertain. The resultant significant imbalance between oil quantities supplied and demanded led to sharp rises in world oil spot market prices. In response to this market evidence, OPEC countries in turn raised their official oil prices. Thus, as world oil stockpiles became (temporarily) full, the real price of oil rose sharply. Subsequently, the worldwide recession with its associated reduction in energy demand, the higher oil prices with their contribution to energy demand reduction, and the high stockpiles created a market softness which would have led to OPEC production cutbacks were it not for the war between Iran and Iraq. The oil supply reductions from that war, together with some oil supply increases by others, have thus far brought world oil markets roughly into balance, at a range of prices more narrow than in the recent past.

Real world oil prices have now reached a level where, absent a new supply crisis, they are likely to remain roughly stable. This is due in part to, and associated with, both the improved conservation induced by higher prices and the beginnings of significant energy supply augmentation caused by higher prices, as well as recent policy decisions in countries such as the United States. Given this supply and demand response, calculations which take account of OPEC self-interest and the dynamics of total energy supply and demand suggest that absent a further major oil supply disruption (for example, 5 million barrels per day for six months or more), real world oil price jumps as sharp as those of the past may be behind us. Thus, current real prices provide a useful directional guide for policy thinking and analysis about the steady state. Due consideration should also be given to oil price externalities and related issues such as interruption policies; the oil price relationship to national security and macroeconomic consequences of oil supply interruptions; environmental protection; research, development, and demonstration strategies; and equity issues.

ENERGY POLICY PRINCIPLES

In this section, factors affecting energy policy formulation are first discussed, along with some related policy issues. Major forces affecting energy supply and demand outcomes are then considered, and the section concludes with a review of the IEA's sets of principles on national energy policy in general and coal action in particular.

Energy Policy Formulation

The overall objective of energy policy, in my judgment, ought to be to achieve a supply/demand balance in energy services at market-clearing prices, such that in the broadest sense a least cost solution over time results for energy consumers.

Energy consumers do not want a barrel of oil. They want energy services: warm houses, transportation, crop drying, process heat, petrochemical molecules. The energy system delivers these services through a production process that includes the transformation of fuels. Energy policy should support an efficient (least cost) approach to that production process, considering all alternatives and their future value as well as present cost. Sant has addressed this question in some detail.[1] In policy planning it is not the past or present cost of alternatives that is important but what these alternatives will cost over the periods of their use in the future. To insure supply responsiveness, signals must be clear, unambiguous, and appropriate to the level of response needed. Interferences with such responses need to be eliminated to the greatest extent possible.

Energy demand must be equally influenced by clear signals that allow energy consumers to shift among levels of use, among fuels, and among alternative use technologies. Countries already have made many energy service reductions normally thought of as energy conservation in the face of higher prices, yet major opportunities in this area through increased efficiency of energy use lie ahead. These increases take time to achieve. It takes ten years to replace the stock of cars in the United States with more efficient cars; major changes in building efficiencies through retrofit and new construction take thirty years for heavy penetration. Increased efficiency of industrial energy use, an area of significant progress already, takes time. These are capital investment processes, just as is the bringing on of new primary energy supplies, and they must compete for funds with other investment needs.

Price signals are the key to much of energy policy. Governments must insure that energy producers and consumers are able to respond to energy service needs and availability by expanding supply and reducing demand as the scarcity value of energy (the cost of the next increment of energy services) changes. But price must be broadly interpreted to include nonmarket factors, which may be expressed in dollars and cents, in policy, or in regulations. Typical nonmarket factors that must be treated in this way include protection of the environment, national security, and macroeconomic consequences of energy policies and potential external developments. For example, it is accepted by many economists and government policymakers that assured domestic energy supply (or increased efficiency of use) has national security value both directly and in terms of balance of payments and inflation benefits. Thus a case can be made for some volume of domestic supply increase or demand reduction through increased efficiency (say, 2 million barrels per day of import replacement) at prices above the world oil price. Such volumes should be obtained incrementally, of course, least cost source first. For example, synthetic fuels development and demonstration is appropriate if it can reduce costs of such supplies to competitive levels, but large-scale production should be avoided as long as less costly opportunities exist for obtaining an incremental 2 million barrels per day of domestic supply or energy use efficiency.

Governments are also recognizing the close link between energy prices, economic growth, and inflation. Thus, energy policy is no longer seen as a separate political matter but as a key issue in overall economic policy. Countries have come to realize that higher prices paid to oil exporters for imported energy must reduce the real goods and services available in the domestic economy. This adjustment can no longer be hidden by compensatory stimulation of domestic economies by monetary expansion. The inflationary and payments balance consequences have been seen to be unacceptable.

Some Related Policy Issues

Lead times significantly affect energy supply expansion, oil substitution, and demand reduction. Decisions must be made, policies set in place, capital mobilized, and physical activity undertaken. Without each step in the process, subsequent steps become difficult or impossible, and calculated supply/demand balances based on anticipated increased supply and dampened demand growth fail to materialize. Thus, the promptness of national policy actions is critical.

Institutional interferences to energy balance adjustment must be removed for that adjustment to be realized. Many regulations have ceased to serve their purpose and should be modified. Electric utilities, for example, can help to improve the efficiency of their customers' energy use through insulation financing, conversion to decentralized electricity, and cogeneration and other options, if they are permitted to enter these markets. Utilities can charge for services that save energy on customer utility bills, thus making financing of such improvements easy for customers who can pay, at least in part, out of the energy savings they obtain. Many current subsidies to particular forms of energy conceal their true relative cost and distort signals for fuel switching, new capacity additions, or efficiency increases. These should be examined and modified.

Policy must deal with dynamic as well as static issues. In addition to long-run supply/demand balance adjustments, major questions about supply interruption policy remain unanswered. Emergency stockpiling actions by government are not yet fully satisfactory in the United States. Nor has enough recognition of the speed and efficiency of private stockpiling taken place. Policies have not been put in place to stimulate continued investment in private stockpiles, such as government assurances that private stockpilers will be permitted to recoup the costs and risks of such action by being permitted to receive the benefits (financial and physical) of such stockpiles in an emergency. Some governments are just beginning to recognize the inadequacy of their rationing plans and to consider such "heresies" as market clearing by price in emergencies, with due regard to equity.[2]

Uncertainty is a major obstacle to orderly adjustment and institutional planning. Some irreducible uncertainty is inherent in the unknown character of future energy supply, demand, and price responses to changing states of the world. Some uncertainty is also inherent in imperfect analysis. Some governments, however, are compounding this uncertainty several times over by either refusing to adopt clear policies or refusing to accept an overt reliance on the market process in many energy decision areas. The result of such uncertainty is to compel caution and to further delay adjustment actions, increasing associated price instability and import dependence.

The tradeoff between efficiency and equity has been raised as a "dilemma" in caring societies. This concern has sometimes been used as a political device to paralyze energy policy action in areas in which costs and benefits are unevenly distributed. For example, until most recently the United States refused to adopt straightforward energy pricing policies to stimulate supply adjustments and use efficiency improvements. Despite well-established policy mechanisms for equity adjustment (income transfers, tax changes, food stamp equivalents) to maintain income for the poor, energy policy was held hostage to the unachievable demand that it, rather than social policy, achieve equity while dealing with energy issues.

One economic role of government in democratic societies that has been well recognized is based on the "public goods," or "lighthouse," argument. When potential public benefits exceed public costs in an area in which the private sector cannot or will not act unaided (usually because circumstances make it impossible for private actors to act collectively or to individually capture enough of the benefits to justify the costs of an action), there is a case for government action. Thus, governments build lighthouses where individual ship owners might not receive sufficient individual benefits to afford to. A similar case for public goods may be made for certain areas of public health and safety. In energy policy this lesson has often been forgotten. Governments sometimes take actions highly beneficial to special interests and detrimental to the general interest, based on small but vocal constituencies in areas neither perceived as a threat nor opposed by the majority. Thus, for example, tobacco driers receive energy allocation priorities on the grounds that they are in agriculture (with "food" implied). Groups receive protection against price increases on the grounds that they don't want to pay more. Despite these abuses, there is a clear role for government in protecting basic human needs of groups unable (not unwilling) to care for themselves. There is a clear role in high-cost, high-risk, high-social-return research, development, and demonstration. There is a regulatory role to prevent abuses of market power or to deal with egregious market imperfections. There is a clear role in protection of the environment. But this does not extend to political distortions of the market process such as the requirement in the United States for "best available control technology" for coal use, which penalizes the use of low-sulfur coal relative to high-sulfur coal. A clearer focus on such distinctions is needed by policymakers.

The role of politics in social compromise is essential. Democracies function through the consent of the governed. A bad law, widely ignored, is worse than no law at all. The challenge of governance is to elicit, express, and effectively deal with policy challenges in a way that will carry along the governed while meeting the physical requirements of a situation. This requires planning and policymaking prior to a crisis situation. At times when lack of foresight or the suddenness of events prevents such policymaking-in-advance, it requires a damage-limiting strategy coupled with public information and the calming of an inflammatory atmosphere. We live in an age when circumstances are much less forgiving of policy errors and delays than in the past. We can no longer accept a negative process in which a minority tries to achieve its objectives by obstruction and delay. Instead, we must move to a more overt conflict-resolution process—for example, to resolve environmental, economic, and energy tradeoffs in a way consistent with the broad public interest.

Major Forces

The three major forces affecting energy supply and demand outcomes are price, real economic growth, and government policies. A series of global studies over time have shown the influences of each, including the Workshop on Alternative Energy Strategies (WAES) study,[3] which considered world energy strategies to the year 2000, taking account of these factors in some detail. Puzzling outcomes from policy often can be explained by considering all of these variables together, not one or two in isolation. For example, despite the 1973–74 oil embargo, driving in the United States increased in each of the next four years except 1974. This apparent ineffectiveness of some US gasoline policies can be explained by parallel price developments (in this case, policy-induced). Gasoline prices were politically controlled by the US Congress in such a way that the real price of gasoline fell from the third quarter 1974 peak of 37 cents per gallon (regular leaded gasoline, 1967 dollars) to 33.7 cents per gallon as recently as the fourth quarter of 1978, a level very similar to that in the first quarter of 1967 (33.2 cents per gallon). Naturally, driving rose·in response to this mistaken signal. After real prices were permitted to rise in early 1979, gasoline consumption began to fall at a 6–7 percent annual rate. The connection is clear and unambiguous, even after adjusting for changes in real economic growth over the period.

A Strategic Translation of Energy Policy Principles—The International Energy Agency

The 1980 IEA Governing Board has adopted one set of principles for national energy policy and another for action on coal, both of which serve as useful frameworks for examining national energy policies. The former set of principles includes:

1. Reducing or limiting oil imports through conservation of energy, expansion of indigenous energy sources, and oil substitution.
2. Attending to environmental, safety, and security concerns and improving the speed and efficiency of public procedures for conflict resolution between these concerns and energy requirements.
3. Allowing domestic energy prices to reach a level that encourages conservation and alternative energy sources.
4. Strongly reinforcing energy conservation, with increased resources, to limit the ratio of energy-demand growth to economic growth; to eliminate inefficiencies; and to encourage substitution through (a) pricing policies, (b) energy efficiency standards, and (c) increased investment in energy-saving equipment and techniques.
5. Progressively replacing oil in electrical, district heating, industrial, and other sectors by (a) discouraging construction of new, exclusively oil-fired power stations, (b) encouraging conversion from oil-fired capacity, (c) encouraging

refinery conversion to avoid an excess of heavy fuel oil, and (d) reducing heavy fuel oil use as primary energy in low-efficiency sectors.

6. Actively promoting steam coal use and trade through (a) rapid phase-in within electrical power and industrial sectors, (b) improved market stability and reliability, and (c) corrected potential infrastructure bottlenecks.
7. Expanding natural gas availability and premium use.
8. Expanding nuclear capacity consistent with national safety, environmental, and security standards, and with weapons proliferation prevention, including (a) adequate fuel supplies at equitable prices and (b) adequate facilities and techniques for nuclear electricity generation, dealing with spent fuel, waste management, and overall handling of the back end of the fuel cycle.
9. Increasing research, development and demonstration (RD&D), including collaborative programs, and emphasizing (a) near-term renewable resources with the fullest possible financial support for energy RD&D, (b) international collaboration, (c) energy/technology development incentives, and (d) consistency between research and development (R&D) policies and energy policies.
10. Establishing a favorable investment climate by minimizing energy and other policy uncertainties (see item 2), by providing government incentives to give priority to exploration, including offshore and frontier areas, and by encouraging rates of exploration and development capacity consistent with optimum economic development of resources.
11. Providing alternatives to oil consumption increases for handling supply or conservation shortfalls.
12. Cooperating appropriately with developed or developing countries or international organizations.

The IEA's principles for coal action are similarly specific and include:

1. Precluding (with limited, stated exceptions) new or replacement oil-fired, base-load power plants.
2. Limiting existing oil-fired, base-load power plants to middle- or peak-load requirements.
3. Limiting oil use in dual-fired power plants.
4. Facilitating timely construction of coal-fired power plants and supporting facilities by improved siting and licensing procedures.
5. Encouraging negotiations and long-term arrangements between electric utilities and coal producers.
6. Encouraging the substitution of coal for oil in the production of industrial steam and process heat.

7. Encouraging the use of large coal-fired boilers for industrial parks, district heating and power cogeneration.
8. Ensuring R&D on coal combustion and related environmental considerations.
9. Encouraging coal gasification and liquefaction technology commercialization, including demonstration.
10. Ensuring that government royalties, severance taxes, transportation tariffs, and so forth do not adversely affect coal-mining viability.
11. Training labor and improving community infrastructure and other services where necessary to increase production.
12. Ensuring timely and effective leasing and licensing of government lands.
13. Ensuring that environmental, safety, and health regulations for mining take account of existing technologies.
14. Encouraging efficient, economic, environmentally acceptable coal transporation systems with adequate capacity and flexibility to handle expected increases in coal trade volumes.

The IEA principles embody many of the policy principles mentioned earlier. In some cases they go further, to stimulate actions in advance of market developments. These steps may be justifiable in order to reduce lead times for more protracted measures, including R&D and infrastructure development, when major public benefits can be gained as a result. However, principles 5 through 9 should be focused on obstacle removal, not on replacing the market process and least cost approaches.

In the light of the principles, what choices have been made by the IEA countries? The next section of this paper examines this question.

NATIONAL ENERGY POLICY CHOICES

National energy policy choices entail a number of considerations, including prices, conservation policies, supply policies, and RD&D policies. These considerations are reviewed in this section.

Energy Prices

In the past five years, countries have made economic adjustments to higher oil prices, although higher inflation, caused in part by energy price rises, economic adjustment policies, and other factors complicated this adjustment. The Iranian revolution and associated oil price rise added further complications and insured that the expected recession would take place, as monetary policies in countries with unacceptable inflation rates and payment balances were forced to contract economic activity.

Energy policy responses of individual IEA countries to the events of 1973–74 differed widely. Some countries allowed domestic energy prices to rise, while others, most notably the United States, maintained price control schemes.

Yet even the United States finally adopted rules to phase out such controls, albeit in the context of pre-1980 world price levels. The US action reflected a compromise: although prices seen by energy users were allowed to rise over time, effective prices seen by oil and gas suppliers were maintained below world price levels through taxation. Further, it remains to be seen whether full decontrol will take place as scheduled. Natural gas price control phase-out formulas based on 1979 conditions need to be modified if world energy prices remain high, to avoid a "gap" jump when phase-out ends. The IEA recommends[4] that phased decontrol of crude oil by 1981 and of natural gas by 1985 should not be interrupted and should be accelerated if possible. The price of gasoline has not yet been deregulated. The IEA indicates that to avoid a repetition of long gasoline lines resulting from price controls, it is essential to bring demand and supply into better balance, and decontrolling gasoline prices is one way to do this. In general, energy prices—especially oil prices—still need to be adjusted to reflect replacement cost.

In Canada, oil import subsidies and wellhead oil price restraint continues to keep the internal price well below world market levels. As of January, 1980, domestic prices were about half of December, 1979, world market levels. If adjustments continued at that pace, a growing disparity between domestic and international prices would occur. Proposals to ameliorate that situation continue to be made. Natural gas continues to be controlled to domestic oil price levels on a heat value basis—a basis which fails to reflect the relative availability and premium nature of natural gas. The IEA points out that if self-sufficiency is not attained and maintained in Canada, public sector budgets will have to take into account permanent, and possibly very expensive, subsidies; demand will be greater than otherwise and supply less, and this will be reflected in the balance of payments position. Energy isolation cannot successfully work in the medium to long run.

In the Federal Republic of Germany, except for coal and electricity prices, energy prices are determined primarily by the market system, albeit with some energy product taxes. Electricity tariffs were improved in 1979, leveling regressive tariffs and eliminating the surcharge on interruptible heat pumps.

In Italy, most energy prices are still controlled and, except for those in the industrial sector, have not kept pace with inflation in the past few years. Electricity tariffs have decreased in real terms over the past few years; about two-thirds of residential consumption is subsidized to levels below average costs of production and distribution on social policy grounds, although limitations in this subsidization were expected to start in August, 1980.

In the United Kingdom, oil prices continue to reflect market levels, except for past domestic natural gas prices. This condition is intended to be corrected, with real domestic gas prices rising 10 percent per year over the next three years.

In Norway, electricity prices were unchanged in 1979; the government had this matter under study in 1980. In Sweden, energy prices remain low, compared with other European countries, through government control. In Denmark, prices are market-determined, although some detailed fiscal measures may not be optimum from a fuel-switching point of view. In June, 1979, Denmark heavily increased taxes on petroleum products and electricity to stimulate demand restraint.

In Japan, market prices play an important role in conservation policies; during 1979 the gasoline tax was raised by about 25 percent, the aviation fuel tax was doubled, and kerosene prices were deregulated. In Australia, oil prices are being brought to world market levels; domestic fuel prices are close to cost because of consumer-owned and cooperatively owned mines. Natural gas prices may be regulated by the states in Australia, as are electricity tariffs.

On balance there has been significant improvement in the IEA countries, in their moving toward market pricing of energy, although much remains to be done in some countries, and vigilance is required to avoid regression in others. Similarly, there has been some progress in shifting away from regressive electricity tariffs.

Energy Conservation Policies

The recent IEA report[5] finds that Denmark and Sweden have developed the strongest and most comprehensive energy conservation programs, with an absolute reduction in oil consumption expected through the 1980s. The Netherlands also has a well-defined program with adequate funding for incentives, although it emphasizes a heavy switch to oil to preserve the Groningen gas reserves. The United States, Canada, and New Zealand are developing comprehensive energy demand management programs that include building codes, loan schemes, fuel switching, and reduction of oil use under utility boilers, as well as stronger incentives for conservation. Canada has adopted a demand growth target of 2 percent annually for primary energy from 1975 to 1990 and thus is expanding its conservation programs. The IEA reports that although Japan, the Federal Republic of Germany, and the United Kingdom are reinforcing their conservation and fuel-switching programs, there is scope for improvement, particularly for fuel-switching incentives in Japan and the Federal Republic of Germany and for conservation of transportation energy in the United Kingdom. Italy is beginning preparation of a comprehensive program, but with limited public funding and staffing. Japan, Greece, Ireland, Austria, and Belgium still appear to have a long way to go to put programs in place at the level of other IEA countries.

On balance it appears that much more could be done to increase energy efficiency. Many countries still have a distance to go in achieving appropriate levels of internal prices for conservation stimulus. Many still approach conservation as "using less" rather than as increasing energy use efficiency through capital investment. Little evidence of integrated tradeoffs between conservation and supply at national policy levels exists. In many cases conservation programs are still seen as a thing apart, as evident from the higher cost policies followed in supply areas, while less costly conservation opportunities remain neglected.

Energy Supply Policies

The United Kingdom and Canada have relatively strong and well-structured supply programs with substantial indigenous resources. The UK government, after remedying the previous government's depletion policy to correct the substantial decline in exploratory drilling activity since 1973, appears likely to follow a more modest development policy to avoid a sharp decline in the North Sea fields in the late 1980s. Canada's synthetic fuel programs are impressive, although a decline from existing fields and the high cost of new frontier production make the outlook guarded.

Australia and Norway are self-sufficient and export large surpluses: oil, gas, and hydro from Norway and coal, liquified petroleum gas (LPG), and uranium from Australia. Increased emphasis on coal exports in Australia and coal use in Norway is needed.

The United States, the Netherlands, and New Zealand have large indigenous resources and good supply programs. New Zealand is switching away from imported oil and toward new sources of domestic gas.

In the United States the gas outlook has improved due to the phased deregulation of natural gas prices. Alaskan and outer continental shelf oil can reduce the speed of decline of established production, although leasing schedules and incentives do not yet fully stimulate potential exploration and development. Coal production is hampered by the regulatory regime, by leasing practices, and also by limitations on means of financing infrastructure development. Further nuclear delays are likely as a result of the Three Mile Island accident and subsequent Nuclear Regulatory Commission (NRC) decisions. Increased training and safety costs imposed on the industry as a result, taken together with regulatory delays (in part due to public opposition), make it likely that new nuclear electricity plant costs are now higher than those for new coal-fired plants, current environmental protection costs included.

In the Netherlands, a substantial increase in oil imports is planned. The IEA recommends review of the Dutch policy of switching from gas to oil in electricity generation, and it also recommends that the Netherlands accelerate their coal and nuclear decision processes.

Denmark is aggressively diversifying its fuel sources by increasing oil and gas production and steam coal imports. Germany is focusing on the use of indigenous coal resources and expanded coal imports. For both countries, nuclear power is at a standstill. Japan has launched a major alternative supply plan including coal, liquified natural gas (LNG), and nuclear and synthetic fuels, to reduce its oil import dependency from 75 percent to 50 percent by 1990. This will require major efforts.

On balance most countries are doing moderately well in supply policies if one considers only traditional fuels. A glaring weakness, however, is the absence of major supply policies for renewable energy sources. Although some limited first steps are taking place, reliance seems largely on the market and on private sector development for solar heating and cooling, solar photovoltaics, biomass, and other renewable sources of energy services. Some tax and related incentives exist. Activity is mainly concentrated on RD&D in these areas. These issues are dealt with below.

Research, Development, and Demonstration Policies

By and large, RD&D programs in IEA's member countries devote far too much funding to electricity production technologies. The IEA reports that almost three-quarters of such budgets are concerned with electricity generation. Since virtually all the electricity technologies involve the production of heat, it is appropriate to shift effort significantly toward substitutes for liquid fuels.

Further, budgets are still overwhelmingly oriented toward the development of supply technologies. Even if historically understandable, this situation must be rectified. Among IEA countries, only the governments of Austria, Ireland, the Netherlands, New Zealand, Norway, and Sweden devoted 10 percent or more of their 1979 energy RD&D budget to conservation. The United States, Italy, Japan, and the Federal Republic of Germany were among the lowest, with less than 6 percent devoted to this work. New energy sources (solar, wind, ocean, biomass, and geothermal), were somewhat better represented. Austria, Denmark, Greece, Ireland, New Zealand, Sweden, Switzerland, and the United States devoted 15 percent or more of their RD&D budget to these areas. Belgium, Japan, the Federal Republic of Germany, and the United Kingdom surprisingly spent less than 5 percent.

Despite public resistance to and the increased costs of nuclear power, RD&D spending was still heavily distorted in nuclear's direction. More than 50 percent of energy RD&D spending was in this category for all IEA countries except Austria, Denmark, Greece, Ireland, New Zealand, Norway, Sweden, and the United States.

Italy spent 86.9 percent of its 1979 government energy RD&D funds on nuclear; Japan, 84.3 percent; Canada, 75.2 percent; the Federal Republic of Germany, 64.7 percent; and the United Kingdom, 64.5 percent. Except for Belgium and Italy, these countries spent more of their budget on nuclear nonbreeder research than on breeder reactors and fusion.

The industry energy RD&D budget situation in 1979 was somewhat better. Over 25 percent of this spending was for conservation RD&D in Japan, New Zealand, Spain, and the United States, while Italy, the Federal Republic of Germany, and the United Kingdom spent less than 10 percent in this area. In Italy and Spain, over 20 percent of such budgets were for new energy source RD&D. New Zealand, Norway, the United Kingdom, and the United States still lag badly, with less than a 3 percent share of such RD&D spending. Industrial nuclear RD&D spending is above 20 percent in Italy, Japan, the Federal Republic of Germany, the Netherlands, Norway, the United Kingdom, and the United States, with 71.8

percent in the United Kingdom, 60.2 percent in the Federal Republic of Germany, 55.7 percent in the Netherlands, and 51.2 percent in Norway.

Thus, a major reexamination of energy RD&D spending seems in order in most IEA countries. It cannot be argued that the distribution of industrial RD&D spending meets some sort of market test, since the major actor in the nuclear market has been the government, in most IEA countries. The IEA concludes that more work is called for in assessing the benefits and costs of various lines of technology development; that increases in RD&D spending still may not be an adequate recognition of the contribution technology development can make to reduce dependence on oil; that the importance of conservation technology is not reflected in the level of its funding in relation to supply; that more attention to the use of waste heat from electricity generation as a substitute for liquid fuels is needed; and that more progress is possible in fossil fuel technology commercialization.

MAJOR STRATEGIES FOR THE FUTURE

In conclusion, the following are strategies which, in my judgment, should be adopted to improve energy production and use:

1. Increased emphasis is necessary on supply interruption policies, including market-clearing policies, and in determining such policies, social welfare policy must be treated separately, not by avoiding market-clearing policies. This area has recently come to the fore as a result of the Iranian revolution and the Iran/Iraq war. National strategies have not yet been decided in many countries, and in some countries those that have been selected seem unworkable. The role of private stockpiles has also been largely ignored in some countries. Yet such interruptions seem increasingly likely, with possibly serious economic consequences.
2. While the role of price signals to stimulate increases in energy supply and end-use efficiency has improved, more must be done.
3. More internal coordination of national energy policies is necessary to permit cost tradeoffs in policy and in the market between alternative ways of providing energy services, including tradeoffs between supply and conservation (demand efficiency).
4. Governments should act promptly to reduce and eliminate uncertainties caused by unclear or ambiguous policies.
5. Specific incentives should be provided for conversion from

oil to other forms of energy, when there is a special national value to such actions and that value is not captured by the price system.

6. During the bridging period to increased use of renewable resources, increased availability of conventional fuels must be assured through accelerated leasing; advanced action to permit warranted infrastructure development; the expansion of coal production, trade, and use; and the removal of inappropriate obstacles that prevent adequate supply response to market price signals. This must take place with due regard to the environment and to occupational health and safety. Nevertheless, market distortions for political purposes should not be permitted under the rubric of environmental protection.

7. Significant enhancement of market acceptance processes for conservation technologies and new energy sources is essential. This should include incentives to compensate for distortions in price signals due to subsidies of other fuels (when such subsidies cannot be corrected), significant stimulus for barrier removal to the entry of these technologies and sources, and expanded education and information programs.

8. A significant RD&D effort for conservation technologies and new energy sources is essential. Major shifts in recent RD&D spending patterns are required.

9. The energy policy review process of the IEA should be strongly supported and enhanced, as should its results publication system, with results widely disseminated within national borders. This review process must increase its emphasis on policies of mid-range and long-range adjustment, while maintaining and enhancing shorter term review content. National governments should be encouraged to cooperate strongly in this process by appropriate means.

NOTES

1. R. W. Sant, "The Least-Cost Energy Strategy" (Arlington, VA: Energy Productivity Center, Mellon Institute, 1979).

2. Aspen Institute for Humanistic Studies, "Petroleum Interruptions and National Security" (New York: Aspen Institute, 1980).

3. Workshop on Alternative Energy Strategies (WAES), *Energy: Global Prospects 1985–2000* (New York: McGraw-Hill, 1977).

4. International Energy Agency (IEA), "Energy Policies and Programmes of IEA Countries—1979 Review" (Paris: Organisation for Economic Co-operation and Development [OECD], 1980).

5. Ibid.

The references listed below are also useful sources of information for the topics discussed in the paper:

Aspen Institute for Humanistic Studies, "Decentralized Electricity and Cogeneration Options" (New York: Aspen Institute, 1979).

International Energy Agency, "Energy Conservation in Industry in IEA Countries" (Paris: OECD, 1979).

———, "Energy Conservation—Results and Prospects," OECD *Observer* (Paris: OECD, November 1979).

———, "Energy Research, Development and Demonstration in the IEA Countries—1979 Review of National Programmes" (Paris: OECD, 1980).

———, "Outlook for the Eighties—Summary of 1979 Review" (Paris OECD, 1980).

G. J. Mangone, ed., *Energy Policies of the World* 1–3 (New York: Elsevier, 1979, 1977, 1979).

J. Sawhill, ed., *Energy Conservation and Public Policy* (Englewood Cliffs, NJ: Prentice-Hall, 1980).

R. H. Shackson and H. J. Leach, "Using Fuel Economy and Synthetic Fuels to Compete with OPEC Oil" (Arlington, VA: Energy Productivity Center, Mellon Institute, 1980).

D. Sternlight, "Petroleum Supply/Demand Balances: 'Notional Gaps' and Policy Consequences," in World Petroleum Congress, *Proceedings of the Tenth World Petroleum Conference* 2.

———, "Scenario Analysis, Policy Relevance, and Model Utility," in IGT Symposium Papers, *Energy Modeling II—The Interface Between Model Builder and Decision Maker* (Chicago: Institute of Gas Technology [IGT], 1979).

World Bank, "Energy in the Developing Countries" (Washington, DC: World Bank, 1980).

World Coal Study (WOCOL), *Coal—Bridge to the Future* (Cambridge, MA: Ballinger, 1980).

CHAPTER FOURTEEN

Alternative Energy Policies for Developing Countries: A Case Study of India

Amulya Kumar N. Reddy
Professor of Chemistry
Indian Institute of Science

Both industrialized countries and oil-importing developing countries are in a serious energy crisis, but there the similarity ends because of the fundamental differences in the natures of the two crises.[1] The developed countries have already attained extravagant standards of living, and their problem is whether they can resolve their energy crises without cutting down their growth rates and reducing their profligacy in the consumption of goods and services. In contrast, developing countries have such pathetic standards of living that their problem is whether the rapid growth so vital to their development can be achieved without becoming constrained by a lack of energy. Thus, the challenge in developed countries is to prevent energy shortages from eroding affluence, but the battle in oil-importing countries is to remove poverty despite escalating oil prices. Since there is not much in common between the contexts and dilemmas of these two categories of nations, it is inevitable that their energy solutions and policies will be quite different and must be separately considered.

Such separate consideration is given here by choosing India as an example of an oil-importing developing country with a low per capita Gross National Product (GNP) and presenting the nature and dimensions of its "energy crisis." Apart from a description of its present and past energy production by source and energy consumption by sector, detailed projections to the year 2000 are given of the likely energy demand in the household, agricultural, industrial, and transportation sectors. These Reference Level Forecasts of the Indian Planning Commission's Working Group on Energy Policy are made on the assumption that *no* deliberate policy-based measures will be initiated to alter current trends.

The Reference Level Forecasts are shown to constitute a set of energy demands which are so prohibitively high relative to India's energy resources that it becomes imperative to intervene by implementing alternative policies for improved energy productivity and production. Such alternative policies are reviewed here by considering in detail the household, agricultural, industrial, and transportation sectors. For each sector, two categories of alternative policy options are discussed: first, soft-option forecasts associated with technological fixes, and then, development-oriented forecasts based on fundamental changes involving need-based consumption and renewable energy sources. It is shown that while the soft-option forecasts only result in marginal reductions in energy demand, the development-oriented forecasts lead to a tractable problem in which energy is an integral aspect of development.

INDIA'S PATTERN OF ENERGY CONSUMPTION

The 1978 contribution by source and consumption by sector of inanimate energy in India is presented in Table 14.1 and Figure 14.1.* The most important feature of the table distinguishing it from a corresponding matrix for a developed country is the prepon-

Table 14-1. **1978 Energy Consumption Matrix for India, by Sector and Source**[a]

Source	House-hold	Indus-try	Trans-port	Agri-culture	Other	Total
Coal	4.00	50.50	12.40	—	1.90	68.80
Electricity	7.70	53.90	2.60	11.95	8.25	84.40
Oil	28.76	9.00	78.18	19.37	5.79	141.10
Commercial energy	40.46	113.40	93.18	31.32	15.94	294.30
Noncommercial energy	200.00	50.00	—	—	—	250.00
Total	240.46	163.40	·93.18	31.32	15.94	544.30

[a]In million tonnes coal replacement (MTCR).

---

*This paper follows the Indian practice of using coal *replacement* units (which take into account conversion efficiencies) rather than coal *equivalent* units. The conversions for these terms are as follows: 1 million tonnes coal = 1 million tonnes coal replacement (MTCR) = 1 million tonnes coal equivalent (MTCE); 1 billion kilowatt-hours (kWh) = 1 MTCR; 1 million tonnes oil = 6.5 MTCR = 2 MTCE; 1 million tonnes firewood = 0.95 MTCR = 0.95 MTCE; where a tonne equals 1 metric ton—i.e., 1,000 kilograms (kg) or 2,205 pounds avoirdupois.

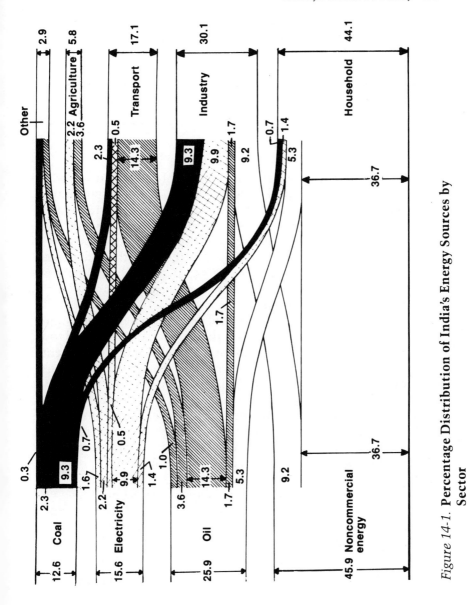

Figure 14-1. Percentage Distribution of India's Energy Sources by Sector

derance of noncommercial energy (firewood, animal wastes, and agro-residues). Another distinguishing feature is the importance of the household sector, which in India consumes a little less than half the total energy, as shown in Table 14.2 comparing Indian and US energy consumption by sector.

A better idea of the differences between a developing country like India and the developed countries can be obtained by present-

Table 14-2. **Energy Consumption in India and the United States, by Sector as Percentage of National Total**

Country	House-hold	Indus-try	Trans-port	Agri-culture	Other
India	44	30	17	6	3
United States	14	30	25	4	27

ing India's matrix in original units as in Table 14-3. With a 1978 population of 634 million, the Indian per capita annual consumption figures were 108.5 kg of coal, 133.1 kWh of electricity, and 0.034 tonnes of oil. For comparison, the corresponding US per capita figures were 2.7 tonnes of coal, 10,403.7 kWh of electricity, and 3.7 tonnes of oil; that is, 25, 78, and 109 times the Indian per capita consumption. Furthermore, in 42, 16, and 14 days respectively the United States consumed the coal, electricity, and oil that India consumed in a whole year.

Despite the low values of total and per capita energy consumption in India, the rapid post-1973 escalation in oil prices has led to a situation where in 1978–79 the country, which imports 60 percent of its oil requirements, spent 35 percent of its total export earnings on these oil imports. This already impossible situation can only worsen—in fact, it already has. Hence, major policy interventions are imperative. In order to define these policy interventions, an analysis by sector of India's energy problem has to be undertaken.

The task of doing this analysis has been greatly facilitated by the elaborate and systematic report of the Working Group on Energy Policy (WGEP) of the Indian government's Planning Commission.[2] The WGEP report has done yeoman service in providing analysis by sector of (1) present energy sources and energy-utilizing activities, (2) energy trends over the past twenty-five years or more, and (3) detailed Reference Level Forecasts (RLFs)

Table 14-3. **1978 Energy Consumption Matrix for India, by Sector and Source**[a]

Source	Unit	House-hold	Indus-try	Trans-port	Agri-culture	Other	Total
Coal	x 10^6 T[b]	4.00	50.50	12.40	—	1.90	68.80
Electricity	x 10^9 kwh	7.70	53.90	2.60	11.95	8.25	84.40
Oil	x 10^6 T	4.42	1.38	12.02	2.98	0.89	21.71

[a]In original units.
[b]Tonnes (One tonne = 1 metric ton, or 1000 kg).

which project the likely level of energy demand up to the year 2000 *if present trends continue—that is, if no deliberate policies are implemented to alter these trends.* The WGEP has also given some suggestions, mostly but not wholly by way of technical fixes, for moderating the demands indicated by the RLFs.

Therefore, the first part of this paper, which describes the current situation, past trends, and RLFs by sector, is largely based on the WGEP report. The second part of the paper, however, constitutes a distinct departure from the WGEP report by emphasizing development-oriented energy policies that drastically reduce the RLFs of sectoral energy demand.

CONSUMPTION PATTERNS, TRENDS, AND FORECASTS BY SECTOR

Consumption patterns, trends, and forecasts are given below for each of the Indian economy's four major sectors. As previously indicated, these include the household, industrial, transportation, and agricultural sectors.

Household Sector

The household sector is the largest consumer of total energy in the economy, accounting for about half of this total. A matrix of its energy sources and energy-consuming activities is presented in Table 14-4 and Figure 14-2.

It can be seen that most of the energy for this sector (for example, 84% in 1975) comes from noncommercial sources. Firewood, agro-wastes, and dung cakes contribute 65 percent, 15 percent, and 20 percent to these noncommercial sources of energy. The commercial sources of energy used in the household sector are kerosene, electricity, coal, and liquified petroleum gas (LPG). The 1975 share of these various energy sources in the household sector's commercial energy consumption was 68 percent, 16 percent, 10 percent, and 6 percent respectively, whereas their share in the total household energy consumption was 10.9 percent, 2.9 percent, 1.3 percent, and 0.7 percent respectively.

Energy is used in the household sector almost wholly for cooking (89 percent) and lighting (11 percent). The differences between the rural and urban patterns of consumption of the various fuels is shown in Figure 14-3. In rural areas, 99 percent of cooking energy sources are noncommercial, and 87 percent of lighting is done with kerosene, in contrast to urban areas, where noncommercial energy sources, kerosene, LPG, and soft coke ac-

Table 14-4. **1975 Energy Consumption Matrix for the Household Sector, by Activity and Source**[a]

Source	Cooking	Lighting	Total
NCE[b]	193.87	—	193.87
Electricity	—	6.79	6.79
Kerosene	8.21	16.88	25.09
LPG	1.70	—	1.70
Soft coke	2.91	—	2.91
Others	—	0.91	0.91
Total	206.69	24.58	231.27

[a]In MTCR.
[b]Noncommercial energy.

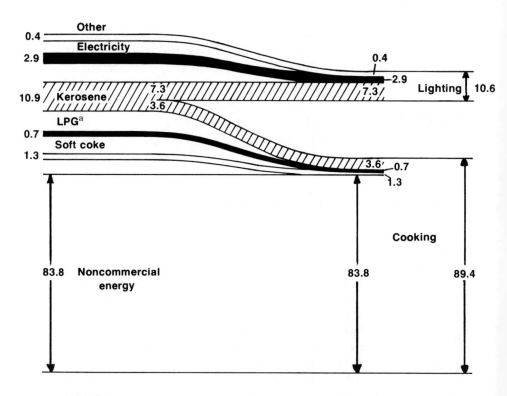

[a]Liquified petroleum gas.

Figure 14-2. **Percentage Distribution of Energy Sources by Household Sector Activity**

count for 75 percent, 16 percent, 4 percent, and 5 percent respectively of the cooking energy, and electricity and kerosene both play major roles (62 percent and 38 percent respectively) as energy sources for lighting.

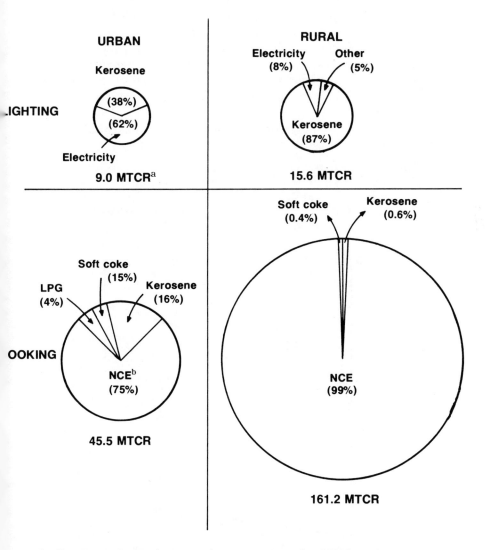

^aMillion Tonnes Coal Replacement, where a tonne is equal to 1,000 kg, or 1 metric ton.
^bNoncommercial energy.

Figure 14-3. **Urban-Rural Differences in Sources of Household Sector Energy**

The share of noncommercial energy is about 90 percent in rural areas, whereas it is only about 63 percent in urban areas. Furthermore, the contribution of electricity to the household sector's energy consumption is less than 1 percent in rural areas compared with about 10 percent in towns and cities.

cooking energy sources are noncommercial, and 87 percent of lighting is done with kerosene, in contrast to urban areas, where noncommercial energy sources, kerosene, LPG, and soft coke account for 75 percent, 16 percent, 4 percent, and 5 percent respectively of the cooking energy, and electricity and kerosene both play major roles (62 percent and 38 percent respectively) as energy sources for lighting.

The trends of energy consumption in the household sector over the past twenty-five years have been as follows:

- The share of noncommercial energy has been declining slightly, though its absolute magnitude has been increasing, as shown in Figure 14-4.
- The consumption of all commercial energy sources has been increasing, as shown in Figure 14-5.

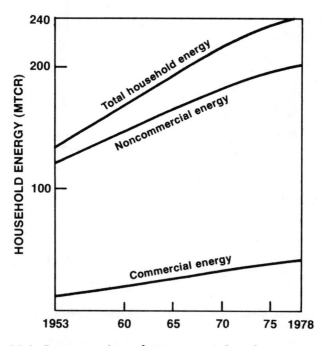

Figure 14-4. **Consumption of Commercial and Noncommercial Energy in Household Sector**

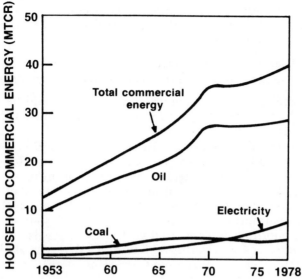

Figure 14-5. **Consumption of Commercial Energy in Household Sector**

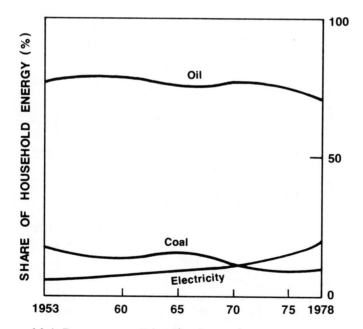

Figure 14-6. **Percentage Distribution of Commercial Energy Sources in Household Sector**

• There have not been any major intersource shifts, but there has been a small tendency for the share of electricity to increase, as shown in Figure 14-6.

Based largely on these trends, the WGEP made RLFs for the household sector. The RLF for the year 2000 shows several features:

• The percentage of noncommercial energy would be reduced from the present 80 percent to about 50 percent, as shown in Figure 14-7.
• Even then, the forecasted firewood consumption would be about 106 million tonnes per year, which is slightly less than the present consumption (This assumes that the present rate of firewood consumption can be sustained for another two decades even though forests are being denuded at a fast rate.)
• The consumption of all commercial energy sources will increase enormously compared with present levels—electricity by a factor of 4.9, reaching 35.8 billion kWh/year, as shown in Figure 14-8; coal by a factor of 6, reaching 24 million tonnes per year, as shown in Figure 14-9; kerosene by a factor of 4.1, reaching 12.95 million tonnes per year, as shown in Figure

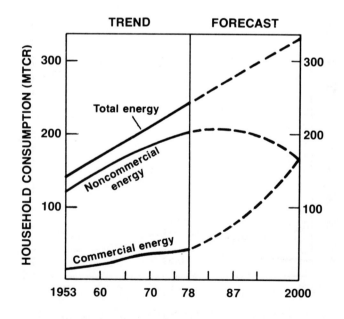

Figure 14-7. **Forecast for Household Energy Consumption**

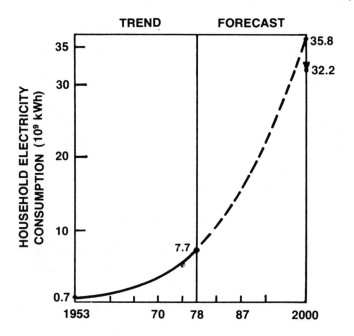

Figure 14-8. **Forecast for Household Electricity Consumption**

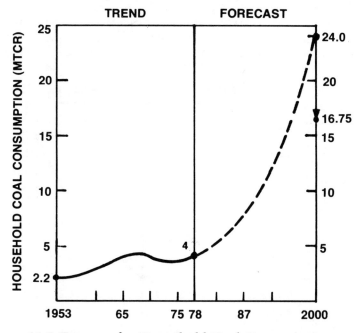

Figure 14-9. **Forecast for Household Coal Consumption**

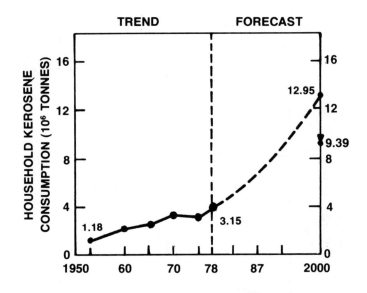

Figure 14-10. **Forecast for Household Kerosene Consumption**

14-10; and LPG by a factor of 10.6, reaching 3.3 million tonnes per year, as shown in Figure 14-11.
• The source mix in 2000 would be as indicated in Figure 14-12: noncommercial energy—50 percent, kerosene—26 percent, electricity—11 percent, coal—7 percent, and LPG—7 percent.

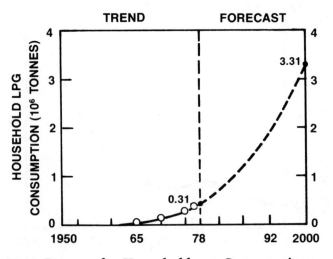

Figure 14-11. **Forecast for Household LPG Consumption**

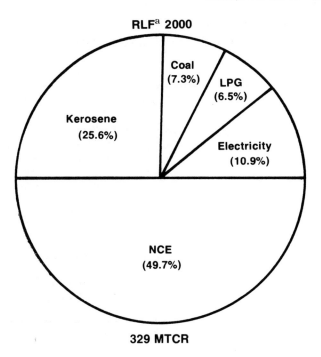

RLF^a 2000

Coal
(7.3%)

LPG
(6.5%)

Kerosene
(25.6%)

Electricity
(10.9%)

NCE
(49.7%)

329 MTCR

^aReference Level Forecast.

Figure 14-12. **Percentage Distribution of Household Energy Sources According to Forecast for 2000**

As shown by the WGEP, the RLF figures can be reduced through technical measures such as (a) increasing the fuel efficiency of lighting by 10 percent over the next two decades, (b) reducing kerosene and soft coke consumption by 3.0 and 7.3 million tonnes respectively and supplanting these fuels with an extra 50 million tonnes of firewood, and (c) increasing the efficiency of kerosene cooking stoves by 5 percent and of noncommercial-source cooking stoves by 10 percent. The reductions that could be achieved by these measures are presented in Table 14-5.

Industrial Sector

The industrial sector is the second largest consumer of total energy in the Indian economy, accounting for about 30 percent of this total. The matrix estimating commercial energy sources and their consumption in the industrial sector is presented in Table 14-6 and Figure 14-13.

Table 14-5. **Possible Reductions in the 2000 RLF for Household Commercial Energy Consumption**

Source	Unit	RLF	Change	Measure(s)	Reduced figure
Electricity	x 10⁹ kwh	35.8	−3.6	Improve lighting efficiency by 10%	32.2
Kerosene	x 10⁶ T	12.95	−2.99	Reduce rural trend toward kerosene cooking	9.39
			−0.57	Improve cooking efficiency by 10%	
LPG	x 10⁶ T	3.31	—	—	3.31
Coal	x 10⁶ T	24.0	−7.25	Limit increase of soft coke production	16.75
Commercial energy	MTCR	165.5	−52.0	Above measures	113.5
Noncommercial energy	MTCR	163.5	+31.2	To substitute for kerosene & soft coke	194.70

Table 14-6. **1977-78 Energy Consumption Matrix for the Industrial Sector, by Industry and Source**[a]

Industry	Coal	Electri-city	Oil for heating	Oil for power	NCE	Total
Steel	21.079	5.481	—	0.003	—	26.563
Cement	3.861	2.346	3.152	0.004	—	9.363
Fertilizers	1.768	3.889	1.448	—	—	7.105
Textiles	2.584	5.255	2.130	0.121	—	10.090
Bricks	5.345	—	—	—	~10	15.345
Aluminum	—	3.552	—	—	—	3.552
Sugar	—	0.795	—	0.024	~30	30.819
Others	16.034	28.033	1.789	0.231	~10	56.087
Total	50.671	49.351	8.519	0.383	~50	158.924

[a]In MTCR.

It is evident that energy for industry comes predominantly from coal (32 percent), electricity (31 percent), and noncommercial energy (31.5 percent). Oil, which contributes 5.5 percent of the energy for industry, is used mainly as fuel oil for heating, but about 3.6 percent of the total oil consumed by industry is in the form of

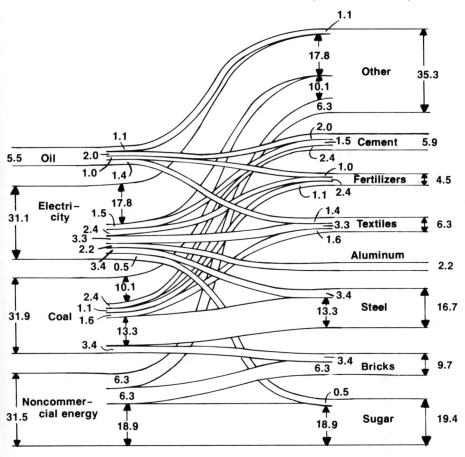

Figure 14-13. **Percentage Distribution of Energy Sources for Various Industries**

diesel oil for captive power generation. The main sources of noncommercial energy for the industrial sector are sugar cane bagasse (about 60 percent), firewood, and vegetable wastes such as rice husks. If noncommercial sources are excluded, the contributions of the commercial energy sources—namely, coal, electricity, and oil—are about 47 percent, 45 percent, and 8 percent respectively.

Over half the commercial energy is used by six industries: steel (24.4 percent), textiles (9.3 percent), cement (8.6 percent), fertilizers (6.5 percent), brick burning (4.9 percent), and aluminum (3.3 percent). With regard to noncommercial energy, the bagasse is used almost completely as a captive energy source for the sugar and jaggery (country sweetener) industries, and about half the firewood is used for brick making.

No data are available on the end uses of industrial energy, but a rough estimate is as follows: process heat, including the use of coal as a reductant—54 percent excluding noncommercial energy and 69 percent including noncommercial energy; and mechanical power—46 percent excluding noncommercial energy and 31 percent including noncommercial energy.

The trends in the industrial consumption of commercial energy are shown in Figure 14-14 and indicate the following:

- Commercial energy consumption in the industrial sector increased at the compound rate of 6.7 percent per year between 1953 and 1978, the rates for coal, electricity, and oil being 5.3 percent, 10.0 percent, and 3.7 percent respectively. Thus, the growth rate for electricity consumption in industry has been 1.9 and 2.7 times the growth rates for coal and oil.
- Throughout the twenty-five-year period, coal has continued meet about half the energy needs of the industrial sector.
- The intensity of commercial energy use in the industrial sector —that is, the ratio of the quantity of energy used to the units of

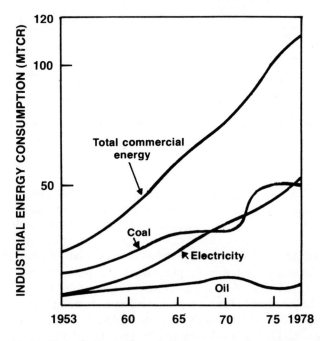

Figure 14-14. **Trends in Industrial Consumption of Commercial Energy**

value added—has been steadily increasing during the twenty-five-year period, as shown in Figure 14–15. But this increase is mainly attributable to the increase in intensities of electricity and coal consumption rather than to that of oil consumption, for the latter has changed little.

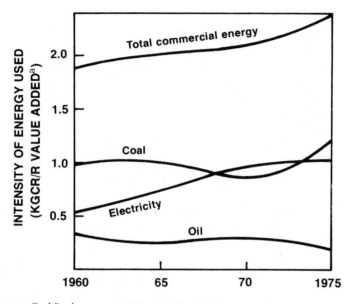

^aKilograms Coal Replacement per Rupees Value Added.

Figure 14-15. **Past Intensity of Commercial Energy Use in the Industrial Sector**

The RLF for commercial energy consumption in the industrial sector is shown in Figure 14-17. The following are some of its salient features:

The RLFs for the industrial sector have been made primarily on the basis of assumptions regarding (1) the growth of value added in the industrial sector and (2) the intensities of electricity, coal, and oil consumption, as shown in Figure 14-16. In the case of coal and oil, the intensities have been considered for industries other than steel; for the steel industry, the coal requirements have been computed using a norm of 2.88 tonnes of coking coal per tonne of steel.

- The RLF for total commercial energy in the year 2000 is 636 MTCR, which is 5.6 times the 1978 consumption by the industrial sector.
- The RLF for electricity in 2000 is 350 billion kWh, or 6.5 times

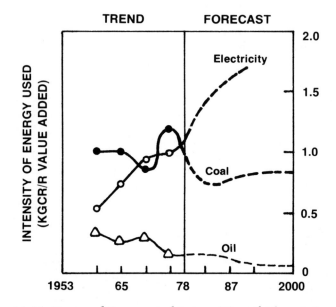

Figure 14-16. **Past and Forecasted Intensities of Electricity, Coal, and Oil Use in the Industrial Sector**

the current industrial consumption; for coal, it is 273 million tonnes, or 5.5 times the current industrial consumption; and for oil, it is 6.2 million tonnes of furnace oil, or 1.4 times the current industrial consumption.

To achieve a steel production of 40 million tonnes (i.e., four times the present production), 115.25 million tonnes of coking coal are required—5.8 times the present coal consumption of the steel industry and 15 percent more than the country's *total* coal consumption at present.

The WGEP has suggested two ways in which the RLFs for the industrial sector can be reduced. First, instead of permitting the intensity of electricity consumption to rise continuously from 1.02 in 1975 to 1.72 in 2000, the WGEP has suggested that it should be pegged at 1.40 from 1982 onward. This can be achieved by "better utilization of capacity, more efficient use of electricity" and a deemphasis of electricity-intensive industries like aluminum, ferro-silicon, and so on. Second, the intensity coefficients of oil and coal can be reduced from the RLF assumption of 0.90 to a value of 0.70 by better fuel efficiencies and by a switch from furnace oil to coal for heating in industries. The reductions attainable by these suggestions are shown in Table 14-7.

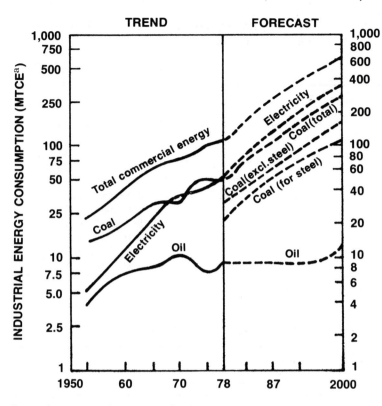

aCoal equivalent units are the same as coal replacement units for electricity and coal, but for oil, 1 x 10⁶ tonnes oil, = 2.0 MTCE = 6.5 MTCR.

Figure 14-17. **Past and Forecasted Commercial Energy Consumption in the Industrial Sector**

Transportation Sector

The transportation sector is the third largest consumer of total energy in the Indian economy, accounting for about 17 percent of this total in 1975. Table 14-8 and Figure 14-18 give estimates of the interrelationships of commercial energy sources and transport modes. As shown in the latter, the contributions of coal, oil, and electricity to the energy for transportation are 1.3 percent, 83.9 percent, and 2.8 percent respectively, and total transport energy is shared among the railways, roads, airlines, and shipping modes according to the following percentages: 24.6 percent, 64.7 percent, 9.7 percent, and 1.0 percent respectively.

This, however, is not a complete picture. Animal energy plays

Table 14-7. **Possible Reductions in the 2000 RLF for Industrial Energy Consumption**

Source	Unit	RLF	Change	Measure(s)[a]	Reduced figure
Electricity	x 10⁹ kwh	350	−66	Better capacity utilization More efficient use of electricity Deemphasis of electricity-intensive industries	284
Coal	x 10⁶ T	158	−35	(a) Better fuel efficiency	123
Oil	x 10⁶ T	6.2	−2.96	(b) Switch from furnace oil to coal for indus-	3.24
	MTCR	40.3	−19.24	trial heating	21.06

[a]Measures (a) and (b) both apply to coal and oil.

Table 14-8. **1978-79 Commercial Energy Consumption Matrix for the Transportation Sector, by Transport Mode and Source[a]**

Source	Railways	Road	Air	Ships	Total
Coal	12.4	—	—	—	12.4
Oil	7.9	60.3	9	1	78.2
Electricity	2.6	—	—	—	2.6
Total	22.9	60.3	9	1	93.2

[a]In MTCR.

an important part in the transport scene via bullock carts. There are approximately 13 million bullock carts in the country with load-bearing capacities of about 0.5 to 1.0 tonne which are used for short-haul freight traffic over distances of 10 to 25 kilometers. It has been estimated that bullock carts transport about 10 billion tonne-kilometers (BTKM) per year—freight traffic which would require about 0.5 million tonnes of diesel fuel or 4.4 MTCR if moved by trucks. When animal energy is also included, the matrix of energy sources and transport modes must be modified as in Table 14.9. The contributions of coal, oil, electricity, and animal energy are then 12 percent, 80 percent, 3 percent, and 5 percent respectively; the shares of railways, road, air, and ships become 24 percent, 66 percent, 9 percent, and 1 percent respectively.

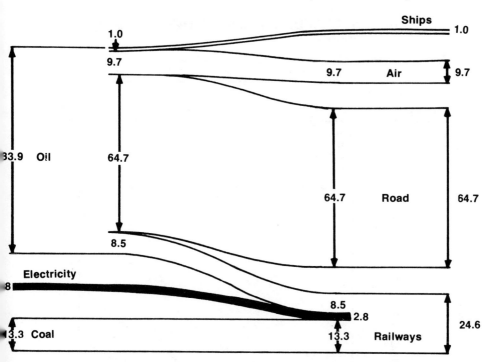

Figure 14-18. **Percentage Distribution of Energy Sources by Transport Mode**

Table 14-9. **1978-79 Energy Consumption Matrix for the Transportation Sector, Including Animal Energy[a]**

Source	Railways	Road	Air	Ships	Total	(%)
Commercial energy	22.9	60.3	9	1	93.2	95.5%
Animal energy	—	4.4	—	—	4.4	4.5%
Total	22.9	64.7	9	1	97.6	100.0%
(%)	23.5%	66.3%	9.2%	1%	100.0%	

[a]In MTCR.

Altogether, the transportation sector has moved goods and passengers as shown in Table 14-10.

The trends for passenger traffic are shown in Figures 14-19 and 14-20, from which the following can be seen:

• The passenger traffic today in billions of passenger-kilometers (BPKM) is five times what it was in 1953.

Table 14-10. **1977-78 Distribution of Freight and Passenger Traffic**

	Railways	Conventional road traffic[a]	All road traffic[b]	Air
Goods[c]	163	77	87	—
Passengers[d]	177	250	?	3.4

[a]Excluding bullock carts.
[b]Including bullock carts.
[c]In billion tonne-kilometers (BTKM).
[d]In billion passenger-kilometers (BPKM).

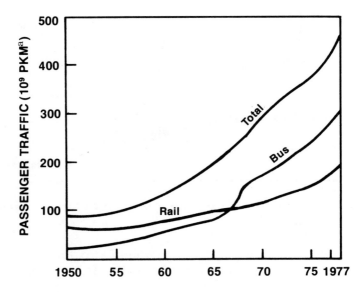

[a]Passenger-Kilometers.

Figure 14-19. **Trends in Passenger Traffic**

- The shares of railways and buses are currently 38.5 percent and 61.5 percent respectively, whereas they were 74.5 percent and 25.5 percent twenty-five years ago. A massive shift in favor of buses and away from rail has taken place.

   The trends for freight traffic are similar (see Figures 14-21 and 14-22), except that freight by trucks has not yet exceeded railway freight. However, the following can be noted:

- The freight traffic in BTKM is now approximately five times what it was in 1950.
- Railways, which carried 88 percent of the freight in 1950, now carry only 67 percent of the total freight.

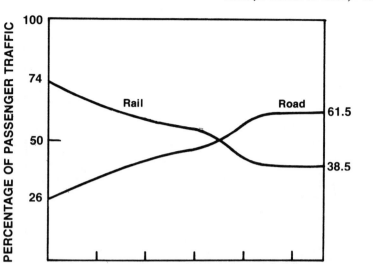

Figure 14-20. **Percentage Distribution of Passenger Traffic**

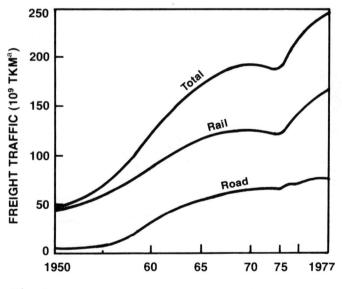

[a]Tonne-Kilometers.

Figure 14-21. **Trends in Freight Traffic**

These trends have had major impacts on the patterns of energy use, as may be seen from Figures 14-23 and 14-24. Figure 14-23

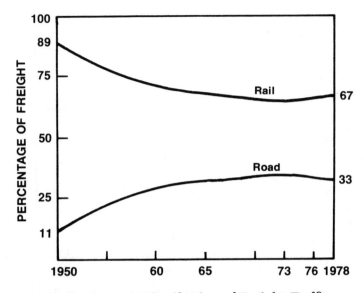

Figure 14-22. **Percentage Distribution of Freight Traffic**

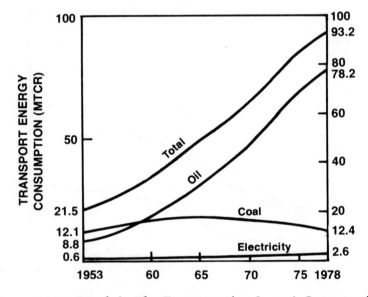

Figure 14-23. **Trends in The Transportation Sector's Consumption of Commercial Energy**

shows that the quantities of coal and electricity used have changed little but that the amount of oil used in the transportation sector has increased dramatically. This trend is revealed more clearly in Figure 14-24, which shows the different sources' percentage contributions to total energy used for transport. From this it can be seen that the share of oil has approximately doubled, going from 41 percent in 1953 to 84 percent in 1978, in contrast to coal, which at 13 percent in 1978 was reduced to about one-quarter of its 1953 level of 56 percent.

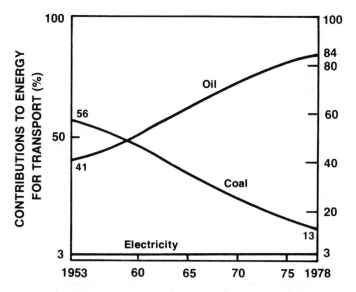

Figure 14-24. **Percentage Distribution of Energy Sources Used for Transport**

The WGEP has done separate RLFs for rail and road transport. Based on past trends and the rate of change of railway transport services to Gross Domestic Product (GDP), the railway transport RLF is as in Figure 14-25 for passenger traffic and as in Figure 14-26 for freight traffic. These forecasts can be combined to give forecasts of the total gross tonne-kilometers for railway transport, as shown in Figure 14-27. It is then assumed that this total railway traffic will be shared between steam, diesel, and electric traction as shown in Figure 14-28. The total forecasted energy consumption for railways can therefore be worked out, as shown in Figure 14-29.

The RLF for road transport (Figures 14-25 and 14-26) has also been based on past trends and the elasticity of road transport services to GDP. This RLF involves projections regarding the sharing of passenger and freight traffic, as shown in Figures 14-30 and 14-31.

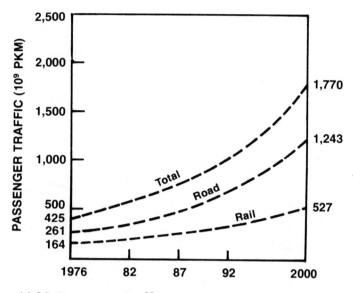

Figure 14-25. **Passenger Traffic Forecast**

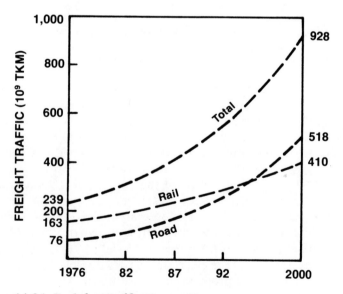

Figure 14-26. **Freight Traffic Forecast**

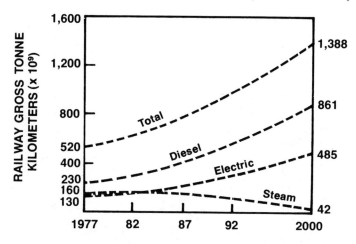

Figure 14-27. **Forecast for Railway Transport**

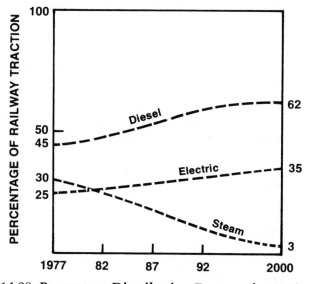

Figure 14-28. **Percentage Distribution Forecast for Railway Transport**

With regard to passenger traffic by road, assumptions about the share of buses on the one hand and cars, taxis, and two- and three-wheelers must also be made, as in Figure 14-32. Based upon these trends and assumptions, the forecasts of the transportation sector's fuel consumption are as in Figure 14-33.

The WGEP has suggested several measures for reducing the

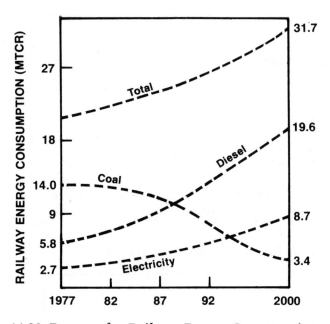

Figure 14-29. **Forecast for Railway Energy Consumption**

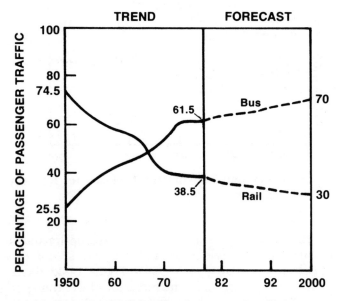

Figure 14-30. **Percentage Distribution Forecast for Passenger Transport Modes**

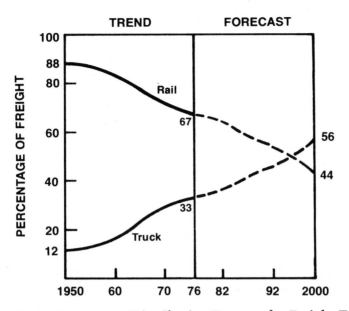

Figure 14-31. **Percentage Distribution Forecast for Freight Transport Modes**

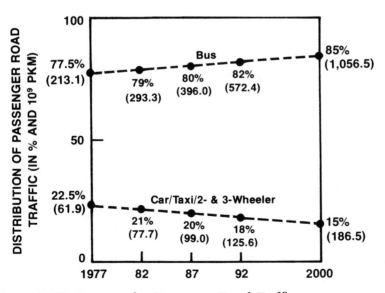

Figure 14-32. **Forecast for Passenger Road Traffic**

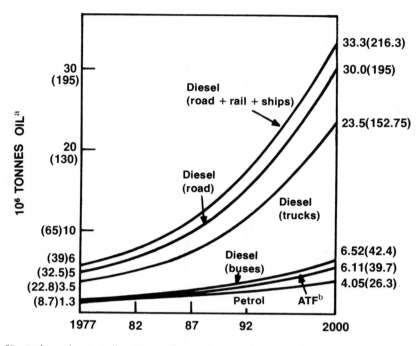

[a] Equivalent values in Million Tonnes Coal Replacement (MTCR) are shown in parentheses. 1 x 10⁶ tonnes oil = 6.5 x 10⁶ tonnes coal replacement.
[b] Alternative Transportation Fuel.

Figure 14-33. **Forecasts for Fuel Consumption in the Transportation Sector**

energy demands indicated by the RLFs. In particular, the following suggestions have been made:

- Improve the fuel efficiency of road transport vehicles—7.5 percent by the year 2000 in the case of gasoline-fueled vehicles and 10 percent in the case of diesel-fueled vehicles.
- Increase the railways' share of passenger and freight traffic compared with the share shown in the RLF—in particular, by attaining a 25 percent increase over the RLF values for intercity passenger and freight rail traffic and by equally distributing this increased rail traffic between diesel and electric traction.
- Reduce the elasticity of traffic demand to GDP—a 6 percent reduction in traffic by 2000.

These suggestions would lead to the reductions in the RLF values shown in Table 14-11.

Table 14-11. **Possible Reductions in the 2000 RLF for Transport Energy Consumption**

Source	Unit	RLF	Change	Measure(s)[a]	Reduced figure
Electricity	x 10⁹ kwh	8.73	+1.75	Increased shift to rail	10.48
Coal	x 10⁶ T	3.4	0	—	3.4
Motor gas	x 10⁶ T	4.05	−0.89	(a) Increased shift from road to rail	3.16
Diesel	x 10⁶ T	33.28	−9.79	(b) Greater vehicles	23.49
ATF[b]	x 10⁶ T	6.11	−0.37	fuel efficiency	5.74
FO[c]	x 10⁶ T	0.40	−0.02	(c) 6% lower volume	0.38
Oil	x 10⁶ T	43.84	−11.07	of transport	32.77

[a]Measures (a), (b), and (c) apply collectively to the petroleum-derived sources.
[b]Alternative transportation fuel.
[c]Fuel oil.

Agricultural Sector

Even though agriculture supports about 80 percent of India's population, it accounts for only about 6 percent of the total inanimate energy and 11 percent of the total commercial energy used in the economy.

Energy is used in the agricultural sector for operations such as land preparation, water lifting, harvesting, and so forth. Traditional agriculture formerly was carried out with only animate sources of energy—in other words, with human beings and draft animals—but electricity and oil have begun to play an increasingly important role during the past three to four decades. The agricultural sector's energy matrix for 1978 is shown in Table 14-12 and Figure 14-34.

Table 14-12. **1978 Energy Consumption Matrix for the Agricultural Sector, by Activity and Source[a]**

Source	Pumpsets	Tractors	Total
Electricity	11.95	—	11.95[b]
Diesel oil	14.32	5.04	19.37[c]
Total	26.27	5.04	31.32

[a]In MTCR.
[b]Equivalently, this total would be 11.95 billion kilowatt-hours.
[c]Equivalently, this total would be 2.92 million tonnes of oil, of which 2.16 million is attributable to pumpsets and 0.76 million to tractors.

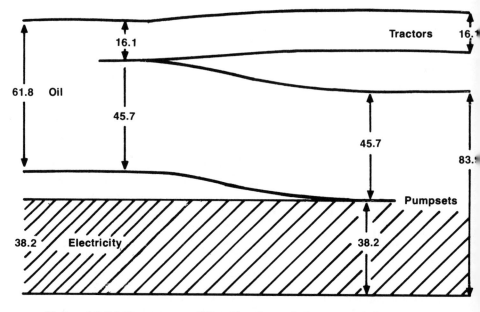

Figure 14-34. **Percentage Distribution of Commercial Energy in the Agricultural Sector**

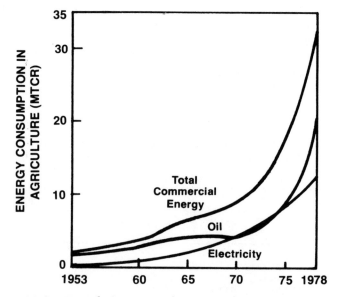

Figure 14-35. **Trends in Agricultural Energy Consumption**

The total inanimate energy used in the agricultural sector is primarily derived from two sources: oil (diesel fuel), 61.8 percent; and electricity, 38.2 percent. Of this total energy, water lifting with pumpsets consumes 83.9 percent with the remaining 16.1 percent going for diesel tractors.

The trends in this sector's consumption of electricity and oil are shown in Figure 14-35, from which it can be seen that during the past five years oil consumption has been increasing at a faster pace than electricity. This is primarily because oil is used for both pumpsets and tractors, and, as shown in Figure 14-36, although the growth rate in the number of electrical and diesel pumpsets has been approximately the same over the past three decades, the growth rate in the number of tractors has increased significantly over the past five years. As a result of the increased number of these machines, agricultural oil consumption has grown substantially, as can be seen in Figure 14-37.

The RLF has been made on the basis of these trends and other

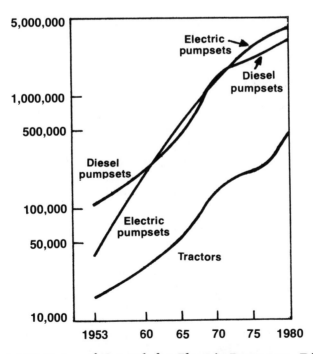

Figure 14-36. **Rates of Growth for Electric Pumpsets, Diesel Pumpsets, and Tractors**

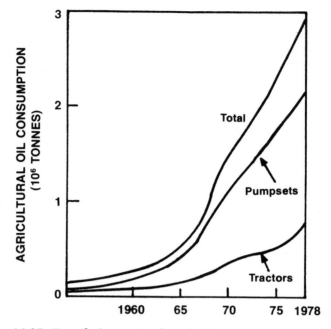

Figure 14-37. **Trends in Agricultural Oil Consumption**

factors such as the area sown, the nature of the crops, the ground water potential, the extent of irrigation and agricultural mechanization, and so forth. The RLF for numbers of pumpsets and tractors is shown in Figure 14-38 and the corresponding oil and electricity consumption in Figures 14-39 and 14-40 respectively. The numbers of electric pumpsets, diesel pumpsets, and tractors in the year 2000 are respectively forecast to be 3.1, 1.6, and 5.3 times the present numbers and the associated electricity and oil consumption represents an increase by a factor of 1.4 and 3.0 respectively.

Some reductions in the RLF values are possible by improving the efficiency of tractors and electric and diesel pumpsets. Through these improvements it is anticipated that a 15 percent savings in the forecasted energy consumption of electric and diesel pumpsets can be realized, with a 10 percent savings in the case of tractors. In addition, animal power can be utilized more fully, and thus a 15 percent savings of diesel fuel can be achieved. Notwithstanding these measures, the overall reduction in oil and electricity consumption would be only 26 percent and 15 percent respectively, as shown in Figures 14-39 and 14-40.

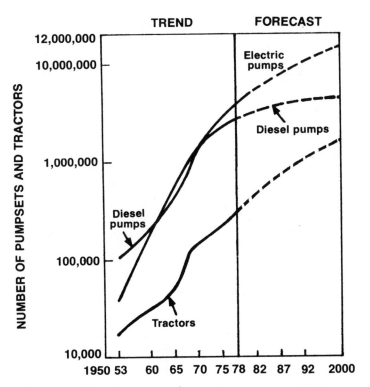

Figure 14-38. **Forecast for Numbers of Pumpsets and Tractors**

IMPLICATIONS OF THE REFERENCE LEVEL FORECASTS FOR 2000

The RLFs for the household, industrial, transportation, and agricultural sectors can now be assembled into a total energy RLF for 2000. The result is shown in Table 14-13.

A better idea of the implications of such a forecast can be obtained by comparing it with present consumption levels. Table 14-14 shows that the year 2000 RLF corresponds to about 5.6, 4.5, 3.4, and 2.7 times the 1978–79 consumption of electricity, coal, oil, and total commercial energy respectively.

Such a forecast constitutes a frightening prospect, because even the present levels of consumption are being met only with the greatest difficulty. In fact, there is currently a crisis with respect to every single energy source—electricity, coal, oil, and noncommercial energy. This crisis has been well documented;[3] hence, only the briefest reference will be made here.

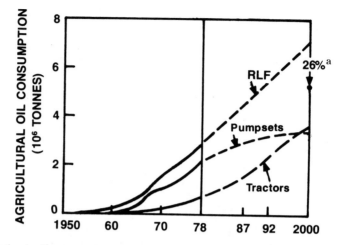

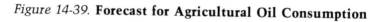

^aReduction in oil consumption possible with improved mechanical efficiency and use of animal power.

Figure 14-39. **Forecast for Agricultural Oil Consumption**

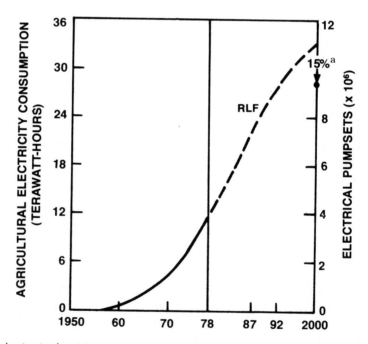

^aReduction in electricity consumption possible with improved mechanical efficiency.

Figure 14-40. **Forecast for Agricultural Electricity Consumption**

Table 14-13. **Reference Level Forecast for 2000, by Sector and Source**[a]

Source	House-hold	Indus-try	Trans-port	Agri-culture	Others	Total	CE[b]	CE & NCE[c]
Electricity	35.8	350.0	8.7	33.0	43.5	471.0	37.3%	32.1%
Coal	24.0	273.5	3.4	—	7.1	308.0	24.4%	21.0%
Oil	105.7	40.3	285.0	46.2	5.1	482.3	38.2%	32.9%
Commer-cial energy	165.5	663.8	297.1	79.2	55.7	1261.3	100.0%	86.1%
(%)	13.1%	52.6%	23.6%	6.3%	4.4%	100.0%		
Noncommer-cial energy	163.5	40.9	—	—	—	204.4		13.9%
Total	329.0	704.7	297.1	79.2	55.7	1465.7		100.0%
(%)	22.4%	48.1%	20.3%	5.4%	3.8%	100.0%		

[a]In MTCR.
[b]Source as a percentage of total commercial energy.
[c]Source as a percentage of total commercial and noncommercial energy.

Table 14-14. **Reference Level Forecast of Consumption in 2000 Compared with 1978 Consumption**

Source	Unit	1978 consumption	2000 RLF	2000 RLF ÷ 1978 consumption
Electricity	x 10⁹ kwh	84.2	471.0	5.59
Coal*	x 10⁶ T	68.8	308.0	4.48
Oil*	x 10⁶ T	21.7	74.2	3.42
Total	MTCR	544.3	1,465.7	2.69

*Excludes coal and oil used for (a) electricity generation and (b) nonenergy purposes, e.g., as feedstock.

In 1978–79, for instance, there was about 10 percent too little electricity to meet demand. This led to production losses in the factory sector which have been valued at 25 billion rupees (i.e., about 3.125 billion dollars) and to underutilization of capacity in many electricity-intensive industries—the capacity utilization in the aluminum, fertilizer, and caustic soda industries was only 69 percent, 72 percent, and 77 percent.

In 1978–79, demand for coal was 115 million tonnes but actual production was 11 percent less, or only 102 million tonnes. This created an "unprecedented scarcity" of coal[4] which had repercussions throughout the economy. Since about 30 percent of the country's coal output is utilized by thermal power stations (which produce about 55 percent of the total power), a shortfall in coal production leads to a reduction in electricity generation. But the coal mines themselves require electricity for their maintenance and operation, for example, for pumping out water. Hence, a drop in power generation due to a coal shortage causes a further drop in coal production—a vicious circle.

Another 22 percent of the total coal output is consumed by steel plants, which also require about 8 percent of the total electricity generated. Two coal-related factors thus combine to decrease steel production: (1) a direct shortage of coal for reductant purposes and (2) an indirect shortage of coal as fuel for thermal power plants supplying electricity to the steel plants. Decreased steel production leads to cutbacks in wagon manufacturing, and a decreased availability of wagons lowers the dispatches of coal from coal mines—another vicious circle.

Finally, about 12 percent of the total coal output is used by the railways, since in India 30 percent of rail traction comes from coal-fired steam locomotives. Hence, a drop in coal supplies produces a decrease in the rail transport of coal from mines to users; in other words, reduced coal dispatches—yet another vicious circle involving coal.

The inability of the railways to meet coal freight traffic demands creates a situation in which 8 percent of the coal is transported by trucks consuming diesel fuel. And, seeking to make up for the electrical power shortage, diesel sets generating captive power are used in industry and diesel pumpsets are used in agriculture. The net result is a rapid rise in diesel consumption, even though two-thirds of the country's oil is imported and oil prices are escalating.

Thus, the whole industrialized sector has been affected by the critical situation with respect to electricity, coal, and oil.

The nonindustrialized rural sector has also been affected. On the one hand, the poor progress in the electrification of rural *households* (as distinct from the electrification of villages) is leading to increasing consumption of kerosene for lighting. And on the other hand, since the only economically accessible cooking fuels for poor households are firewood, dung cakes, and agro-wastes, there is a rising demand for noncommercial sources of energy for rural cooking in the face of increasing competition for these same sources from brick kilns and from the kitchens of the urban poor. Furthermore this whole situation is being aggravated by diminishing firewood resources—the so-called other energy crisis.

In conclusion, therefore, there is a highly critical situation with respect to electricity, coal, oil, and noncommercial energy. In such a context, the outstanding role of the RLFs for 2000 is that they warn what will happen if current trends continue. Alternative realizable futures must be worked out.

A DEVELOPMENT-ORIENTED APPROACH TO ENERGY PLANNING

The conventional "wisdom" in such predicaments is to accept the sanctity of energy demand forecasts and the political economy underlying them and to prospect desperately for energy resources. That such an approach is neither feasible nor moral has been argued in detail elsewhere.[5,6] Suffice it to say here that this conventional growth-obsessed approach is based on (1) turning a blind eye to the nature of energy demand and the societal end uses of energy—that is, what energy is used for and which sections of stratified countries within a stratified world are the beneficiaries of this energy; (2) concentrating solely on the supply aspects of energy; (3) equating development with growth and viewing energy as an engine (oil-burning, of course!) of growth; and (4) considering energy as a sustainer and amplifier of global and national inequalities rather than a mechanism for realizing more just and equitable international and national economic, social, and political orders.

The alternative is a *development-oriented* approach to energy planning. In this approach, energy is turned into an instrument of development which is viewed as a socioeconomic process directed towards (1) satisfaction of basic human needs, *starting with the needs of the neediest*, (2) endogenous self-reliance, and (3) harmony with the environment.

The first step in such a development-oriented approach to energy is to examine whether the basic energy needs of the population, and particularly its deprived sections, are being met, and if not, to plan for this overridingly important goal. The next step is to scrutinize the various technological alternatives for satisfying these energy needs and to select those technologies which promote self-reliance in an environmentally sustainable manner without seriously retarding the pace of satisfaction of basic needs.

All this constitutes a major exercise of a kind infrequently discussed and rarely conducted. In the context of a task of such magnitude, all that can be attempted here is to take a few indicative steps to reveal the type of policies that need to be implemented as an integral part of development-oriented energy planning.

DEVELOPMENT-ORIENTED FORECASTS FOR 2000

In this section, development-oriented suggestions are made for each of the four sectors previously discussed. By way of introduction, it should be noted that many of the suggestions in this section have been briefly mentioned but not elaborated upon or quantified in the WGEP report.

Household Sector

From a development point of view, the important features of the RLFs up to 2000 for the household sector are as follows:

- Of the 128.3 million *rural* households (constituting 71 percent of the country's projected population of 921 million), 70 percent will still be cooking with noncommercial energy sources, which have low thermal efficiency, and lighting their houses with kerosene, which has a pathetic luminous efficiency level.
- The poorest 10 percent of the 51.7 million *urban* households will continue to use noncommercial energy for cooking and kerosene for lighting.
- Along with kerosene's continued use for lighting, its use for cooking can be expected to extend to 22 percent and 50 percent respectively of the rural and urban households, raising the total demand for kerosene to 13 million tonnes—several times the 1978 consumption level of 3.9 million tonnes.
- The availability of LPG to rural and urban households will be extended from a 1975 level of 0 percent and 5 percent respectively to 3 percent and 25 percent respectively in 2000. This will result in a 2000 consumption level of 3.3 million tonnes, compared with the 1978 consumption of 0.4 million tonnes.

Quite apart from the fact that this forecasted pattern may not even be feasible because of the problems associated with the availability and sustainability of noncommercial energy and kerosene, this pattern is totally inconsistent with development objectives, for a convenient cooking fuel and adequate illumination must be considered as basic developmental needs. A development-oriented energy plan for the household sector would aim at totally different goals that are primarily concerned with improving the quality of life in homes. The following is one possible set of goals:

(1) Every urban *and* rural household must be electrified so that all homes can be illuminated electrically.
(2) Every *rural* household must be provided with biogas cooking fuel so that (a) the pressure on forests can be reduced, (b) animal wastes can be diverted to fertilizer applications,

(c) agricultural residues can be used for more appropriate ends, and (d) the enormous time spent on gathering these noncommercial fuels[7] can be devoted to more lucrative or meaningful pursuits.

(3) The tendency in towns and cities has been to move irreversibly from firewood and dung cakes to charcoal to kerosene to LPG; hence, urban households must be supplied with a fuel as convenient as LPG—that is, a gaseous fuel. Sewage gas, which is now being wasted, is an obvious energy source but can meet only part of the cooking fuel requirements; hence, it has to be supplemented. Two possible supplements are (a) wood gas generated from forests grown for energy purposes around towns and cities and (b) in arid regions, coal gas produced from soft coke or coal.

(4) Kerosene must be *completely* replaced as a cooking fuel and illuminant.

(5) LPG must be *totally* diverted to socially more advantageous end uses such as feedstock for the fertilizer or petrochemical industries.

Some quantitative implications of these goals can be worked out by assuming a period of transition—for example, a ten-year

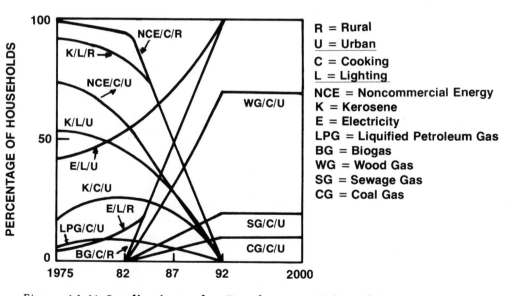

Figure 14-41. **Implications of a Development-Oriented Energy Plan for the Household Sector with Ten-Year Transition Period Assumed**

transition, as shown in Figure 14-41. The resulting estimates are shown in detail in Tables 14-15 through 14-18, and the aggregated requirements are shown in Figure 14-42. It turns out that the development-oriented forecast results in markedly lower requirements for kerosene and LPG compared with the RLF, and the former also produces a marginally lower requirement for electricity by 2000, although this requirement is 3 percent and 36 percent greater in 1987 and 1992 respectively because of the more rapid drive for household electrification. In addition, kerosene (40 percent of which is imported today) would be completely replaced in the development-oriented forecast.

Table 14-15. **Development-Oriented Forecast for Rural Lighting**

		1975	1982	1987	1992	2000
Total no. of households	(x 10⁶)	90.1	100.9	108.6	116.1	128.3
Electricity						
% of HH[a]		4.4	12.9	50.	100.0	100.0
No. of HH	(x 10⁶)	4.0	13.0	54.3	116.1	128.3
Ann. reqt.[b]	(x 10⁹ kwh)	1.2	3.9	8.1	17.4	19.2
Ann. reqt.	(MTCR)	1.2	3.9	8.1	17.4	19.2
Kerosene						
% of HH		91.4	84.0	50.0	—	—
No. of HH	(x 10⁶)	82.4	84.8	54.3	—	—
Ann. reqt.	(x 10⁶ T)	1.55	1.60	1.02	—	—
Ann. reqt.	(MTCR)	12.9	13.3	8.5	—	—
Other sources						
% of HH		4.2	3.1	—	—	—
No. of HH	(x 10⁶)	3.7	3.1	—	—	—

[a]Households.
[b]Annual requirement.

Industrial Sector

The RLFs for the industrial sector are based on growth trends and intensities of energy consumption—the latter figured in kilograms coal replacement per rupees value added. As stressed previously, the most striking features of the RLFs are the enormous demands for energy supplies involved, in particular for coal and electricity, when electricity and coal supplies and, by extension, the railway sector are already under severe strain.

A development-oriented approach should first try to define the type of society which is envisaged for the country. In particular, it should clearly indicate whether India can replicate the path of industrialization and consumption-obsessed lifestyles made possible in developed countries by the now bygone era of cheap energy. It should then sketch the pattern of industrialization essential for building the desired type of society.

What is required, therefore, is a specification of the end uses of the economy's industrial products and the quantities of these

Table 14-16. **Development-Oriented Forecast for Urban Lighting**

		1975	1982	1987	1992	2000
Total no. of households	(x 10⁶)	26.1	32.1	36.9	42.0	51.7
Electricity						
% of HH		42.1	53.0	71.5	100.0	100.0
No. of HH	(x 10⁶)	11.0	17.0	26.4	42.0	51.7
Ann. reqt.	(x 10⁹ kwh)	3.3	5.1	7.9	12.6	15.4
Ann. reqt.	(MTCR)	3.3	5.1	7.9	12.6	15.4
Kerosene						
% of HH		53.6	45.2	28.5	—	—
No. of HH	(x 10⁶)	14.1	14.5	10.5	—	—
Ann. reqt.	(x 10⁶ T)	0.40	0.41	0.30	—	—
Ann. reqt.	(MTCR)	3.31	3.40	2.46	—	—
Other sources						
% of HH		4.3	1.8	—	—	—
No. of HH	(x 10⁶)	1.0	0.6	—	—	—

Table 14-17. **Development-Oriented Forecast for Rural Cooking**

		1975	1982	1987	1992	2000
Total no. of households	(x 10⁶)	90.1	100.9	108.6	116.1	128.3
NCE						
% of HH		99.0	94.5	50.0	0	0
No. of HH	(x 10⁶)	89.2	95.4	54.3	0	0
Ann. reqt.	(x 10⁶ T)	194.0	207.0	102.0	0	0
Ann. reqt.	(MTCR)	184.0	197.0	97.0	0	0
Biogas						
% of HH		0	0	50.0	100.0	100.0
No. of HH	(x 10⁶)	0	0	54.3	116.1	128.3
Ann. reqt.	(x 10⁹ m³)	—	—	17.7	37.8	41.8
Ann. reqt.	(MTCR)	—	—	18.9	40.5	44.7
Other sources						
% of HH		1.0	5.5	—	—	—
No. of HH	(x 10⁶)	0.9	5.5	—	—	—

products necessary to sustain these end uses. It is in this context that alternative ways of achieving these end uses must be considered. In other words, the product mix of the economy must be thoroughly scrutinized and rigorously justified. The questions of production for what and production for whom must be answered in detail. In parallel with formulating a perspective on these funda-

Table 14-18. **Development-Oriented Forecast for Urban Cooking**

		1975	*1982*	*1987*	*1992*	*2000*
Total no. of households	(x 10⁶)	26.1	32.1	36.9	42.0	51.7
Kerosene						
% of HH		17.2	26.5	20.0	—	—
No. of HH	(x 10⁶)	4.5	8.5	7.4	—	—
Ann. reqt.	(x 10⁶ T)	0.9	1.6	1.4	—	—
Ann. reqt.	(MTCR)	7.5	13.3	11.6	—	—
NCE						
% of HH		73.2	58.1	28.5	—	—
No. of HH	(x 10⁶)	19.1	18.7	10.5	—	—
Ann. reqt.	(x 10⁶ T)	35.2	34.4	19.3	—	—
Ann. reqt.	(MTCR)	33.4	32.6	18.4	—	—
LPG						
% of HH		5.0	8.4	5.0	—	—
No. of HH	(x 10⁶)	1.3	2.7	1.8	—	—
Ann. reqt.	(x 10⁶ T)	0.27	0.55	0.37	—	—
Ann. reqt.	(MTCR)	2.24	4.57	3.07	—	—
Sewage gas						
% of HH		—	—	10.0	20.0	20.0
No. of HH	(x 10⁶)	—	—	3.7	8.4	10.3
Ann. reqt.	(x 10⁹ m³)	—	—	1.20	2.74	3.37
Ann. reqt.	(MTCR)	—	—	1.28	2.93	3.61
Producer gas						
% of HH		—	—	26.5	70.0	70.0
No. of HH	(x 10⁶)	—	—	9.8	29.4	36.2
Ann. reqt.	(x 10⁶ T firewood)	—	—	5.1	15.2	18.7
Ann. reqt.	(MTCR)	—	—	4.8	14.4	17.8
Coal gas						
% of HH		—	—	10.0	10.0	10.0
No. of HH	(x 10⁶)	—	—	3.7	4.2	5.2
Ann. reqt.	(MTCR)	—	—	6.5	7.4	9.1
Soft coke						
% of HH		4.6	7.0	—	—	—
No. of HH	(x 10⁶)	1.2	2.2	—	—	—
Ann. reqt.	(x 10⁶ T)	2.1	3.9	—	—	—
Ann. reqt.	(MTCR)	1.4	2.6	—	—	—

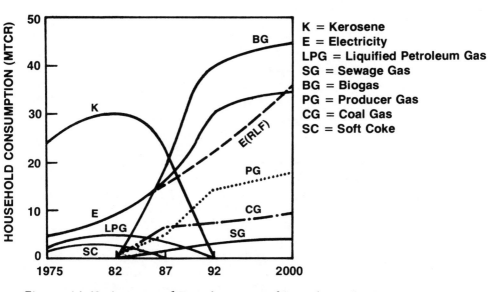

Figure 14-42. **Aggregated Requirements of Development-Oriented Energy Plan for Household Sector**

mental matters, there should be a detailed analysis of the current product mix of Indian industry from the standpoint of developmental needs.

The crucial industries needing scrutiny are steel, brick, textile, cement, fertilizer, and aluminum production. Of these, the most critical from the standpoint of energy is the steel industry. It currently consumes 21 million tonnes of coal and 5.5 billion kWh of electricity (23 percent and 6.5 percent respectively) and accounts for 14 percent of the freight tonnage carried by the railways. Since the shortages of coal, electrical power, and transport are on this order, a reduction in the steel industry's energy consumption and in steel demand would have beneficial impacts on the coal, electricity, and transport situations and perhaps could break the vicious circles mentioned previously. Thus, conservation in two senses—in the industry itself and in the use of its products—has to be a central objective of development-oriented energy planning for the industrial sector.

With regard to energy conservation in steel production, it must be noted that Japan uses half the energy per tonne of steel that India does, showing that there is ample scope for improvement in the industry. A detailed review is required, but the coal demand by the steel industry projected in the RLF for the year 2000 could be reduced at least 25 percent to 86.4 million tonnes. For energy conservation in the use of steel, a thorough study has to be made of classes of situations in which steel is used today when it is not needed at all (e.g., for single-story reinforced concrete constructions) and in which it is used to a greater extent than necessary because of poor design, as in machinery and equipment.

Similar conservation possibilities must be explored in the building industry. The brick and cement industries account for about 16 percent of the total energy used in the industrial sector; in addition, cement alone accounts for 6 percent and 3 percent of the freight (in tonne-kilometers) carried by rail and road respectively, with average leads of 717 and 286 kilometers in 1977–78. Yet, there are categories of buildings which do not require bricks and cement at all.

The products of the highly energy-intensive aluminum industry also merit scrutiny. A large fraction of aluminum is used for electrical conductors where it plays the key role of replacing copper, but only 3 percent of the aluminum product is used for cooking vessels, even though a widespread use of aluminum cooking pots in rural areas would drastically decrease fuel consumption.

Considerable study is required before the energy savings from all these conservation measures can be quantified so that precise development-oriented forecasts can be made. All that can be done in this paper is to consider the 25 percent reduction in the coal needs of the steel industry and to adopt the energy savings suggested by the WGEP, as shown previously in Table 14-17.

Transportation Sector

For the transportation sector, the two important features of the RLFs to 2000 are (1) a major increase in the *magnitude of goods transported*— the RLF for total rail and road freight traffic in 2000 is expected to be four times the 1976 level—and (2) an accentuation of the already major *shift from rail to road* so that the railroad's present share of 67 percent of goods transported and 39 percent of passenger traffic will become 44 percent and 30 percent respectively, leading to road transport consuming 30 million tonnes of diesel fuel in 2000 (or 40 percent of the total oil consumption projected for that year) compared with 4.3 million tonnes (or 23 percent of total oil consumption) in 1976. These features are related to some important developmental issues.

First, freight traffic currently has an unduly large volume—a volume which has been amplified by the uneven development of industry and agriculture. This unevenness results in large distances between surplus and deficit regions and between production and consumption centers. Table 14-19 gives an idea of the present state of affairs by showing that many commodities are moved in large quantities over large distances. The two well-known examples of this are food grains being carried from the north of the country to the south and cement being carried in the opposite direction.

Table 14-19. **Flows of Interregional Freight Traffic in 1977-78, by Commodity**

Commodity	Originating freight quantity (x 10⁶ T)	Freight traffic volume (x 10⁶ TKM)	Average lead (km)
Coal	69.3	46,332	669
Food grains	23.7	22,681	957
Iron ore	17.2	9,052	526
Mineral oil	17.9	9.676	541
Cement	15.4	9,452	614
Limestone & dolomite	6.9	2,516	365
Iron & steel	15.2	12,778	841
Fertilizers	11.6	9,213	794
Fruits & vegetables	7.9	4,206	532
Building materials	7.9	1,909	242
Stones & marbles	6.7	1,940	290
Wood & timber	6.6	3,723	564
Provisions & household items	6.9	3,826	554
Others	61.8	43,465	703
Total	275.0	180,769	657

Source: "Report of the National Transport Policy Committee," Government of India, Planning Commission (New Delhi: 1980).

A more uniform industrial and agricultural development with greater self-sufficiency of states and regions would result in much lower volumes of freight traffic (in tonne-kilometers) and therefore smaller leads (in kilometers) for the same quantity of freight (in tonnes). But as freight traffic decreases, energy consumption also decreases. Hence, the greater the self-sufficiency of districts, states, and regions, the lower the energy consumption for the transport of goods. The RLF for a fourfold increase in freight traffic thus is implicitly based on a perpetuation of the present unevenness of industrial and agicultural development.

Second, the RLF for the shift from rail to road implies a continued increase in the average lead distance for road transport and a continued decrease in the proportion of short-haul road traffic. Today, the average lead distance of road transport is 354 kilometers (km), or less than half of that for railways, which is 808 km. But this average lead conceals the fact that a few high-value commodities (particularly machinery and equipment) have comparatively longer leads of over 500 km, and 30 percent of interregional traffic moves over an average lead of 700 km. Furthermore, some low-value bulk commodities are now moving by road, coal being the most notorious example with 8 percent (69 million tonnes) of the total coal transported being carried by trucks. A major steel plant with a capacity of 1.5 million tonnes per year of steel was forced to transport by truck about 1 million tonnes of coal per year—about one-third of its total requirements—at an additional cost of 60 rupees (approximately 7.5 dollars) per tonne.

There is, however, a break-even distance below which road transport is more economical and above which rail transport is preferable. As shown in Table 14-20, the present leads for most commodities are *beyond* the break-even distances. This means that the diseconomies of moving goods by road beyond the economical distances are being borne by society—in fact, through increased oil import bills. Equally important is the fact that these break-even distances decrease with increasing fuel prices. This means that as diesel fuel costs increase, the freight traffic must shift correspondingly from road to rail. As the RLF warns, however, if there was an increasing shift in the opposite direction—that is, from rail to road—against a background of rising diesel fuel costs, it would mean that society would increasingly subsidize the trucking industry, 99.5 percent of which is owned by the private sector.

Table 14-20. **Originating Freight Quantities, Average Leads, and Break-Even Distances in 1977-78, by Commodity**

Commodity	Originating freight quantity (x 10⁶ T)	Average lead (km) Rail	Road	Break-even distance (km)ᵃ
Coal	69.26	691	408	106
Food grains	23.65	1278	277	130
Iron ore	17.23	529	96	106
Iron & steel	15.21	1101	371	128
Cement	15.20	717	286	116
Fertilizers	11.21	1039	267	107

Source: "Report of the National Transport Policy Committee," Government of India, Planning Commission (New Delhi: 1980).
ᵃBased on a 50% increase over 1979 diesel fuel prices.

A development-oriented approach to the transport sector must therefore address itself to two basic issues: (1) whether the present and projected volumes of goods and people transported are absolutely unavoidable or whether these transport needs can be satisfied in other ways that are not transport-dependent, and (2) whether the mix of transport modes (that obtains at any juncture or is forecasted) is the most optimum from the standpoint of resource use. These two issues have distinct considerations.

The first issue concerns fundamental matters such as the evenness of industrial and agricultural development; the nature of the goods being transported and which sections of society are consumers of these goods; and the purposes for which people travel and whether these purposes can be achieved by other means such as better communication. Unfortunately, the currently available information base is far too inadequate to deal with these matters, and they must therefore be deferred.

The second issue, namely, the optimal mix of transport modes, has been the main preoccupation of India's National Transport Policy Committee (NTPC), which published its comprehensive report in May, 1980. This report[8] is based on the excellent study of Rail India Technical and Economic Services (RITES) on traffic flows by commodity, resource costs, and break-even distances for various transport modes. Using break-even distances calculated on the basis of a 50 percent increase in diesel fuel prices beyond the 1979 value, and assuming that 75 percent of the traffic moving by road *beyond* break-even distances will be shifted to rail, a modal split of 75:25 for rail and road was suggested. The NTPC report also made careful projections of freight and passenger traffic by several methods and came out with the estimates of 650 BTKM and 1320 BPKM respectively, figures which correspond to 30 percent and 25 percent reductions in the RLF levels. Finally, with the aid of energy norms, the NTPC report forecasts 14 million tonnes of diesel fuel in 2000 compared with the RLF value of 30 million tonnes, a reduction of about 50 percent.

These are indeed drastic reductions—but it is possible to go much further. For instance, instead of about $20 per barrel, as apparently assumed in the NTPC report, estimates can be made on the basis of $50 or even $100 per barrel. These estimates will push the break-even points even lower. Further, one can assume that in such crisis situations, 90 percent or even 95 percent of traffic potentially allocable to rail can in fact be shifted away from road to rail. Under such circumstances, the oil consumption may become as low as 10 million tonnes, which is one-third the RLF value.

When road traffic becomes overwhelmingly short-haul in character, the advantages of liquid fuels are increasingly eroded, and the World War II solution of running trucks and buses with charcoal—that is, with producer gas—becomes more and more valid. This option would reduce diesel fuel consumption even further. For instance, assuming that road transport accounted for only 10 percent of the projected freight traffic in 2000 and that all this road transport was fueled by producer gas, it would turn out that 21.65 million tonnes of wood or 0.87 million hectares of forest grown for energy would be required to replace completely the 2.95 million tonnes of diesel fuel which would have been used for the same transport task.

Agricultural Sector

The RLFs to 2000 for the agricultural sector concern themselves exclusively with the numbers of pumpsets and tractors and their associated energy needs. If present trends continue, there will be an alarming increase in diesel fuel and electricity consumption, as shown in Figures 14-39 and 14-40, due to the respective 305 percent, 163 percent, and 528 percent increases in the number of electric pumpsets, diesel pumpsets, and tractors, as shown in Figure 14-38. The 2000 RLF for diesel fuel consumption by pumpsets and tractors is 7.1 million tonnes, which is three times the present consumption in the agricultural sector and one-third the 1978 total consumption. Before wondering where this oil will come from and what its costs will be, it is necessary to identify the present and prospective beneficiaries of these energy inputs.

It is well known that there are grave inequalities in India's land ownership pattern, as in most developing countries. For instance, in Punjab (one of the most agriculturally advanced states in India) the pattern of land ownership is as shown in Figure 14-43.

It is no coincidence, therefore, that there appears to be strong correlation between the number of pumpsets in a state and the number of large land holdings. Thus, in Punjab, Haryana, and Uttar Pradesh—the "green revolution" states—the number of pumpsets (P) appears to be correlated to the number of holdings above 5 hectares (N > 5) as follows:

P = 1.1041 + 0.3658 (N > 5)
Correlation coefficient = 0.9768

In the states of Bihar, Gujarat, Karnataka, and Maharashtra, the

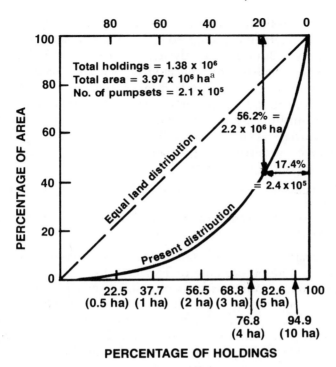

[a]Hectares.

Figure 14-43. **Land Ownership Pattern in the State of Punjab**

coefficient is 0.9598 for the correlation between pumpsets and holdings above 10 hectares; in Andhra, Orissa, and Rajasthan, 0.9859 for holdings above 20 hectares. Similarly, there seems to be a correlation between tractors and large holdings.

This search for correlations has been stimulated by a lack of awareness about the size of land tracts held by pumpset owners and tractor owners. But it seems self-evident that those who own large holdings are predominantly the ones who can afford electric or diesel pumpsets and tractors. The percentage of such large landowners may be small, but they own a large percentage of the total land. In Punjab, for instance, the percentage of land holdings above 5 hectares is 17.4 percent, but these holdings account for 56.2 percent of the total land and thus produce a major portion of the agricultural output.

This is not the context for considering whether energy planning for agriculture should cater solely to these large landowners—as has been the trend—or whether their contributions to society are substantial enough to justify their consumption of oil and electricity. But it is certainly sensible from the standpoint of a need-oriented, self-reliant, and environmentally sustainable development strategy to insure that (a) the consumption by landowners of scarce and, particularly, imported nonrenewable energy sources is minimized and (b) the use of locally available renewable energy sources is maximized. Thus, the main thrusts of a development-oriented energy plan for the agricultural sector would involve the following:

(1) the cessation of the use of oil as diesel fuel for pumpsets and tractors, and

(2) the stabilization at the present level of the use of *centrally produced* electricity for pumpsets.

Fortunately, there are technological options for achieving these goals. Diesel—a liquid fuel with an extensive distribution system—is most convenient for mobile power (e.g., vehicles), but even this convenience applies primarily to long-distance movements. For stationary equipment (e.g., pumpsets) and for mobile equipment (e.g., tractors) operating over short ranges of, say, 15 kilometers from a base, the advantages of liquid fuels are not fully utilized, and solid or gaseous fuels are almost as convenient. In fact, both producer gas (wood gas or charcoal gas) and biogas can be used for pumpsets and for tractors.*

The objective therefore includes three targets. They are:

(1) to run pumpsets and tractors only with biogas or producer gas,

(2) to utilize animal energy more efficiently and effectively for plowing and water lifting, and

(3) to limit electrical pumpsets powered by centrally produced electricity to the numbers that have already been connected to centralized generating stations via the grids.

Assuming a transition period of, say, ten years, the quantitative implications of these targets can be estimated. For instance, with the pattern of increase of water-lifting devices shown in Figure 14-44, the associated consumption of diesel and electricity would be reduced as in Figure 14-45. Similarly, by running tractors with producer gas or biogas instead of diesel fuel, the diesel consumption would be reduced as in Figure 14-46 without reducing

---

*Little attention has been devoted in this paper to alcohol, primarily because of the grave shortage of good agricultural land in the country. If nonagricultural land can be used for alcohol, it must be considered a serious candidate.

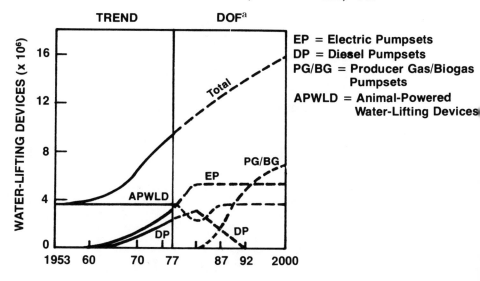

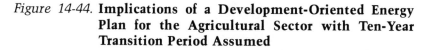

ᵃDevelopment-oriented forecast.

Figure 14-44. **Implications of a Development-Oriented Energy Plan for the Agricultural Sector with Ten-Year Transition Period Assumed**

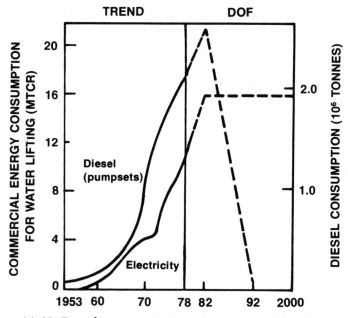

Figure 14-45. **Development-Oriented Forecast for Commercial Energy Consumed by Water Lifting**

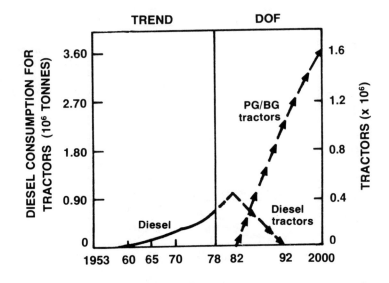

Figure 14-46. **Development-Oriented Forecast for Diesel Fuel Consumed by Tractors**

the RLF number of tractors. In short, the RLF diesel fuel consumption of 7.1 million tonnes can be completely avoided.

Thus, the case of the agricultural sector is parallel to that of the other sectors. There, too, the development-oriented approach leads to major reductions in the consumption of commercial energy sources without affecting the energy-utilizing tasks envisaged by the RLF.

Comparison of Development-Oriented and Reference Level Forecasts for 2000

It is now possible to synthesize the development-oriented forecasts (DOFs) for the various sectors into a composite picture that can be compared with the RLFs. This comparison is presented in Table 14-21 in MTCR units. The DOF values for commercial energy and for total energy (which includes noncommercial energy) are 63 percent and 60 percent respectively of the RLF values; that is, less than two-thirds of the RLFs. Table 14-22 provides the same information in original units and also gives the DOFs as percentages of the RLFs.

Table 14-21. **Comparison of Energy Consumption Levels in Development-Oriented and Reference Level Forecasts for 2000, by Sector and Source**[a]

Source		House-hold	Indus-try	Trans-port	Agri-culture	Other	Total
Electricity	-DOF	34.6	284.0[b]	12.8[e]	16.2	43.5[c]	391.1
	-RLF	35.8	350.0	8.7	33.0	43.5	471.0
Coal	-DOF	9.1	209.4[d]	3.2[e]	—	7.1	228.8
	-RLF	24.0	273.5	3.4	—	7.1	308.0
Oil	-DOF	—	21.1[b]	152.8[e]	—	5.2	179.1
	-RLF	105.7	40.3	285.0	46.2	5.2	482.3
Commercial	-DOF	43.7	514.5	168.8	16.2	55.7	799.0
energy	-RLF	1665.5	663.8	297.1	79.2	55.7	1261.4
Noncommer-	-DOF	17.8	—	—[f]	44.1[g]	—	61.9
cial energy	-RLF	163.5	—	—	—	—	163.5
Total	-DOF	61.5	514.5	168.8	60.3	55.7	860.9
	-RLF	329.0	663.8	297.1	79.2	55.7	1424.9

[a]In MTCR.
[b]From WGEP Report.
[c]RLF values.
[d]Apart from reductions in the WGEP Report, the coking coal requirements in the steel industry have been reduced as specified in this paper.
[e]From National Transport Policy Committee Report.
[f]Possibility of using producer gas (generated from wood) not considered here.
[g]It is assumed that pumpsets run on producer gas (generated from wood) and tractors on biogas.

Table 14-22. **Aggregate Comparison of Energy Consumption Levels in Development-Oriented and Reference Level Forecasts for 2000**[a]

Source	Unit	DOF	RLF	DOF as % of RLF
Electricity	x 10⁹ kwh	391.1	471.0	83%
Coal	x 10⁶ T	228.8	308.0	74%
Oil	x 10⁶ T	24.4	74.2	33%
NCE	x 10⁶ T	65.2	172.1	38%

[a]In original units.

Three major points emerge from these comparisons. These are detailed immediately below.

(1) The DOFs for electricity and coal are *not* significantly lower than the RLFs; in fact, the reductions are only 17 percent and 26 percent respectively. But this need not be a cause for disappointment. In the case of both electricity and coal, the industrial sector

accounts for the bulk of these energy sources' consumption—in fact, 73 percent and 92 percent in the DOF and 74 percent and 89 percent in the RLF—and the industrial sector was the one sector in the development-oriented approach which could not be explored and quantified because of inadequate information. It is well known, however, that there are enormous possibilities for energy conservation in this sector. Within particular industries, there is tremendous scope for reducing energy consumption by minimizing wastage, designing more efficient energy-utilizing equipment, exploiting cogeneration possibilities, using renewable sources (e.g., solar energy for heating or preheating), and so forth. And of equal or greater importance is the potential for conservation in the use of products; that is, utilizing industrial products to advance genuine development and not for conspicuous consumption or unnecessary purposes. Until these conservation opportunities are investigated throughly and quantified, the DOFs for industry given in Table 14-21 are quasi-RLFs for all practical purposes. (In fact, all the soft-option technical fixes listed at the end of the discussions of the four sectors' RLFs result only in marginal reductions of these RLFs.)

(2) In contrast to electricity and coal, the DOFs for oil and noncommercial energy, which are only about one-third the RLFs, demonstrate what can be achieved with a development-oriented approach to energy. The most striking feature of the DOF for oil is that it turns out to be less than the current consumption. In fact, as pointed out above, if road transport can be almost wholly restricted to small leads—for example, less than about 150 kilometers—and if buses and trucks for these small distances can be run on renewable fuels—for example, producer gas from wood or charcoal—oil consumption for energy purposes can be largely eliminated or at least brought within the reach of indigenous production, which is currently about 12 million tonnes. If essential to do so, oil can then be used for nonenergy purposes, for example, as feedstock. Thus, the development-oriented forecast can become the basis of a ten-year, 1982–92 plan for drastically reducing or even removing India's dependence on imported oil. These imports have been estimated to constitute about 60 percent of the country's export earnings in 1979–80, and as this percentage is bound to increase with increasing oil prices, it seems that a development-oriented approach is the only hope.

(3) Fortunately, this developmental orientation also shows that the key to a safe energy future is the provision of an alternative cooking fuel (e.g., biogas) to the poorest households. At present,

these approximately 95 million households (constituting 80 percent of India's population) burn about 195 million tonnes of noncommercial energy, including about 130 million tonnes of firewood, in cooking stoves that operate at efficiencies of less than 10 percent. Once the cooking fuel requirements of these households are satisfied, the firewood they would have used becomes available for alternative uses. Now, with respect to these alternative uses (e.g., wood gas for urban cooking, producer gas for pumpsets, etc.), it turns out that (a) they demand much less energy than cooking (since every household cooks, but only about 12 percent of the rural households would have pumpsets according to the RLF for 2000), and (b) they can be achieved with greater efficiency than cooking. The combined effect of these two factors is that despite the provision of more sophisticated cooking fuels to the poor, the net consumption of firewood becomes only one-third the RLF demand in 2000. In fact, the 2000 DOF for firewood is only about *half* the present consumption. Thus, by eliminating the poor households' need to burn firewood as cooking fuel, the crucial condition for the success of energy forestry programs—a major reduction in pressure on the country's firewood resources—would be created.

DEVELOPMENT-ORIENTED ENERGY POLICIES

It is now necessary to list the policies that must be adopted and implemented to transform the development-oriented approach into a reality. These are listed below for each sector.

Household Sector

Among other things, the major development-oriented policies for the household sector would include the following:

- the electrification of every rural and urban household so that every home can have electric lights;
- the supply of electricity to households on a first priority basis either from grids linked to centralized electricity-generating stations or from decentralized generators, depending upon the relative economic and social costs and benefits;
- the provision to all rural households of biogas cooking fuel produced from livestock and other fermentable wastes converted either at large village-size or sub-village-size biogas plants in compact settlements or at family-size plants in settlements with low housing densities;

- the supply to all urban households of a gaseous cooking fuel which is a mix of sewage gas (produced from city and town wastes), wood gas (generated from forests grown for energy purposes in green belts around cities and towns), and coal gas—the composition of the mix to be dependent upon these resources' local availability;
- the complete halt of the supply of kerosene to homes for cooking and lighting, with the supply cessation coming into force after the households have been provided with electricity and gaseous cooking fuel; and
- the total diversion of LPG to the fertilizer and petrochemical industries, but, as above, only after providing an alternative cooking fuel to households.

In addition, soft-option policies are necessary. These would include:

- the improvement of lighting devices' efficiency,
- the manufacture, supply, and installation of these improved lighting devices,
- the improved design of cooking stoves, especially for gaseous cooking fuels, and
- the manufacture, supply, and installation of these improved cooking stoves.

Industrial Sector

A precise listing of development-oriented policies for the industrial sector is not possible because of inadequate analysis of this sector. As stressed previously, a detailed exercise has to be carried out to scrutinize the product mix of industry, to identify the end uses of these industrial products, to examine whether these end uses can be achieved with less energy-intensive products, and to determine the volumes of industrial production needed for end uses essential to development. In parallel, it is important to collate the studies that have already been carried out on measures for energy conservation within industries, particularly in the production of steel, bricks, textiles, cement, fertilizers, and aluminum, and to formulate policies for the reduction of energy consumption in industry.

Pending the completion of the studies mentioned above, some immediate steps are required. These include:

- regulating the use of furnace oil and conversion to coal as a heating source,
- improving capacity utilization in industries, and

• reducing the intensity of electricity use—that is, the quantity of electricity used per units of value added.

The steel industry is in a special category—it is a vital factor in the vicious cycles aggravating the present energy crisis in the country. Hence, policies must be enunciated to achieve the following:

• a substantial reduction in the coal requirements of the steel industry so that low coke rates like those achieved in Japan can be realized here, and
• a reduced level of power consumed by the steel industry.

Transportation Sector

The most important policy determining energy consumption in the transportation sector concerns the industrial and agricultural sectors. This is the policy of promoting and realizing a far greater self-sufficiency by different regions and much more evenness of industrial and agricultural development than obtains today. Centralized production for national markets may have been in the private interests of enterprises in the erstwhile era of cheap energy, but the energy price that has to be paid for such a national reach is too high to be continued except where the region's resource endowments are incompatible with intraregional production. In other words, there must be a deliberate policy to reduce to the essential minimum the transport of goods and passengers.

Corresponding to this irreducible minimum of goods and passenger traffic, the next most important cluster of policies revolves around the optimum mix of transport modes. Some of the crucial policies pertaining to this issue are as follows:

• Given transport modes should be restricted to the regimes of break-even distances within which they are the most economical. In particular, truck and bus transport must be compelled to keep within ranges where they are optimum from a national point of view.
• All the transport modes, even those operating optimally within their given distance regimes, should use the socially most desirable fuel—in particular, electric traction for the railways and renewable sources (e.g., producer gas) for road transport.
• Vastly increased investments for railways should be made available, especially for multiple tracking and electrification.
• Coastal shipping should be developed to take advantage of the peninsular character of India.

- The major rivers should be utilized for inland water transport.
- Oil-derived fuels should be priced realistically.

In addition, it is necessary to pursue softer option policies for:

- improving the fuel efficiency of transport vehicles,
- developing alternative renewable fuels, and
- utilizing animal-drawn and pedal-powered vehicles in conditions where they are most appropriate.

Agricultural Sector

For the agricultural sector, the main development-oriented policies are as follows:

- the establishment, in rural areas, of forests grown for energy purposes;
- the utilization of these forests to provide alternative renewable fuels (e.g., producer gas) for pumpsets;
- the allocation of only biomass-derived fuels (i.e., wood, charcoal, methanol, or ethanol) for tractors;
- the prohibition of diesel fuel use for pumpsets and tractors after the above alternative fuels are supplied;
- after alternative renewable fuels or locally produced electricity is made available, the ban on further connections of pumpsets to grids bringing electricity from centralized generating stations; and
- the utilization of animal- and pedal-powered devices and of solar and wind energy in those agricultural operations where they are economical.

It is also essential to follow policies that emphasize conservation—for example, conservation in the use of water for agriculture and in the use of agricultural equipment such as pumpsets and tractors.

CONCLUSIONS

Notwithstanding the promise held out by the development-oriented forecasts and the attractiveness of the policies necessary for the realization of these forecasts, there may be major obstacles to their acceptance.

An internal obstacle may arise from those groups within the country that have been responsible for the past and present trends in energy consumption. In the case of oil consumption, for instance, the main groups are those using diesel fuel, furnace oil, and kerosene, which account for 35 percent, 23 percent, and 14 percent respectively of the total oil used. Taken together, these groups use over two-thirds of the total oil consumed. The diesel consumers consist of three main categories—road transport operators, pump-set owners, and industries that have set up captive power generation facilities. Furnace oil consumption is mainly due to industrial heating and to thermal power plants, and as pointed out previously, kerosene is used mainly in the household sector for lighting and cooking.

Many of these groups, however, consume oil only because other energy sources are either not available or too scarce. For instance, households use kerosene because of the lack of electricity for lighting and the lack of convenient fuels for cooking; industries use furnace oil because of coal availability problems and set up captive power generation units because of electricity shortages; and landowners use diesel pumpsets because of power shortages or the absence of electrical connections. All these groups would resist a reduction in oil supply *only if alternative fuels were not made available.* Since, however, the development-oriented policies envisage not only the supply of these alternative fuels but also increased energy consumption by all these consumers, it is unlikely that the consumers will offer major resistance.

The real resistance, therefore, is likely to emerge from road transport operators because development-oriented policies demand a massive shift of long-haul goods and passenger traffic away from them to the railways. Whether they can be made to surrender this traffic is an issue which will be settled in the political arena, but it is likely that in the conflict between national and road transport interests, the former will prevail in the escalating crisis.

Obstacles external to the country are also likely because, however tempting the development-oriented policies, there is the stupendous problem of the transition period. The shift to alternative energy sources and policies will take a finite time—in fact, the development-oriented forecasts envisage a ten-year period from 1982 to 1992. *Before and during this period, the country has to survive with present trends and energy systems, and in particular with an oil demand which cannot be reduced significantly for at least a decade.* How will the country's energy needs, and especially its oil requirements, be met during this transitional period? Will the industrialized countries help the oil-importing countries through the transition?

If international organizations, and particularly industrialized countries, become an obstacle to the vital transitional oil supplies of the oil-importing developing countries, the repercussions of inadequate oil over the next decade will be catastrophic. The transition to a development-oriented energy future will be totally distorted or even thwarted. Development itself may be undermined, and there will be little choice for oil-importing developing countries except to struggle in all possible ways for the oil needed, first, to survive, and then, to achieve the transition to an oil-free future.

It is in this context that there is a great deal of justified suspicion that the real aim of international advice regarding energy conservation in developing countries is to insure that, in a situation of dwindling oil supplies, these poor countries reduce their oil demand and thus guarantee for the industrialized countries a greater share to sustain their energy extravagance. An equally strong suspicion centers around the developed world's proliferation of interest in the Third World's rural energy problems. Is it the purpose of this interest to accumulate information to be used at the bargaining tables of international forums to point out to developing countries that their real priorities must be oriented towards rural energy problems (e.g., firewood stoves) rather than towards insuring sufficient oil supplies for survival during a period of transition? The answer to all this apparently altruistic concern is to note that the greatest contribution which developed countries can make to help solve global and national energy problems is to demonstrate in their own societies that reductions in the consumption of energy, particularly oil, can be achieved without impairment of the quality of life.

In conclusion, it must be stressed that these factors—(1) the rural firewood crisis due to the excessive use of noncommercial energy by the poor, (2) the national diesel fuel crisis due to the major encroachment of road transport into the domain of more energy-efficient transport modes such as the railways, and (3) the international oil crisis due to the reckless consumption of oil by the developed countries—all illustrate a well-known environmental principle:[9] whenever grave economic inequalities exist, there is an irrational and wasteful use of natural resources at both extremes of the economic spectrum. The poor undermine the resource base in order to survive—they cannot avoid this—and the rich exhaust the resource base in order to preserve their affluence—they do not want to change this. The resolution of these crises lies above all in the reduction of inequalities within and between nations and is therefore the crux of national and international development. Thus, the solution of the energy problem lies in a commitment to a need-based development approach. The slogan must be "Look after the needs of the neediest, and energy will look after itself!"

Acknowledgments

The author learned a great deal through his association with T. L. Sankar, T. R. Satish Chandran, and N. B. Prasad in the Planning Commission's Working Group on Energy Policy, and he therefore wishes to express his gratitude to them. The information shortage which exists in the field of energy could not have been tackled without the generous help of T. R. Satish Chandran, S. N. Roy, M. de P. Miranda, and above all Narottam Shah and his Centre for Monitoring Indian Economy—to all these friends must be extended the author's sincere thanks. The author must also thank K. S. Jagadish and C. R. Prasad for willingly acting as sounding boards through the long process of writing this paper. Finally, it must be acknowledged that this paper could not have been written without the data contained in the reports of the Working Group on Energy Policy and the National Transport Policy Committee.

NOTES

1. A. K. N. Reddy, "Energy Options for the Third World—Part I," *Energy Management* II No. 4 (1978), pp. 225–31; and "Energy Options for the Third World—Part II," *Bulletin of Atomic Scientists* (1978), pp. 28–33.

2. "Report of the Working Group on Energy Policy," Government of India, Planning Commission (New Delhi: 1979).

3. N. Shah, "Power, Coal and Oil: Review of 1978–79 and Prospects," Centre for Monitoring Indian Economy (Worli, Bombay: 1979).

4. Ibid.

5. A. K. N. Reddy and K. Prasad, "Technological Alternatives and the Indian Energy Crisis," *Economic and Political Weekly* XII (1977), pp. 1465–1502.

6. A. K. N. Reddy, as cited in note 1 above.

7. N. H. Ravindranath, H. I. Somashekar, R. Ramesh, A. Reddy, K. Venkatram, and A. K. N. Reddy, "The Design of a Rural Energy Centre for Pura Village: Part I—Its Present Pattern of Energy Consumption," *Employment Expansion in Indian Agriculture* (Bangkok: International Labour Organisation [ILO], 1979).

8. "Report of the National Transport Policy Committee," Government of India, Planning Commission (New Delhi: 1980).

9. A. K. N. Reddy, "Technology, Development and the Environment—a Re-appraisal" (Nairobi: United Nations Environment Programme [UNEP], 1979).

CHAPTER FIFTEEN

Selected Comments

KERRY McHUGH

It has been clear in looking at the papers and listening to the speakers that the energy problem has diverse aspects both within countries, where it spans different social groups, and internationally. And we have available to us a range of responses, from synfuels to ethanols to energy use efficiency. In all this diversity of problem and response, we are faced with choices, as Dr. Alonso said in his introduction to this session. In making those choices, the issue that is perhaps common to them all is the fact that we really are talking about making the best use of what is obviously and rapidly becoming a very limited range of resources. These limitations exist whether the resource is oil or its possible substitutes. Professor Reddy indicated that technology is an instrument in overcoming these limitations, and I put it to the Symposium that the other major instrument is the one that Dr. Sternlight mentioned throughout his paper: the question of pricing and the need to ensure its appropriateness.

When energy choices are made, such decisions ought to be informed by as certain a knowledge as possible about the real cost of these rapidly depleting resources. This is true, I think, whether or not in the end the decision comes down to value judgments of one sort of another. Realistic pricing is absolutely vital to that process, and it cannot be regarded as a soft path or a soft option. Particularly over the last couple of years, we have seen examples throughout the industrialized world of the hard choices faced by politicians in deciding to adopt appropriate and realistic pricing policies for energy—the Canadian experience is a good example and the current Australian elections provide another example. Nevertheless, unless a realistic pricing policy is adopted, it seems to me that our people will be fooled with "soft options" which in fact aren't soft and with paths which in the end lead us into a situation of endless difficulties.

MARCELO ALONSO

I think it's very good that you emphasize the matter of choice. If I may add an additional remark: at the same time you have to know how to choose, and that is a very crucial point.

INGI R. HELGASON

I wish to thank the two speakers in this session for their excellent presentations. Professor Reddy's report from India was for me a startling one, indicating the seriousness of India's situation with regard to power, as well as the relevance of international relations to the problems of India.

As a legal adviser to Iceland's Ministry for Industry and Power, especially in the field of utilization of our hydro and geothermal power, I am representing the ministry at the Symposium. In that official capacity, and also as chairman of the Central Bank of Iceland's board of directors, I came here more to listen than to submit scientific papers or give speeches. Let us hope it will be the other way around in Symposium II.

Professor Sadli said at the conclusion of his paper that the sun never sets for these energy conferences. This comment reminded me of the geographical, or should I say astronomical, fact that in my country during the summer the sun does not go down. I can read Professor Sadli's report in bright daylight at midnight next June. You all know that we in Iceland constitute a small nation of only 240,000 souls wandering about in a vast volcanic country— we could all live on one street in New York, but we are sitting on the top of the greatest furnace there is. Every five years since 1874, Iceland has had volcanic eruptions. This big furnace is a great natural resource. We dig some holes in the ground and we get boiling water up. We heat all our houses in Reykjavik, a city of 130,000 people and the capital of Iceland, with geothermal water. It is in this field, the field of geothermal power—a renewable energy source—that Iceland probably could contribute something to the Symposia discussions of the energy problem.

ALVIN M. WEINBERG

I was a member of the Symposium's program committee, but I fear that I failed to exercise my rights on the program committee by allowing the Symposium to reach its end without once having heard from a person who deeply believes that nuclear energy (1) is needed and (2) can be fixed. This is a very serious deficiency which I hope will be corrected in the future Symposia.

Why do I believe it is so important to correct this deficiency? Because the fundamental problem over the next fifteen or twenty years, as so many of the speakers have stated, is oil. An obvious strategy that removes oil as a bone of contention in international politics is the one proposed by Henry Linden, the head of the Gas Research Institute, who suggested removing oil from the non-transport sectors. The one country in the western world that is aiming at exactly this is France. And as far as I know, no one at this Symposium has talked about the problem in France. France today uses oil for 67 percent of all of its energy. It is planning by 1990—just ten years from now—to have 30 percent oil, 30 percent coal and gas, 30 percent nuclear, 5 percent hydro, 5 percent solar, and a considerable amount of conservation. If the western countries of the world emulated France and reduced their call on world oil from 50 million barrels per day to 30 million barrels per day, as in France, then this oil would be available for Professor Reddy's India and the rest of the underdeveloped countries. I do not see, in heaven's name, why we don't take seriously this pattern being worked out in France. I submit that there is a good chance that by 1990, just ten years from now, when we have another conference—perhaps not in Knoxville but of the same character as this Symposium—we will all look back and say, "Why didn't we do what France did?"

In short, I believe that nuclear energy is an essential part of the picture, and that despite the underlying assumption—supported by the climate of opinion which I sense around this table—that nuclear is beyond repair, it is not beyond repair. Nuclear power for energy can be fixed, and it must be fixed.

DAVID STERNLIGHT

Dr. Weinberg and I have had this discussion many times before. In the context of my talk this morning, I think it is an egregious blunder for countries to put most of their research and development eggs in the nuclear basket, and I think a more balanced mix is urgently needed.

MILTON KLEIN

I asked for the floor to address questions to our speakers, who presented very interesting and fine papers. But I also must take this opportunity to comment on Mr. Weinberg's comments. I certainly concur that the failure to address the need for nuclear power is a mistake and that nuclear energy is one of the essential elements in the solution to our energy problems. So, I can only subscribe to what Alvin Weinberg said. Having recently returned from the International Energy Agency (IEA), I would also say that this is clearly the view as seen from the IEA. The industrial countries have a tremendous responsibility to get off oil. The problems of these countries, serious as they are, are easier than the problems of the developing countries, and we in the industrial world bear a particular responsibility to move as rapidly away from oil as possible.

I would also like to comment on David Sternlight's remark about electricity production technologies, since while at the IEA I essentially wrote the report to which he referred. I think his point needs to be put in perspective: there was no intention in our review, our assessment, to say that significant efforts devoted toward electricity technology were unwarranted; quite the contrary. What we were trying to point out was that there is not enough effort being put into other aspects of energy research and development—particularly liquid fuels. So one must not read that out of that context.

Now to my questions. First, I had a question for Dr. Sternlight which relates to a remark he made in his speech implying, I thought, that he felt that oil prices were likely to remain stable in real terms for an extended period. I want to know whether I heard him right or what his view is on that.

Then a question for Professor Reddy. I agree with him that there is a need to look at details, and so I am going to ask a couple of detailed questions relating to his remarks. I have heard stories— and probably others have heard them as well—that the need for improved cooking efficiency in rural areas of developing countries has met with cultural problems because of the difficulty of getting individuals to adapt to new techniques in these functional areas. I welcome Professor Reddy's comments on that. In addition, I would like to ask whether he looked at the capital investment needed to completely replace the cooking utensils—the cooking stoves—in the country, and how those capital needs compete with the need for capital for other purposes in the development of India's society.

DAVID STERNLIGHT

Because I work for an oil company and because of the antitrust laws I have to choose my words very carefully when talking about prices or future prices. So I hope that you will forgive me for having to speak more carefully than I would wish as a professional economist.

I did not say what you thought you heard. What I did say was that the real price of oil has now reached the point where we are seeing significant conservation and the beginnings of significant development of energy supplies that we've not seen in the past, such that, absent a major further disruption in world oil supplies—on the order of a further 5 million barrels a day for six months or more—the sharp, real price increases of the past may well be behind us. That's about as strongly as I can put it. If I were a prudent man trying to decide what real price I would use to make tradeoffs and policy analyses, I might well choose today's real prices as a reasonable analytic assumption for such perspective analysis.

AMULYA KUMAR N. REDDY

There are two questions: one involves the acceptability of alternative cooking fuels; and the other is concerned with capital cost. I think it has been the experience of everyone working in rural areas of developing countries that whenever more convenient cooking fuels have been provided, there has been very little resistance to it. With regard to the capital costs, they are on roughly the same order as rural electrification. To be more specific, in a rural energy center being implemented in South India for a village with a population of 360 living in 60 households, the center's capital costs are about $10,000 for that village. But when we look at these capital costs, what we must compare them with is really the serious fiscal hemorrhage taking place due to oil imports.

YUMI AKIMOTO

After the two excellent presentations on alternative policies for improving energy productivity and production in developed and developing countries, I would like to bring you a little news from Japan.

On October 1, 1980, just two weeks prior to this Symposium, Japan inaugurated a new government developmental organization called the New Energy Development Organization (NEDO). This organization is devoted to (1) the development of alternative energy sources including coal liquefaction and the generation of electricity by solar energy; (2) geothermal source development; and (3) overseas exploration for coal. Our government has allowed a NEDO budget of $88 million for this first fiscal year—a budget which will soon grow to $180 million. This organization is similar to the Power Reactors and Nuclear Fuel Development Corporation which was organized ten years ago and is presently playing an important role in Japan's development of advanced reactors and its establishment of fuel cycle technologies, such as uranium enrichment and reprocessing.

As Dr. Sternlight correctly stated, in 1977 Japan relied on imported oil for 75 percent of its total energy. Together with the effort to expand Japan's nuclear energy capacity and to save energy through conservation, as was promised at the Tokyo Summit last year, the purpose of this new organization is to cut this oil reliance to 50 percent by 1990. Taking all efforts into account, we expect that the new energy sources will be able to cover 6.5 percent of energy consumption (a percentage which is practically zero at present). We also expect to increase the reliance on nuclear energy from its present 5 percent to 11 percent within the coming ten years, and we envision another 15 percent of effective met demand from energy conservation. However, even with these measures, we will still have to increase the percentage of coal utilization from the present 14 percent to 18 percent, and this is nothing but a switch from imported oil to imported coal.

Another point: because of the difference in its energy consumption structure, energy saving in Japan will have to proceed in a somewhat different way from other industrialized nations. In Japan, the household sector's share of total energy use is only 13 percent, in contrast to the 30–32 percent share in European countries and the United States. This is partly due to our living quarters, which some European Community officials bluntly referred to as "rabbit huts." Even now, most private houses lack central heating, and the ones lucky enough to have it use it very modestly—for example, they cut the heat off before going to bed and so on.

So the elasticity for energy conservation in the household sector is not expected to be high. On the other hand, 57 percent of Japan's total energy use is attributable to its industrial sector, in contrast to 30–40 percent in the United States and Europe. We may look for a considerable increase in energy efficiency, as Dr. Tominaga explained yesterday [see chap. 8]. However, after exhausting almost all the easy countermeasures, we have reached the point where the cost/benefit performance of further energy conservation is not necessarily high enough. But in view of the instability of oil prices and supply, most Japanese industries are still enthusiastic about energy conservation. As Professor Sadli, Professor Reddy, and Dr. Nwosu have pointed out, reducing oil consumption will help not only the concerned country but also those who will need more energy in the future. We are seriously tackling this program.

Regarding Dr. Weinberg's point, I also feel that throughout this Symposium the role of nuclear energy has been discussed too little in contrast to its importance as an alternative to oil, at least for the short and medium term. As a matter of fact, due to the oil price rise in the last year, all the Japanese utilities had to increase their electricity prices by 50 to 70 percent. Among them, the utility that is the most heavily involved in nuclear energy had the most modest price increase, and even so, that company was the first to recover from the financial loss it incurred from the oil crisis. This company's reliance on nuclear energy is 20 percent in the daytime, increasing to 40 percent at night when electricity consumption decreases.

If I may be permitted to express my personal feeling, I believe, as many Japanese do, that the world is transitional and existence relative. There is no absolute evil or absolute good; instead, evil and goodness are nothing but different sides of the same coin. And it is mankind that chooses to put one side or the other up.

In August of 1945 I was about 50 miles from Hiroshima when I saw the first atomic bomb explode over that city. I crossed the city two weeks later, and still I could find many wounded people wandering about seeking their lost relatives in this flattened and decimated city. Though I was raised in a Christian home, the crucial question which caught me was, "If God really exists, why should people die in such a cruel way? Why did God give such terrific power to mankind?" Eleven years later when I was assigned to be on one of the first crews engaged in the peaceful use of nuclear energy, I felt extremely happy because I thought I had finally found the answer to my old question, and I had the opportunity to prove this answer through my life work. That answer is, "God gave mankind the key to nuclear energy because He trusted us to have the wisdom to control and convert it for the benefit of mankind."

As Professor Rose correctly stated yesterday, solar and nuclear are all God gave us for our future. Indeed, solar energy also originates from nuclear reactions, and there is no reason to look on only one side of the coin for solar and only the other side for nuclear.

Controlling nuclear energy is not an easy task. We have to solve various social and political problems, and we need to maintain nuclear's safety strictly. But as one nuclear engineer, I believe in a future where every nation can enjoy the benefit of peaceful nuclear power depending upon their need and willingness. Above all, to escape from the present vicious cycle that resulted from overreliance on petroleum, the only way left before us is to call for all practical possibilities of alternative technologies and to organize such efforts into global collaboration.

MÅNS LÖNNROTH

I come from a country [Sweden] with by far the biggest nuclear power program in the world on a per capita basis, but I would like to start off by disagreeing with Alvin Weinberg and Milton Klein, and I would like to thank the program and organizing committee for not letting yet another meeting get bogged down in endless arguments about the pros and cons of nuclear power. I think we have had too many of those arguments already. Also, I agree with David Sternlight that there is too much money spent on nuclear research and development compared with the R&D in other areas, and furthermore, I think too much social and political energy is spent on nuclear energy.

However, one thing I *would* have liked to see in a meeting like this is more focus on the difference between manifestoes and realistic assessments of the supply future. I would like to cite a few examples and ask David Sternlight a few questions on this. Take the Exxon "World Energy Outlook" study, for instance. It seems to me to be relatively realistic with respect to nuclear and coal, but frankly, I think it is rather optimistic when it comes to the role of synthetic fuels in the United States. And I wonder whether David Sternlight would agree with me when I say that the Exxon synfuel projections are much more a manifesto of what they would like to do than an assessment of what is realistic, given the political constraints in the United States.

Second, regarding the coal questions and the role of coal that David also brought up: if one looks at the International Energy Agency's steam coal study and the World Coal Study, one finds that a major role in the world of coal exporting is supposed to be played by the United States. I wonder if David would agree with me that the role given the United States in the world coal market in these two international studies is also rather optimistic.

Third, when it comes to the energy exports of other countries, much has been said about the absorptive capacity problems of less developed countries such as Saudi Arabia, Iran, and so on concerning their sensible use of income from oil exports. Much less has been said about the same problems in developed countries, but they also exist. Take Norway, for instance. Norway has significant oil exports, and the net effect of the trade balances due to these oil exports is that the Norwegian currency is revalued at a higher level, with the result that the industrial products of Norway are priced out of the marketplace. So Norway's economy is becoming significantly changed and restructured in a difficult direction due to its large oil exports. I think that Jane Carter would agree that there are similar problems in the United Kingdom with respect to a large oil export in relation to the rest of the economy, and I think, if I understand it correctly, that you will find the same problems in Australia. In other words, if one looks at the way the economies operate in those countries that export energy, whether it is coal or oil, one might also find some other grounds for being more pessimistic about the world coal export scene. Which leads me back to the question that perhaps we have to look much more seriously at conservation programs than we have so far.

DAVID STERNLIGHT

I think that those are excellent points. First of all, the Exxon forecast of synfuels is, in Exxon's own view, an expression of what they believe might be possible under the most favorable of circumstances. So I don't really think that even Exxon would attempt to defend it as a projected forecast; it is rather what they believe to be possible. Exxon's own representations are of the class that says "Let's decide what we're going to do, and let's see what can be done in this direction," rather than saying "Well, here's what's going to happen."

On coal, I do believe that, based on the kind of detailed analysis in the world coal study as well as in the IEA studies, it is possible that the volumes projected for most of the world coal study cases—except perhaps the most stringent one in which there are further disasters in oil, nuclear, and so on—are achievable in a realistic policy planning context rather than the context alluded to a moment ago in terms of synfuels and the Exxon forecasts. This assumes that the United States government will take appropriately long lead-time actions for infrastructure development, particularly in the port area, and that world coal consumers will get together with world coal producers to sign long-term contracts. (Japan is one consuming country that's particularly active in trying to do this through a variety of very innovative mechanisms, including equity participation, sharing, and so on.)

Most enlightened energy analysts do take account of absorptive capacity and a number of other factors that are of a pessimistic-seeming character in making forecasts and then coming up with policy conclusions such as those I referred to. In particular, if you look at the published forecasts of the US Central Intelligence Agency, if you look at the study by the US Office of Technology Assessment called "World Petroleum Availability—1980 to 2000" and any number of other recent studies including the IEA's own recent study, you find that for Norway, the United Kingdom, Mexico, and other countries, likely future oil production is very much a function of national concerns about absorptive capacity and other factors. Similarly, OPEC production from a number of countries is a function not only of engineering and geological considerations but also of absorptive questions and supply/demand balance questions. One must not, however, take this too far. For example, the often-heard statement that oil in the ground is worth more than money in the bank is simply untrue in the realistic context of many producing countries. This would be true if the oil were in the ground in some canton in Switzerland, but it's not: it's in the ground in countries with regimes where it is not at all clear who is going to get the benefit of the stream of revenues coming from that oil ten or twenty years from now. That tends to bias such regimes toward converting the oil into money and investing it safely.

MARCELO ALONSO

I would like to make a comment about the problem of nuclear energy. I don't think that we have to look at nuclear energy as if we were for or against it, in the same way that I don't see the issue as whether we are "for coal" or "against coal." So I would like to see the issue looked at in a slightly different way, the way, I think, that Alvin Weinberg was putting it.

Whenever a decision about energy is to be made—and the major decision at present is to move away from oil as much as possible—one cannot at the beginning rule out the nuclear option. One must consider the nuclear option jointly with many other energy options, and each one is decided not just in terms of cost but by taking other factors into account as well. I think it is a big mistake to make a decision about an energy option just in terms of cost—not only the cost of production but also the cost of capital investment. I think it is a terrible mistake. The decision has to be made in terms of many other factors which you know so well that I don't have to mention them here. I only want to emphasize that cost—or the environment—cannot be the only factor. So one has to blend all the elements and use the best possible judgment in each case, case by case.

ISHRAT H. USMANI

I have two observations and one announcement to make. I would also like to pose a question.

The first observation is that of all forms of energy, electricity is by far the most versatile. Its importance as an essential index of living was first recognized by Lenin when he is supposed to have said (I forget the exact words) that Bolshevism was nothing but Marxism plus electricity! Since then, provision of electric power to the Soviet Union's people has been a cardinal principle of planning in that country. Another great man who realized the benefits of universally available electric power was Franklin D. Roosevelt. As president of the United States, he sponsored a nationwide rural electrification program which advanced loans at a 2 percent interest rate through a special federal agency, the Rural Electrification Administration (REA). The REA was constituted in 1936 and still exists. Through it, the coast-to-coast extension of the national power grid to homesteads, farms, and factories became a great instrument of change in rural America's standard of living.

My second observation is that the vast majority of the human race living in the Third World's rural areas can also be provided with electric power if the developing countries, most of which happen to lie in the solar belt from 35° north to 35° south of the equator, promote rural electrification. This could be done on a decentralized, village-to-village basis by using the locally available renewable sources of energy, for in a great majority of the villages in the Third World, the sun shines every day, the wind blows, the crops grow, and the cattle graze. Thus, that solar energy, wind power, and energy from biomass (agricultural and animal wastes) can be harnessed to produce enough electric power to meet all the basic needs of the small rural communities. I launched this idea when I was the energy advisor to the United Nations Environment

Programme (UNEP), and I am happy to find from his paper [chap. 14] that Amulya Reddy, my friend and colleague at UNEP, has at last seen the benefits of using renewable energy sources for rural electrification in India.

I´am happy to announce that just before coming here, I received a telex message that the world's first Rural Energy Center (REC) sponsored by UNEP and located in Sri Lanka has been successfully commissioned, with firm electric power generated by integrating the output from variable sources of energy, including solar photovoltaic cells, wind turbines, and a biogas-fired generator.

My question is: To eradicate the pangs of poverty from most of the Third World, why can't an international effort be launched to undertake rural electrification based on renewable energy sources by establishing a network of RECs before the close of the twentieth century?

EDWARD LUMSDAINE

I just have a very brief remark concerning Dr. Usmani's comment on the Sri Lankan village. Rural electrification seems to work and seems to be quite permanent, but from many years' experience with an Egyptian village, I and a group of Egyptian coinvestigaters found that introducing technologies to electrify a village by using alternate energy sources like wind and photovoltaics—well, we still don't know whether it works or not. Just the operation and maintenance problems are quite serious. We have been addressing this for about two years now, and although I am sure that if they ran a power line into a village it would work, just as it would elsewhere in the world, we still have some doubts about the long-term operation of this electricity-generating system in the Egyptian village of Basaisa, which is like the one in Sri Lanka.

MIGUEL S. USSHER

If we scratch a little under the surface of the two excellent presentations at this session, we find a big difference in their philosophies. The first one is free enterprise, free pricing. The second one is participation by the government in the economy on a half government-owned basis. And there is still one that was not mentioned here: an entirely socialized economy.

According to Dr. Sternlight's presentation, three policies for energy are (1) pricing, (2) economic growth, and (3) political decisions. With time, with less energy or fewer chances for cheap energy, I think it will be the other way around. Very likely, political decisions will weigh more.

Now, with respect to Professor Reddy's presentation, I would like to know: What is the pricing policy in India to implement the programs under way? For evidently the country must charge a price that will not kill the economy but at the same time will provoke a shift in consumption from certain types of energy to others that are considered more convenient.

DAVID STERNLIGHT

When I recited those major factors, they were not in any priority order. The influence of each varies with the country, the circumstance, and the policy issue and energy form you are considering.

AMULYA KUMAR N. REDDY

The Rural Electrification Corporation of India's government is responsible for the electrification of villages, and, if my memory is right, it gives an interest rate of about 6 percent over twenty years in order to promote rural electrification.

The pricing that is especially crucial to the type of development-oriented forecast I gave really concerns diesel fuel, because in India the cost of diesel is about half that of gasoline, and the maintenance of these diesel prices has been very much at the behest of the trucking lobby, which is an extremely powerful lobby in India. So a market mechanism is not operating here. This is one more illustration of my point that government intervention is absolutely necessary to stimulate equity.

MOHAMMAD SADLI

To practically do away with kerosene consumption in India, what instruments do you have in mind? For example, can the market take care of itself, apart from what you said about the influences of the lobby, and so on? There is also this phenomenon of substitutability between diesel fuel and kerosene. If you have to price kerosene rather low in the intermediate period for social reasons, then you cannot raise the price of diesel fuel and vice versa. Do you have to impose a tax—a levy—on diesel and kerosene above their normal commercial acquisition cost?

AMULYA KUMAR N. REDDY

The bulk of the kerosene in India (actually, 70 percent of it) is used for rural lighting, and the price of kerosene has been going up, so today there is a powerful demand by rural households for electric lighting. But this demand has not been responded to with *home* electrification programs (which are not equivalent to village electrification programs because on the average only about 14 percent of the homes in electrified villages have obtained domestic connections).

It is in this context that the point just made about the substitutability of kerosene and diesel fuel in trucks is very important. If diesel fuel prices are raised without an associated increase in kerosene prices, the diesel used in trucks will be replaced with the cheaper kerosene. On the other hand, kerosene prices cannot be raised without causing great hardship to the poor for whom kerosene is the main illuminant.

Hence, reducing diesel fuel consumption and preventing the replacement of diesel with kerosene cannot be achieved by a pricing mechanism alone. It is also essential (1) to electrify all homes and make kerosene redundant as an illuminant, and (2) to enforce the zonal and national permit systems now in existence in the country so that trucks are restricted to break-even zones which may be color-coded so that a truck with one zonal color sticks out like a sore thumb if it is in another zone.

IGOR MAKAROV

I want to thank the sponsors and organizers of this Symposium for extending an invitation to a Soviet representative. I am very much impressed by the whole event and would like to commend those responsible.

After what has been already said by the distinguished experts—and unfortunately I am not one of them—there is no need to deliberate on the importance of solving, and solving very quickly, the energy-related problems which, without exaggeration, will have a great impact on the future course of world developments and perhaps on the fate of the world itself. We can go on talking and persuading each other about how urgent the present energy predicament is, but until these nice talks and important meetings—undoubtedly done in good faith—are backed up by our concrete deeds, most of that will be rather futile. We have to be aware that present-day energy problems have political underpinnings and origins with very serious international undertones and implications. First and foremost, the solution lies in these domains. We can come up with mind-boggling results in energy research and development, in backing commercially promising programs in energy-intensive technology, but unless we find a common ground and understanding for effective joint actions, these achievements will not bring their desired results.

I understand that this is no proper place to go into prolonged political statements—to which Russian diplomats are sometimes prone—about the USSR's position on this matter. However, I don't have courage to let such a chance slip by. So in conclusion I would like to add a few more words. It appears to me a bit of an oversimplification to make scapegoats of certain countries, particular big nations, and squarely put blame on them for the world's present energy situation. On the other hand, I cannot quite agree with the sort of doomsday scenario presented by the esteemed Mr. Lovins regarding the energy and environmental future which is precariously lurking somewhere ahead of us. Paraphrasing him, if I may: truly to a significant degree it depends on what kind of hammer we put in whose hands and which nails we choose to strike most effectively.

AMORY B. LOVINS

First, I am surprised to hear characterized by Igor Makarov as a "doomsday scenario" what some others around the table have been characterizing as wild-eyed technological optimism. James Branch Cabell once remarked that the optimist believes we live in the best of all possible worlds; the pessimist fears this is true.

Second, I would remind Ishrat Usmani that institutionalized good intentions tend to go astray after some decades. In recent years about three-quarters of the cheap money available to the rural electric program in this country has been used to bail out the large, and mainly nuclear, construction projects of private utilities which could not finance those projects in the market.

Third, I yield to nobody, including Alvin Weinberg, in my sense of urgency about displacing oil. We only disagree, as we have for years, about how best to do it. Our paper yesterday argued that at the margin, nuclear and indeed coal-electric investments actually *retard* oil displacement by diverting finite resources from well-known measures that can save more oil faster, more cheaply, and more surely. So I think Alvin has been starting on the left-hand end of the spaghetti charts, looking at the imported oil, without paying enough attention to careful and symmetrical comparisons of costs, rates, and difficulties in providing energy services. Electricity is not oil and cannot be very straightforwardly substituted for it in most applications. Only a tenth of the world's oil is used in power stations. In our summer, 1980, article in *Foreign Affairs* ("Nuclear Power and Nuclear Bombs") we argued that nuclear expansion, far from decreasing the already serious risk of war over oil, actually *increases* that risk in three ways: by retarding oil displacement; by creating new international conflicts over uranium, technologies, and fuel cycle services; and by unavoidably spreading nuclear bombs, innocent disguises for bombs, and ambiguous threats that motivate others to acquire bombs. Our paper also argues quantitatively that nuclear displacement of oil is at best extremely slow and limited because of purely logistical and practical constraints.

I think a more likely conclusion if we reconvene in 1990 is that the collapse of the French nuclear program in the 1980s was a natural and predicted consequence of the inability to solve the formidable combination of technical, economic, and political problems incurred by substituting determined technocracy for the salutary discipline of the marketplace. It is that discipline which has largely destroyed nuclear prospects outside the centrally planned economies. We argued that governments need not accept antinuclear rhetoric or sentiments in order to reject nuclear power. They can love nuclear power, provided that they love the market more and accept its verdict in good grace.

Finally, a quick question for David Sternlight. I was impressed in Sweden a few years ago by the number of people who were saying, "In Sweden we will become self-reliant in energy, whether by nuclear or soft technologies or whatever. We are also going to put a great deal of government help into strengthening our export-oriented industries, like steel and pulp and paper. Why? Because we are a trading nation." So I asked, "Well, if you are not going to be importing all that oil, what if you increase your exports—won't the krona become so strong that people can't afford to buy your exports?" So let me slightly reformulate Måns Lönnroth's question. Do you know of any country, David, that has really looked at the structural shifts in the economy? If we do achieve substantial self-reliance in energy, clearly we won't need to export so much, and that may change the whole shape of the national economy.

DAVID STERNLIGHT

I know of no serious studies of the question. I suspect that the answer depends on where you start. If you start by saying "Well, somehow we are going to move into a wonderful world where we don't have to export so much (which I assume is shorthand for 'to pay for the energy'); therefore things will be much better," you reach certain conclusions. Before the initial sharp OPEC price rise we were in that position, and we did not have massively strong currencies in many countries. I believe that the problem with the dollar, for example, is of rather longer standing than the period beginning with the sharp oil price rise in 1974, and if you look at historical relationships, I'm not persuaded at this point of the argument you're making.

AMORY B. LOVINS

But it does imply, for example, that the best kind of Middle Eastern arms control might be American roof insulation.

DAVID STERNLIGHT

I agree with that completely.

MILTON KLEIN

I am sorry to ask for the floor a second time, but I just could not leave unchallenged one statement that Mr. Lovins made—that the money intended for the rural electric cooperatives has been diverted against their interests into the private utilities to pay for nuclear plants. I haven't been associated with the utility industry for very long, but I can't imagine that any funds have been diverted and would like to see the source of that information.

AMORY B. LOVINS

I can answer Milton Klein's comment very shortly. I did not say that the money was "going into the private utilities." I said it was going to bail out their construction projects by buying shares of these projects. I can give you the exact reference later, but it's in the annual reports of the Rural Electrification Administration.

[The following reference was supplied by A. B. Lovins after the conclusion of the Symposium.] The US Senate Committee on Appropriations Hearings, *Agriculture, Rural Development and Related Agencies, I—Justification, FY 1979* (US Government Printing Office, 1978) reports at p. 886 that three-fourths of FY 1977 REA loan guarantee commitments were used for partnerships in large plants being built by private utilities, often to serve remote urban loads. The REA loan program rose from $0.62 billion in FY 1973 to $4.8 billion in FY 1977, and the Federal Power Commission—*Factors Affecting the Electric Power Supply, 1980–85*, Dec. 1976, Executive Summary Recommendations, p. 46—projected a cumulative total of about $20–40 billion over the next decade.

MILTON KLEIN

Certainly some rural cooperatives are participating in nuclear power projects by buying and owning a share of such plants. This is done to meet their power needs on the basis of decisions by these rural cooperatives. There is nothing nefarious about such participation, as implied by Mr. Lovins. There may be some cases in which higher costs or reduced projections of loads have led original sponsors to seek additional participation in a project. But again, if a rural cooperative or any other utility decides to join in, it is a decision based upon the cooperative's needs and not a diversion or "bail-out."

A further comment about Mr. Lovins's statement that three-quarters of the funds for the rural electric program are being used for such purposes. According to my information, about 15–20 percent of rural electric funding in recent years has been related to the purchases of shares in these plants.

AMORY B. LOVINS

My source is noted above.

Mr. Klein fails to note that the dramatically growing raid on cheap (6–8 percent per year) REA money has been assiduously encouraged by many investor-owned utilities unable to finance their projects normally. Certain cooperatives, in turn, have become notoriously unresponsive to their members' wishes and are in practical effect serving as subsidized vehicles, not for bringing electricity to remote farms as in the 1930s, but for powering the heavy industrialization of rural and semiurban areas. This is especially common in the Rocky Mountain West. To pick a single example of some cooperatives' ambitions (though not in this case for buying into private projects), Colorado-Ute, once a small rural cooperative, currently plans to build at least a dozen 400-megawatt coal-fired electric plants to run proposed coal mines, synfuel plants, and the like. Colorado-Ute explains that it doesn't care what the demand will be: it will simply build more and export the surplus power to Arizona and California!

For an excellent history of the REA, see *Lines Across the Land: Rural Electric Cooperatives, the Changing Politics of Energy in Rural America* (Washington, DC: Rural Land and Energy Project, Environmental Policy Institute, 1979).

ALVIN M. WEINBERG

Amory Lovins in arguments of this sort has touched on a fairly central point when he implies that electricity displaces oil only with great difficulty. But Amory has somehow been bamboozled by what my colleagues in the American Physical Society call the "second law efficiency." And the elevation of this idea—that it is necessary somehow to match the quality of the energy source with the quality of the end use—I think is a wicked if not a meretricious dogma. The fact of the matter is that in the United States in 1978, one million houses went electric, two hundred thousand went gas, one hundred thousand went oil, and this resulted from market forces. The market knows nothing about the second law of thermodynamics, but it has led to a sensible result: removing oil from the nontransportation sectors!

AMORY B. LOVINS

I think that had Alvin Weinberg heard our paper yesterday, he would appreciate that we explicitly based our analysis not on any kind of thermodynamic ideology but on neoclassical economics—that is, the lowest private internal cost per unit of service delivered. The historic shift he describes toward electric heat in many countries did not take place in a competitive free market. It took place where marginal consumption did not attract marginal cost but was instead shielded from it by rolled-in pricing*, by massive direct tax subsidies to utilities (in this country at least), and by cross-subsidies to electric heat users from other classes of consumers. Anyhow, the real competitor for heating isn't oil and gas; it's efficiency improvements and renewables.

MÅNS LÖNNROTH

I want to add something to what David Sternlight said previously about coal. If you look at all the various countries in the world and how they have planned their energy supply, you find that during the 1960s and the 1970s, they planned for oil. When the oil crisis came, they all planned for nuclear. Now that nuclear is falling through, they all plan for coal. The all jump on the bandwagon, so what the United States *could* do if the president did this or that or Congress did this or that or the coal mining states did this or that is one thing; what will actually happen is another. I would submit that the World Coal Study and the International Energy Agency's steam coal study are much more manifestoes of what they would like to happen rather than what is likely to happen. I wonder whether David has any comments on this.

DAVID STERNLIGHT

The major purpose of those studies was to identify and call attention to the long lead-time policy decisions that needed to be taken and the bottlenecks and other considerations that needed to be dealt with to allow the market process to make coal more widely available in world trade and use. Rather than being manifestoes both those studies were intended to be precursors of policies. The test then becomes: Have we seen any evidence that some of those policies are now being put into place? We are, in fact, seeing considerable such evidence; we've seen a shift in the US government's policies with more specific attention being paid to some of the early policy decisions that need to be made.

We are also seeing a great deal of activity in the other main area of recommendation—coal consumers and coal producers are getting together to sign long-term agreements. That's the precursor to

---

*I.e., by averaging new costly supplies with old cheap supplies so that the true incremental cost is concealed from the consumer.

much of the physical development that takes place in the energy field or in any other major investment field: when you have a buyer and seller in agreement it's usually easy, if the agreement is complete, to take it somewhere and get financing to build the infrastructure, to develop the mines, to build the ships, and so on. I think the evidence is that we are seeing the kinds of movement that are necessary, but we will have to watch things closely and see how they develop.

It certainly wasn't my intention to suggest that coal is any sort of panacea. One has to compare all alternatives for dealing with the energy supply/demand balance, including increased efficiency of end use and various other means of achieving the goal of providing energy services to people at least cost.

ZYGMUNT KOLENDA[1]

One of the most important problems in nuclear and coal-fired energy systems is environmental protection considered from different standpoints. Great attention should be paid to the problem of human life protection. Present research in this area, especially in a long-term sense, is still very far from accepted conclusions: for example, the influence on human life of carbon dioxide, sulfur dioxide, and hot water from coal-fired power plants has not been examined in a proper way. The "greenhouse effect" is only one factor among others and is not very important; solar wave length distribution changes resulting from atmospheric pollution seem to be much more important from the standpoint of human health. The problem of environmental protection is strictly connected with energy strategies. The as yet unknown influence on human life of present and future energy technologies can stop the world's energy progress.

In addition, it should be noted that the 1974 energy crisis resulted from a political situation, especially from international conflicts between developed and developing countries. The present critical situation should also be considered on a political basis. No new energy technologies can solve the energy problem without a rigorous policy of international agreement among all nations. The present rule of rich countries is particularly important, from different points of view. The "peace" policy should not be confined to military problems; it should also involve the energy situation of the world. God-given natural resources belong to all of the people living in our very inhuman world.

In closing, I should like to applaud Professor Rose's integration [chap. 12] of Session III. His magnificent and human summary gave me extraordinary satisfaction.

JEROME E. DOBSON[2]

A common thread running through all the papers and discussions is the acknowledged need to protect the poorest of the poor from the detrimental effects of increased petroleum prices. There seems to be a rather naive assumption that the world community will do whatever it can to help these people if a reasonable course of action can be found. In reality this moral obligation, which I share with the speakers, will be submerged in a sea of economic, political, and military forces which have little to do with fairness. Equity will be valued only to the extent that it represents true power and that inequity threatens other economic, political, and military goals. I think it would be in the best interest of the less industrialized countries for all of us to take stock of what real power these countries may have in the new era.

NOTES
1. Dr. Zygmunt Kolenda; Professor of Energy Engineering; St. Staszic University of Mining and Metallurgy; Cracow, Poland.

2. Jerome E. Dobson; Leader, Resource Analysis Group; Oak Ridge National Laboratory; Oak Ridge, TN.

For identification of the other discussants, please refer to the list of participants in Appendix II.

CHAPTER SIXTEEN

Summary

Vaclav Smil
Professor of Geography
The University of Manitoba

The last speaker at a symposium devoted to analyzing a single, albeit a very large, topic has a doubly unenviable task. He should hold the attention of a tired and restive audience, and he should not repeat what has been already said several times during the three days of discussions.

So I decided to bring in some unorthodox angles for this adieu performance and will tell you, not necessarily in this order, about straw mushrooms, inflation in the Federal Republic of Germany, Australian eucalyptus forests, and natural insecticides of Amazonia, and, as you might expect, I will quote some Chinese philosophy. And, of course, I will try to weave this together with Dr. Sternlight's and Professor Reddy's presentations.

This entails, as a statistician would put it, a search for some communalities. The first, and to me an essential one as it coincides with my likings and preferences, is the stress both speakers placed, as did some participants before them, on deemphasizing the supply side of energy systems.

Dr. Sternlight did so in relation to the research and development effort in different affluent countries where an inordinate amount of attention is still given to the increased supply of electricity, and especially of nuclear energy, while Professor Reddy urged us to look at the kind, quality, and distribution of energy as opposed to an indiscriminate accent just on gross energy flows. Indeed, this must be a key consideration in system energy studies, because a closer look will quickly show that mere energy use increases by themselves do not provide and guarantee anything.

Higher energy consumption does not strengthen national security. Steady postwar increases of US energy use from about 5.5

tonnes of oil equivalent per capita in 1945 to some 11 tonnes of oil equivalent today (i.e., doubling the relative supply) unfortunately have been paralleled by a steady decline of the country's military might and global prestige, by a decreasing ability to project power in guarding the nation's interest, and by an astonishing erosion of the will to act in international affairs.

Climbing energy use does not guarantee economic security. Between 1972 and 1979—that is, in the wake of sharp price hikes—US oil consumption rose 13 percent, from 16.4 to 18.4 million barrels per day (mmbd), while the Federal Republic of Germany's use advanced just 6 percent, from 2.52 to 2.66 mmbd. Yet during the past three years US consumer prices were climbing at the high rates of 6.5, 7.6, and 11.3 percent per year, in contrast to the Federal Republic of Germany's low inflation rates of 3.7, 2.7, and 4.1 percent.

High per capita energy flows do not bring about higher quality of life. The new field of quality of life (QOL) studies came of age during the 1970s, and, among many other things, it showed persuasively that above certain minimum levels there is precious little, if any, correlation between energy use and enjoyable living. In fact, as Adams so rightly remarked, the opposite it true: higher general energy flows cause an inevitable decline in quality of life for some segments of populations (i.e., those living near the extraction or conversion facilities) and, I would argue, eventually for everybody.

Even huge increases of energy supply do not guarantee industrial modernization. In aggregate terms China is now the world's third largest energy consumer (behind the United States and the Soviet Union) and is ahead of Japan, yet the Japanese national product is, in absolute terms, about 4.2 times bigger. About half of this discrepancy can be ascribed to different product mixes (the advanced technology economy of Japan versus the still largely agricultural and raw material orientation of China), but the other half is due to the pervasive inefficiency of China's fuel conversion (an average efficiency of less than 30 percent nationwide compared with Japan's average of more than 50 percent).

Analogically, higher absolute flows of energy into agriculture are no guarantee of farming modernization. High-yielding variety technology, often called the green revolution, brought much higher direct energy inputs (above all, fuels for pumpsets and for field machinery) and indirect energy inputs (maily synthetic fertilizers and pesticides) into Indian agriculture. Yet, with the lack of infrastructure, education inputs, and equal opportunities for credit, the process mainly favored the rich minority and made poor farmers poorer.

For all of these tasks, other inputs and conditions are no less essential than the requisite energy flows. National security comes no less from energy-intensive weaponry than from national cohesion, a sense of history, and shared memories and concerns. Economic security comes when nations do not live beyond their means. Quality of life arises from strong cultural values, from a sense of belonging, and from preservation of nature's services as opposed to just extraction of its goods. Industrial modernization requires careful, overall, strategic attention to priorities, products, and processes that will do away with traditional energy-GNP elasticities. And agricultural modernization in a poor developing nation cannot achieve widespread and substantial success without stress on equity.

The other key concept shared by the two speakers and myself is even more important: the concept of the possiblity of choices. Although Dr. Sternlight, as one must expect of a chief economist in an oil company, presented us with an image of a world where prices, deregulation, and incentives rule supreme, and while Professor Reddy talked about a development-oriented economy with strong government interventions on behalf of the neediest, both speakers considered a variety of choices. Dr. Sternlight listed and briefly dissected many policies and strategies available to affluent nations in general terms and on an individual level, while Professor Reddy argued that even in a country in such economic and environmental straits as India there are rational alternatives to the mindless perpetuation of the past trends.

When talking about choices it is fitting to introduce a concern—more than that, an imperative: the need to consider energy in an ecosystem framework. Science, like so many things, goes in fashion cycles, and if we were here a decade ago we would have been talking environment. That was the thing—and today energy is nearly everybody's concern. As someone who has been preoccupied and fascinated by both environmental and energy problems, I would submit that our fundamental approach should be to look at energy within an ecosystem framework.

Such an ecosystem approach should be multifaceted, but at least two great lessons and needs emerge immediately. First, we can learn so much because the most complex, most intricate energy systems on this planet are not human systems; they are ecosystems on which ultimately the survival of our civilization rests. This should help us to make better and sustainable choices in managing human affairs. The other thing, of course, is that the preservation of complex ecosystems is *the* ultimate energy problem, in comparison with which any fuel supply disruptions look quite insignificant. If we do not preserve complex ecosystems in the long run, then worries such as "shall we have an extra 4 million barrels of oil per day or not?" will look very trivial.

I want to focus briefly on just two among many useful lessons offered by the study of ecosystems. The first is the importance of "switching" in energy flows. Except in polar regions the surface of the earth is flooded with large quanta of solar energy, but this energy's conversion is switched on or off—attenuated, damped, or oscillated with various intensities depending on a host of factors—and sometimes the switches are truly fascinating.

In a mature grove of *Eucalyptus regnans,* the stately tree of southeastern Australia, all stems are evenly aged with no poles and saplings, a major puzzle as all the trees flower regularly and fruit profusely. The seed rain annually reaches 7.5 million per hectare, and even with low germination rates about half a million young seedlings should come up! But hardly any make it, because the seeds are carried into the soil beyond their germination depth by two species of ants, are dried out by summer heat, are infested by winter fungi, are overcome by competition from other plants, and are grazed by wallabies.

Intricate interplay of switches—climate, soil, plants, fungi, insects, mammals—combine to prevent this particular energy flow, and only when natural fire in a very dry summer destroys these obstacles in a sheet of flame do the seeds germinate in abundance and give rise to a new uniform grove.

Simpler and more prosaic examples are readily available. A tree in your backyard refuses to grow although it has plentiful sunshine, water, and principal nutrients. But still the older leaves are turning yellow and the new ones are barely growing. Here the switch is simple. The plant needs the element which we do not like in acid rain but which gives taste to radishes and, more fundamentally, creates bridges between complex living molecules: sulfur. A gardener applies a tiny amount of chelate, the switch is turned on—and energy flows.

This is, of course, a classical illustration of the old (1840) but dependable Liebig's law: "growth of a plant is dependent on the amount of foodstuff which is presented to it in minimum quantity." (Reality is, as usual, a bit more complex, and nonsteady states and factor interactions make for some exceptions in the law.) And this illustration reminds us that to achieve the optimum flow of desired energy it is far from sufficient to flip on all the obvious major switches, for all too often one must search for a seemingly insignificant but crucial stuck connection—or anticipate a new hidden closure and circumvent it or open it.

From the snail darter who blocks a major hydro station to public distrust of nuclear stations, from lack of water in shale-rich Colorado to peculiar notions of the world held by aging Persian imams, one can see an endless variety of "switches" that keep energy pipelines in a stuck position, regardless of the availability of resources on the one hand and the ability to harness and deliver them on the other. Search for the switches should thus become a critical part of our strategic energy planning, augmenting the necessary, but insufficient, considerations of supply and final uses.

The only other lesson I want to mention here is a grand strategy of how to preserve complexity and stability by not doing anything. We are always urged to do something. But many complex ecosystems show that an absence of a particular energy flow— nonaction if you want—may actually be the best solution. This ecosystemic "wisdom" reminds me of Lao-tse's perceptive lines:

> With a wall all around
> A clay bowl is molded;
> But the use of the bowl
> Will depend on the part
> Of the bowl that is void.
>
> . . .
>
> So advantage is had
> From whatever is there;
> But usefulness rises
> From whatever is not.

We want to have endless energy "advantages," but the usefulness often comes through abstaining from them. The planet's most intricate ecosystem, a tropical rain forest, uses this strategy to maintain its complexity and richness by not allowing any emergent species (i.e., a particular energy flow) to acquire even a small permanent advantage—and we could emulate the pattern.

Examples abound. The western world currently is being swept by a fashion wave of biomass energy, and plans are laid to convert trees and crops into liquid fuels—as if the liquid fuels for the inefficient cars were the only kind of energy we would need. If it appears sensible to use our shrinking and eroding (and in many ways nonrenewable) arable land for food and feed crops rather than for fuel alcohol, the same argument extends even to crop residues.

Cereal straws are considered good candidates for further fuel conversion, but even after cautiously leaving a large part of these wastes on the ground to guard against erosion and to improve the organic content and tilth of the soil, the best use of these by-products is to produce more fine food. It so happens that mushrooms are heterotrophs; they can't photosynthesize; they must have something to feed on. One of the best things for them to feed on is straw, preferably the straw that has been already used as bedding for domestic animals because it is full of nitrogen. And we can achieve some fabulous efficiencies. You take a kilogram of straw and you can get half a kilogram of mushrooms, not even trying hard. If you try hard, you can have 100 percent biological conversion. But even with a moderate efficiency, if only one-quarter of the world's cereal straw were used to cultivate mushrooms—a food with pleasing flavor, fine texture, and adequate protein content—every person on this planet could eat a quarter kilogram of *Agaricus* (the commonly cultivated white mushroom) or *Lentinus* (the Japanese shi-i-take mushroom) per day and receive about one-fifth of his or her protein requirements. (And mushrooms, unlike green plants, have all the essential amino acids!)

Nor is the energy in forests best used by harvesting the trees. Perhaps the best example of leaving things alone is the tropical rain forest. We should not make sugar cane or cassava or fuelwood plantations out of it, but we should leave it as it is, because it is the last remaining huge chunk of variety on this planet. Its very existence is its most important asset, for it contains an as yet largely unexplored richness of biodynamic plants, therapeutic drugs, waxes, oils, and powerful insecticides. The goods that could be harvested by destroying large parts of the forest—gasohol or fuelwood—could not ever remotely compare with the irreplaceable services we are getting from the forest, and with many services that might come to fulfill our as yet unforeseen needs. And, at a fundamental level, biomass is a much better source of chemicals than of energy: being roughly 40 percent oxygenated, it is an excellent feedstock for making about half of the fifty most common chemicals, ranging from acetone to glycerol.

Unfortunately, as Cambridge ecologist Clifford Evans remarks, "human mind has no intuitive understanding of plants," and thus to many people, trees and crops appear to be just a ready raw material for fuel alcohol. Some things cannot be repeated too often, and so every gathering of energy experts—composed, as it usually is, largely of engineers, technocrats, economists, and lawyers—should be reminded that the key energy conversion on this earth is occurring neither inside internal combustion engines nor inside power plant boilers and generators, but within the living cells of green plants. Photosynthesis is the foundation of life not just because the plants provide us with food but, less visibly but no less importantly, because they are providing us with countless irreplaceable and invaluable services. These range from controlling climate, water, and nutrient cycles to making soil and preserving genetic variability. Destruction of tropical rain forests or tall climax grasslands should be of immeasurably higher concern than exhaustion of oil fields and maintenance of ridiculously low prices of gasoline.

At a time when even the most optimistic estimates do not foresee any natural tropical rain forest remaining beyond a few generations, when arable land around the world is disappearing and degrading, and when crops we depend on for our food are becoming genetically uniform and hence uncomfortably vulnerable, our priorities are hardly set right when we worry about maintaining our bloated, inefficient, fossil fuel energy flows. One is forced to say that we are transfixed by the wrong energy crisis.

To maintain the richness of human civilizations we should approach our energy problems as complexifying minimalists rather than as simplifying maximalists. The issue is, as all issues are in the final instance, moral. That means, of course, as the eminent French philosopher and theologian Jacques Ellul begs us, that we should have a point of reference, a point of comparison to anchor our choices and decisions.

And, as Ellul so eloquently argues, humankind or history does not provide the necessary point of reference, as there are many cultures and many possible interpretations. Yet many may not want to share Ellul's point of reference, religious faith—his foundation for the look from the outside of human experience. But there are other links binding us all together. We can, if we work on it, be humble and patient and persevering. This is a great source of hope, because the difficulties facing us call for all of these qualities.

Once again, Lao-tse's insight is illuminative:

As for those who would take the whole world
To tinker as they see fit,
I observe that they never succeed . . .

For indeed there are things
That must move ahead,
While others must lag . . .

So the Wise Man discards
Extreme inclinations
To make sweeping judgments . . .

Closing Address

Lynn R. Coleman
Acting Deputy Secretary
United States Department of Energy

I would like to commend the organizing and program committees of this International Symposium for their efforts in planning this event. We are indebted particularly to Walter Lambert and the members of the Symposium staff for the true southern hospitality they have extended during our visit to Knoxville. I know that after two and a half days of bearing the weight of one of the world's greatest problems, you must be anxious to get on with the active pursuit of the resolves you have made, individually and with each other. There is no more pressing or more urgent problem in the world than energy—not just its production or development, but the awful danger that its absence or shortage poses to world stability, and also the indispensable role of energy in helping fulfill the legitimate economic and social aspirations of most of the world's population residing in less developed countries. But while the breadth of the problem is humbling, it presents great opportunities, particularly for the participants in this Symposium, to contribute to the course of civilization. So this is the topic we have. This is the challenge before us. This is an opportunity we can accept.

Conferences such as this one—and because this one is part of a World's Fair, this one more than others—serve us all by providing the best possible reality test for our plans and dreams. By bringing together a balanced collection of informed and conscientiously shared perspectives, these forums frame mankind's best grasp of what their choices are and, more importantly, of what the implications of those choices are. We cannot know the future, but there's no better test of what it might be than in the reflections of thoughtful men and women with deliberately different perspectives, striving to learn more about themselves and each other.

I am pleased to be here to address you at the conclusion of this conference. I have been asked to summarize and comment upon the course and progress of your deliberations, a chore I embrace with the enthusiasm and complete objectivity of one who has just arrived in Knoxville and who is therefore not at all biased or influenced by listening to what has actually been said. However, although I have not been here during most of your deliberations, I have followed the Symposium with keen interest. I have reviewed your papers, been kept informed of your questions and comments, and discussed your insights with several of my colleagues who have been present here throughout. So I feel prepared to take on the task of comparing your progress to a charge laid down by your chairman, my colleague Dr. John Sawhill, whom we all wish well in his new task as chairman of the US Synthetic Fuels Corporation.

Commenting on what you have done gives me no trouble. I am humbled, however, by the assignment I have assumed of presenting the charge to you for your next Symposium in this remarkable Symposia Series. On Tuesday night, John Sawhill charged you, and his words bear repeating and remembering, that you should "clearly identify and define major energy issues in the context of a global community, in which we seek the betterment of all our citizens. We should focus national, regional, and global attention on these issues and force debate in a cooperative and concerted manner." So John asked you to identify and define the major issues. I shall ask you to do something about them.

But first, let me commend you for being here. Your presence and participation is important in and of itself. I applaud the Symposium organizers for assembling this group. It is a stunning achievement and gives substances to our hopes for the future. I think we all agree that we have arrived at the point in history when the only real solutions are global solutions; when nothing will work unless it is based on international cooperation. Here today is the evidence not that we can cooperate but that we are cooperating. In this chamber have come together producers and consumers, hard technologists and soft technologists (as the terminology goes), optimists and pessimists, the developed countries and the less developed countries. In fact, every available perspective is represented. An example of your success is the cooperative agreement which Minister Altmann of Costa Rica introduced into the record and which appears in Session I. If you keep at it, the future will be as rich as you will it to be. So to make a pedagogical reference, and I understand that some of you are from the academic community, you all get an A for attendance and an A+ for substance. You gave brillant papers and, I am told, "tours de force" on several occasions. Judging from a content analysis of the questions posed, your papers addressed all the relevant issues. You were not always in agreement on the implications of that relevance, but you were in agreement on the dimensions of each issue.

That observation leads me to note perhaps the most remarkable aspect of your agreements. They have taken on a kind of liturgical stability. They were not arrived at lightly, and I do not believe that they were anything other than the rigorous conclusion of careful thinkers, but they do have a ring of familiarity. It strikes me that we have not seen the birth of a foundation from which we can build a rational, effective strategy, but rather we have witnessed the confirmation ceremony for an understanding which will lead us out of the wilderness in which, over some years, we have wandered. That is an exciting achievement for this, the first Symposium in the World's Fair Symposia Series.

The substance of this body of understanding is well captured in your session summaries. There is no purpose to be served in my reciting them to you. I would note that concerning public policy over the new few years, your areas of disagreement are as important as those things on which you agree. As might be expected, there are important disagreements over the appropriate role for nuclear power, the significance of the changes in carbon dioxide in the atmosphere in relation to the use of fossil fuels, and the advisability of reliance on synthetic fuels. These points are familiar to those of us in the United States who have been involved in forging and implementing the US energy policy during the last three years. We learned that with energy, consensus does not come easily, but enough of it must come in a timely fashion to allow governments to act effectively. We may be short of oil, but it seems that time is even shorter.

The disagreements I mentioned are manifestations of a philosophy which looks at the future as having to be protected from high technology. No one at this conference and no prudent government doubts the plain economic and strategic need to move vigorously in order to ensure that we improve the efficiency of energy use. Some see the issue as whether we will move to new technologies and renewable resources through increased use of fossil fuels, or whether we will be able to avoid additional reliance on these sources. In other words, there is a disagreement about whether the increased use of fossil fuels is now necessary. On this, the US government, the Summit countries, and, I think, many others have now taken a position. As we see the world over the next twenty years, we just have no alternative but to rely on continued heavy use of fossil fuels, with coal (through direct burning and synthetics) and natural gas assuming a larger role. The United States has made the decision not only to develop but also to deploy a large synthetic fuel industry, and it is to that end that John Sawhill took office this past week, creating the necessity for my filling his position in the US Department of Energy.

You dispelled a myth in your deliberations here—the myth that the problems we face are only oil problems. I think there is a broad agreement that we do have a highly complex energy problem and that it is a global issue which touches the quality of life throughout the foreseeable future. It is felt acutely in the developed world and even more acutely in the developing world, including, ironically, some of the oil-rich but less developed countries. The problem is here now and cannot safely be ignored. It interacts with food production, environmental quality, and economic development, and, considering the given time frames and the potential for additional disruption of supplies, it cannot be solved by technology alone. That is the energy problem in a nutshell. It must be dealt with politically, socially, and now.

I come to these problems from a background in the law, politics, and government of my country, while most of you come to these problems from a technical background. It falls to those of us in the governmental arena to clear away political obstacles, so that those of you with technical knowledge and vision can have the opportunity to provide real solutions to problems which threaten to block the world's continued development. Overlying all of this is the close interrelationship between energy development and use and the other vital needs of world society. There's only so much capital to go around. Energy has traditionally been extremely expensive to produce. New forms of energy have an obvious tradeoff with our future food supplies and obviously with the quality of the environment in which we all shall exist. Those factors are vital to national and global deliberations on energy.

One lesson to be drawn from our discussions here is a need on the part of the industrialized world to relieve pressure on the supplies of international oil. The industrialized countries are coming to this out of a legitimate concern for their self-interest. Those of us in the industrialized world are emphasizing and shall continue to emphasize greater national productivity of energy and improved efficiency in using the energy we produce.

The decision has been made in the United States that the price of energy can no longer be artificially determined. President Carter made that decision in April of 1979, and we are now moving towards a fair market pricing of energy. The American people are responding to this by taking effective, individual actions to soften the economic impact of increasing energy costs.

If you stayed here this weekend and had time to visit a building materials supply store where insulation or weather stripping is sold, you would find large numbers of Americans buying these and other materials to retrofit their homes, to enable them to use the energy they buy more efficiently. They arrive at that position primarily through economics.

In the United States we have reversed a long-standing trend and are now consuming 8 percent less energy this year than last year. We are also domestically producing a greater share of the energy we consume. Our dependence on imported oil, which peaked at about 50 percent in 1978, is now about 40 percent. We believe that our oil imports this year will be less than 7 million barrels a day, down from a high of a 8.5 million barrels in 1978.

While the United States is starting to use energy more efficiently, it is also increasing national production. Exploration for oil and natural gas in this country is increasing dramatically. The long-standing trend of a decline in domestic production of these two energy sources has been reversed. Coal use is continuing to increase, and a new synthetic fuels industry is being incubated.

These patterns of change in my country are in consort with the theme chosen for this Symposium Series. You have all agreed that there is great opportunity for increased energy productivity, and I interpret this to mean that you seek to enhance the overall efficiency in the use of available energy resources. This agreement will permit us to move forward at the next Symposium, developing strategies and, we hope, articulating tactics for implementing these strategies.

Reaching agreement on the strategies will not be easy, but we must strive to do it. This will require far greater cooperation than the world and its countries have so far been able to bring to bear on this problem. We are not talking about agreements on relatively simple transactions; we are talking about one society agreeing with another society on how to order the lifestyles of their political leaders and of all their people. There is agreement on the perception of a common need to reduce worldwide dependence on petroleum. With this as a base, we should be able to embrace the concerns for equity among societies. It is going to be very difficult to come to agreement on these issues and reduce them to strategies, but we simply have no alternative except to be about it, and to do our best. So my charge to you for your next Symposium is to present your views on strategies and tactics for increasing world energy productivity. Your discussion and debate of these issues will, we hope, lead to a consensus that can be presented to political leaders. This would truly be an outstanding accomplishment.

I congratulate you for being here, and I look forward to your coming back together. You and perhaps other representatives of your countries can now move forward towards coming to grips with more concrete strategies for addressing this most challenging problem. You hold in your hands an opportunity for improving this world.

Appendices

Symposium I Program

The 1982 World's Fair
INTERNATIONAL ENERGY SYMPOSIUM I
Energy Production and Productivity

October 14 - 17, 1980

Symposium Chairperson: **John C. Sawhill**
Deputy Secretary
United States Department of Energy

Tuesday, October 14, 1980

Opening Address
8:00 p.m. - 9:30 p.m.

Speaker: **John C. Sawhill**
Deputy Secretary
United States Department of Energy

Wednesday, October 15, 1980

Session I - World Energy Productivity and Production:
   The Nature of the Problem
8:30 a.m. - 12:00 p.m.

Honorary Chairperson: **S. David Freeman**
Chairman
Board of Directors
Tennessee Valley Authority

Chairperson: **McGeorge Bundy**
Professor of History
New York University

Papers: **Mohammad Sadli**
Professor of Economics
The University of Indonesia

Wolf Häfele
Deputy Director
International Institute for
Applied Systems Analysis

Integrator: **Shem Arungu-Olende**
Senior Technical Officer
United Nations Conference on New
and Renewable Sources of Energy

Thursday, October 16, 1980

Session II - Improving World Energy Productivity and Production: The Role of Technology

8:30 a.m. - 12:00 noon

Honorary Chairperson: **Fernando Altmann Ortiz**
Minister of Energy
Costa Rica

Chairperson: **Lin Hua**
Director of Second Bureau
State Scientific and Technical
Commission
People's Republic of China

Papers: **John M. Deutch**
Arthur C. Cope Professor of
Chemistry
Massachusetts Institute of
Technology

Amory B. and L. Hunter Lovins
Friends of the Earth, Inc.
United States

Integrator: **Hiroo Tominaga**
Professor of Synthetic Chemistry
The University of Tokyo

Session III - Towards an Efficient Energy Future:
           Critical Paths, Conflicts, and Constraints

2:00 p.m. - 5:30 p.m.

Honorary Chairperson:	**Ishrat H. Usmani** Inter-regional Energy Advisor United Nations Division of Natural Resources and Energy
Chairperson:	**Hans H. Landsberg** Senior Fellow Resources for the Future Washington, DC
Papers:	**José Goldemberg** Director of the Institute of Physics The University of São Paulo
	B. C. E. Nwosu Chief Education Officer (Science) Federal Ministry of Education Nigeria
Integrator:	**David J. Rose** Professor of Nuclear Engineering Massachusetts Institute of Technology

Friday, October 17, 1980

Session IV - Alternative Policies for Improved Energy
           Productivity and Production

8:30 a.m. - 12:00 noon

Honorary Chairperson:	**Jane Carter** Undersecretary United Kingdom Department of Energy
Chairperson:	**Marcelo Alonso** Executive Director Florida Institute of Technology Research and Engineering, Inc.
Papers:	**David Sternlight** Chief Economist Atlantic Richfield Company

Amulya Kumar N. Reddy
Professor of Chemistry
Indian Institute of Science

Integrator:
Vaclav Smil
Professor of Geography
The University of Manitoba

Closing Address
12:00 noon - 12:30 p.m.

Speaker:
Lynn R. Coleman
Acting Deputy Secretary
United States Department of Energy

Symposium I Program
Members and Participants

PROGRAM MEMBERS

Marcelo Alonso was employed by the Organization of American States for twenty years, until July, 1980. During that time he was Director of Science and Technology and Executive Secretary of the Inter-American Nuclear Commission. He was, until 1960, Technical Director of the Cuban Nuclear Energy Commission. He recently became the Executive Director of the Florida Institute of Technology Research and Engineering.

Shem Arungu-Olende, a native of Kenya, is a Senior Technical Officer with the United Nations. He is responsible for solar, hydro, and rural energy program development for the UN Conference on New and Renewable Sources of Energy which will be held in Nigeria in 1981. He has worked with the United Nations since 1971 for the Center for Natural Resources as Technical Specialist in Energy and Transportation.

McGeorge Bundy, a former Professor of Government at Harvard University, is now Professor of History at New York University. He has served as Special Assistant to the President for National Security Affairs under Presidents Kennedy and Johnson. He was President of the Ford Foundation for thirteen years prior to taking his current position at New York University. He also serves as Chairman of the General Advisory Committee of the Arms Control and Disarmament Agency, to which he was appointed by President Carter.

Lynn R. Coleman served as Acting Deputy Secretary of the US Department of Energy, replacing John C. Sawhill who joined the US Synthetic Fuels Corporation during the week of Symposium I. Coleman came to this position from a background in law, politics, and government.

John M. Deutch is the Arthur C. Cope Professor of Chemistry at the Massachusetts Institute of Technology. In the past, he has served as the Director of the Office for Energy Research, Acting Assistant Secretary for Energy Technology, and Acting Undersecretary of the US Department of Energy, and as Chairman of the Advisory Panel for the National Science Foundation.

José Goldemberg is the Director of the Institute of Physics at the University of São Paulo and has been a professor of physics in France, Canada, and the United States as well as in his native Brazil. In addition to his expertise in nuclear physics, he has chaired the Energy from Biomass Commission for the Brazilian Ministry of Agriculture. He has many publications in the area of science, technology, and energy problems, and is President of the Brazilian Society for the Advancement of Science.

Wolf Häfele is Deputy Director of the International Institute for Applied Systems Analysis (IIASA) at Laxenburg, Austria, and leads IIASA's research on energy systems, working on problems of energy demand and supply systems. He came to IIASA from the Karlsruhe Nuclear Research Center, where he led the West German Fast Breeder Project. He was responsible for the development of the International Materials Safeguard System for the International Atomic Energy Agency. His research deals with energy demand and strategy.

Hans H. Landsberg was formerly the Director of the Center for Policy Research at Resources for the Future, where he is currently Senior Fellow. In this capacity he has chaired several important energy studies and has published a number of books and articles on energy. He has served as consultant to numerous government agencies and scientific organizations. Before joining Resources for the Future, he served as a consulting economist with Gass, Bell & Associates.

Lin Hua is currently serving as Director of the Second Bureau, State Scientific and Technological Commission of the People's Republic of China. This bureau is responsible for planning and coordinating research and development in natural resources, energy, and materials on the national level. From 1956 to 1960 he was Director of the Bureau of Technology, Ministry of the Chemical Industry. He left that post to become Director and Chief Engineer of Lazhou Complex Chemical Corporation, one of the largest chemical trusts in China. He held that position for seventeen years, until 1978 when he was assigned to his present post.

Amory B. and L. Hunter Lovins work as a team on energy policy. Amory Lovins is a consulting experimental physicist. He has been associated with Friends of the Earth, Inc., a nonprofit conservation lobbying group, since 1971. He is active in energy affairs on both technical and political levels and has published widely on the "soft-path" approach to energy. L. Hunter Lovins is a lawyer and a specialist in energy and environmental education. Previously she cofounded and directed the California Conservation Project. She has also consulted in the field of urban forestry, community energy participation, and alternative energy strategies.

B. C. E. Nwosu is the Chief Education Officer (Science) of the Federal Ministry of Education of Nigeria. He has served as Chief Scientific Officer, Natural Sciences Research Council of Nigeria, and on the Faculty of the University of Nigeria. He holds a doctorate in Nuclear Physics from Ohio State University. His research activities and writings deal with science education policy, public acceptance of nuclear power, and scientific and industrial development in Africa.

Amulya Kumar N. Reddy is on the faculty of the Department of Chemistry at the Indian Institute of Science. He is on temporary leave as a Visiting Senior Research Scientist at Princeton University. He is also Convenor of the Centre for the Application of Science and Technologies in Rural Areas. His research interests include energy problems of developing countries, rural energy consumption, and biomass utilization.

David J. Rose has been a Professor of Nuclear Engineering at the Massachusetts Institute of Technology since 1958. He has served with the United Kingdom Atomic Energy Authority and as Director of the Office of Long Range Planning at Oak Ridge National Laboratory. For several years he has been working with the World Council of Churches in an effort to deal with ethical issues of energy supply and demand.

Mohammad Sadli is Chairman of the Board of Indonesia's national oil company, Perta Mina, in addition to being a member of the Economics faculty of the University of Indonesia. He was Indonesia's Minister for Mines and Energy and in 1974 sat in that capacity as President of the Organization of Petroleum Exporting Countries.

John C. Sawhill was appointed to the US Synthetic Fuels Corporation; prior to that appointment, he served as Deputy Secretary of Energy for the US Department of Energy. He has also served as President of New York University; Administrator of the Federal Energy Administration; Deputy Administrator of the Federal Energy Office; and Associate Director for Natural Resources, Energy, and Science of the Office of Management and Budget. In addition, he has served on boards of directors of a number of major American corporations. Following Symposium I, he became Director of McKinsey and Company, Inc.

Vaclav Smil immigrated to Canada from Czechoslovakia in 1969. He is now a Professor of Geography at the University of Manitoba. His most recent research has dealt with developing countries' energy needs, including energy-related agricultural development and environmental issues in the People's Republic of China. He has published extensively in these areas.

David Sternlight is the Chief Economist of the Atlantic Richfield Company. He has also been Deputy Director of the Secretary of Commerce's Office of Policy Development and prior to that was Director of Economic Planning for Litton Industries. He has served as a full-time consultant to the United Nations and UNESCO in energy and economic statistics and computer systems development, and he has worked in large-scale systems with IBM and the Rand Corporation.

Hiroo Tominaga is a Professor of Synthetic Chemistry and a member of the Graduate Chemical Energy Engineering School at Tokyo University. Before joining the University he worked for the Mitsubishi Oil Company doing research and long-range planning. His current work deals with the chemistry of petroleum and coal.

PROGRAM PARTICIPANTS

Naim Afgan
Boris Kidric Institute of Nuclear Science
University of Belgrade
Yugoslavia

Yumi Akimoto
Director
Mitsubishi Metal Corporation
Japan

Randrianjafisdo Alexandre
Chief Engineer
Ministry of Posts and Telecommunications
Madagascar

Marcelo Alonso
Executive Director
Florida Institute of Technology Research and Engineering
United States

Fernando Altmann Ortiz
Minister of Energy
Costa Rica

Shem Arungu-Olende
Senior Technical Officer
United Nations Conference on New
and Renewable Sources of Energy

Leonard L. Bennett
Head, Economic Studies Section
International Atomic Energy Agency

Jan K. Black
Division of Inter-American Affairs
University of New Mexico
United States

Jay H. Blowers
Executive Director
US Man and the Biosphere Program
United States

McGeorge Bundy
Professor of History
New York University
United States

Jane Carter
Undersecretary
Department of Energy
United Kingdom

Lynn R. Coleman
Acting Deputy Secretary
Department of Energy
United States

Paul Danels
Director
Energy and Urban Environment
National Urban League
United States

John M. Deutch
Arthur C. Cope Professor of Chemistry
Massachusetts Institute of Technology
United States

Joy Dunkerley
Senior Fellow
Resources for the Future
United States

John S. Foster, Jr.
Vice-President, Science and Technology
TRW, Inc.
United States

Herman Franssen
Chief Economist
International Energy Agency
Organisation for Economic Co-operation and Development

S. David Freeman
Chairman of the Board of Directors
Tennessee Valley Authority
United States

John H. Gibbons
Director
Office of Technology Assessment
US Congress
United States

Luc Gillon
Catholic University of Louvain
Belgium

José Goldemberg
Director of the Institute of Physics
The University of São Paulo
Brazil

Wolf Häfele
Deputy Director
International Institute for Applied Systems Analysis

Ingi R. Helgason
Ministry for Industry and Power
Iceland

Pierre Jonon
Director
Production Facilities
Electricité de France

Milton Klein
Assistant to the President
Electric Power Research Institute
United States

Hans H. Landsberg
Senior Fellow
Resources for the Future
United States

Lin Hua
Director of the Second Bureau
State Scientific and Technological Commission
People's Republic of China

Måns Lönnroth
Secretariat for Future Studies
Sweden

Amory B. Lovins
Friends of the Earth, Inc.
United States

L. Hunter Lovins
Friends of the Earth, Inc.
United States

Edward Lumsdaine
Director
Energy, Environment, and Resources Center
The University of Tennessee
United States

Kerry McHugh
Assistant Secretary
Department of National Development and Energy
Australia

Igor Makarov
Science Attaché
Embassy of the Union of Soviet Socialist Republics
Washington, DC

B. C. E. Nwosu
Chief Education Officer (Science)
Federal Ministry of Education
Nigeria

Guy J. Pauker
Rand Corporation
United States

Umberto Ratti
Scientific Counselor
Embassy of Italy
Washington, DC

Amulya Kumar N. Reddy
Professor of Chemistry
Indian Institute of Science

P. C. Roberts
Director of Research
Department of Environment
United Kingdom

David J. Rose
Professor of Nuclear Engineering
Massachusetts Institute of Technology
United States

Mohammad Sadli
Professor of Economics
The University of Indonesia

John C. Sawhill
Deputy Secretary
United States Department of Energy

Luis Sedgwick Baez
Advisor to the Director of Energy
Ministry of Energy and Mines
Venezuela

Mitchell Sharp
Commissioner
The Northern Pipeline Agency
Canada

Allen C. Sheldon
Vice President for Environment and Energy Resources
Aluminum Company of America
United States

Vaclav Smil
Professor of Geography
The University of Manitoba
Canada

Bogumil Staniszewski
Director
Institute for Thermal Engineering
Warsaw Technical University
Poland

David Sternlight
Chief Economist
Atlantic Richfield Company
United States

Sun Wan-Zhu
Deputy Division Chief
State Economic Commission
People's Republic of China

Hiroo Tominaga
Professor of Synthetic Chemistry
The University of Tokyo
Japan

Ishrat H. Usmani
Inter-regional Energy Advisor
United Nations Division of Natural Resources and Energy

Miguel S. Ussher
Director General
Planning Secretariat
Office of the President
Argentina

Johannes Pieter Van Rij
First Secretary for Energy
Delegation for European Economic Communities

Alvin M. Weinberg
Director
Institute for Energy Analysis
United States

C. Segbe Wotorson
Advisor
Ministry of Lands and Mines
Liberia

Xu Zeguang
Engineer, Second Bureau
State Scientific and Technological Commission
People's Republic of China

Symposium I Organizers, Contributors, and Staff

ORGANIZING COMMITTEE

The Organizing Committee is composed of representatives from major energy-related organizations in the Knoxville area. Primary responsibilities of the Organizing Committee are to advise Symposia personnel regarding overall operation in areas of finance, program management, facilities, services and hospitality, and communications.

Norbert J. Ackermann
Technology for Energy Corporation

Robert A. Bohm
The University of Tennessee

William E. Fulkerson
Oak Ridge National Laboratory

Robert F. Hemphill
Tennessee Valley Authority

Walter N. Lambert
Energy Opportunities Consortium

Lillian T. Mashburn
United American Bank

S. H. Roberts, Jr.
Knoxville International
Energy Exposition

Carl O. Thomas
The University of Tennessee

PROGRAM COMMITTEE

The Program Committee is composed of internationally known experts in energy-related areas. The primary responsibility of the Committee, chaired by John H. Gibbons, is to screen participants and identify speakers for the Symposia.

James E. Akins
Former Ambassador to Saudia Arabia

Kenneth E. Boulding
Chairman of the Board
American Association for
the Advancement of Science
Professor Emeritus,
University of Colorado

John H. Gibbons
Director
Office of Technology Assessment
United States Congress

Hans H. Landsberg
Senior Fellow
Resources for the Future

Henry R. Linden
President
Gas Research Institute

Amory B. and L. Hunter Lovins
Friends of the Earth, Inc.

David J. Rose
Professor
Massachusetts Institute of Technology

Alvin M. Weinberg
Director
Institute for Energy Analysis

LOCAL CONTRIBUTORS

Beaty Chevrolet

Butcher Volkswagen

Clayton Motors

Clift Chrysler Plymouth

Hyatt Regency Knoxville

John Banks Buick

Park National Bank

Patty Brothers Datsun

Reeder Chevrolet

Rodgers Cadillac

Snider Motors

South Central Bell

United American Bank

United States Post Office

SYMPOSIUM I STAFF

Management Staff

Robert A. Bohm

Lillian A. Clinard

Richard D. Jacobs

A. Richard Johnson

Nelda T. Kersey

Wallace C. Koehler

Walter N. Lambert

Sheila W. McCullough

Karen L. Stanley

Supporting Staff

1982 World's Fair — Executive Office

S. H. Roberts, Jr., President

Edward S. Keen, Executive Vice President
and Chief Operating Officer

Walter N. Lambert, Executive Vice President
and President of Energy Programs

George M. Siler, Executive Vice President
and Chief Policy Officer

1982 World's Fair — Staff

Vernie Ausman
Ginny Baich
Carole Brailey
Jon Brock
Cookie Crowson
Emmett Edwards
Cathy Farmer
Julian Forrester
Jim Friedrich
Lotta Gradin
Shirley Gray
David Haber
Terri Haws
Leigh Hendry
Diane Irwin

Sheila Killion
Lea Ann King
Marian Kozar
Ken Lett
Carroll Logan
Beverly Mack
Dora McCoury
Jean Miller
John Perry
Jack Proffitt
Barbara Ragland
Jack Rankin
Carrie Suarez
Julia Walker
Kim Wright

Energy Opportunities Consortium

Traci Brakebill
Lisa Bridges
Evette Cobb
Rick Gregory

Katherine M. Lones
Carolyn Vande
John Winebrenner

The University of Tennessee
Energy, Environment, and Resources Center

William Clemons
Georgi Coiro
Debra Durnin
Mary Ellen Edmondson
Mary R. English
Joyce Finney
Nancy Gibson
Helen Hafford
Polly Koehler
Bernie McGraw
Betty Moss

Robert L. Reid
Joe Stines
Carolyn Srite
Rica Swisher
Peggy Taylor
Celina Tenpenny
Joyce Troxler
Gerry Troy
Karen Wallace
Beverly Worman

Department of Energy

John C. Bradburne, Jr., Chief
  Public Presentations
  Office of Public Affairs
William Bibb

Doris Brooks
Julia Redford
Wayne Range

*Union Carbide Corporation
Nuclear Division*

Edward Aebischer
Lee Berry
Carol Grametbauer

Cindy R. Lundy
Fred Mynatt
Collin West

Tennessee Valley Authority

Jean Baldwin
Roger Bolinger
Reid Campbell
Ely Driver
Clara Dunn
L. B. Kennedy
Mary Knurr
Mary Longmire
James McBrearty

Linda Oxendine
Jean Pafford
Virgil Reynolds
Mary Troy
Barbara Turner
James Ward
Arthur B. Wardner
Brown Wright
Thomas Zarger

The University of Tennessee

Robert Bledsoe
Houston Luttrell
O. Kenneth McCullough

Sheldon J. Reaven
Joseph Simpson
William T. Snyder

City of Knoxville

Randy Tyree, Mayor
Jim Humphrey

Anne Russell
John Underwood

Hyatt Regency Knoxville

Paul Sherbakoff
  General Manager
Chester Benton
Pam Caldwell
Steve Caldwell
Willie Cannon
John Cardona
Sam Choy
Steve Dewire

Chuck Dickinson
Tricia Gilbert
Ruth Hauk
Frank Hilbert
Karl Holmes
Buck Mathias
Dan May
Vaughn McCoy
Andrea Michael

Margaret Ogle
Joyce Roth
Jack Simpson
George Stewart
Richard Taber

Mary Lou Wardell
George Weidmann
Lee Wettengel
Eli White
Steve White

Miscellaneous Support

Patti Anderson
Williams E. Arant
E. Ray Asbury
John Banks
Tom Barnard
Helen Blackwell
Homer Butcher
David Cash
John Casillo
Lynn Clapp
Joe Clayton
Lloyd Clift
Helen Collins
Ralph Culvahouse
Eugene M. Daniels
William Ditmore
Henry Dubroff
Gregory B. Ebert
James Elliot
Jack Fagan
Peter Fennelly
Stanley Griffin
Sherri Harrell
Ronnie Huffaker
Tony Igar
Jerry Jenkins
Kristopher Kendrick
Melvin E. Kersey, Jr.
Tom Kinnard
Ann Lambert

Robert F. Lash, M.D.
C. R. Loy
Randy Loy
Barbara Miller
Lee Munz
Bill Nifong
Carol O'Dell
Gloria A. Owens
Kenneth Parry
Winifred Parry
John Patty
Kenneth Rainey
Robert Roach
Leila Roach
Roddy Rodgers
Elaine Kelly Russell
Patsy Scruggs
Karl Seger
Carroll Shetterly
Ed Shouse
Paul Siler
Harrison Snider
Warren Stephenson
Frank Wells
Jim Wells
Don Woods
Mara Yachnin
David Yarnell